Multiple-point geostatistics

Multiple-point geostatistics

Stochastic modeling with training images

Gregoire Mariethoz

Faculty of Geosciences and Environment
University of Lausanne, Switzerland

Jef Caers

Energy Resources Engineering Department
Stanford University, USA

WILEY Blackwell

Registered office: John Wiley & Sons, Ltd, The Atrium, Southern Gate, Chichester,
West Sussex, PO19 8SQ, UK

Editorial offices: 9600 Garsington Road, Oxford, OX4 2DQ, UK
The Atrium, Southern Gate, Chichester, West Sussex, PO19 8SQ, UK
111 River Street, Hoboken, NJ 07030-5774, USA

For details of our global editorial offices, for customer services and for information about how to apply for
permission to reuse the copyright material in this book please see our website at www.wiley.com/wiley-blackwell

The right of the author to be identified as the author of this work has been asserted in accordance with the UK
Copyright, Designs and Patents Act 1988.

Designations used by companies to distinguish their products are often claimed as trademarks. All brand names and
product names used in this book are trade names, service marks, trademarks or registered trademarks of their
respective owners. The publisher is not associated with any product or vendor mentioned in this book. It is sold on
the understanding that the publisher is not engaged in rendering professional services. If professional advice or other
expert assistance is required, the services of a competent professional should be sought.

The contents of this work are intended to further general scientific research, understanding, and discussion only and
are not intended and should not be relied upon as recommending or promoting a specific method, diagnosis, or
treatment by health science practitioners for any particular patient. The publisher and the author make no
representations or warranties with respect to the accuracy or completeness of the contents of this work and
specifically disclaim all warranties, including without limitation any implied warranties of fitness for a particular
purpose. In view of ongoing research, equipment modifications, changes in governmental regulations, and the
constant flow of information relating to the use of medicines, equipment, and devices, the reader is urged to review
and evaluate the information provided in the package insert or instructions for each medicine, equipment, or device
for, among other things, any changes in the instructions or indication of usage and for added warnings and
precautions. Readers should consult with a specialist where appropriate. The fact that an organization or Website is
referred to in this work as a citation and/or a potential source of further information does not mean that the author
or the publisher endorses the information the organization or Website may provide or recommendations it may
make. Further, readers should be aware that Internet Websites listed in this work may have changed or disappeared
between when this work was written and when it is read. No warranty may be created or extended by any
promotional statements for this work. Neither the publisher nor the author shall be liable for any damages arising
herefrom.

Library of Congress Cataloging-in-Publication Data

Mariethoz, Gregoire, author.
   Multiple-point geostatistics : stochastic modeling with training images / Gregoire Mariethoz and Jef Caers.
       pages cm
   Includes index.
Summary: "The topic of this book concerns an area of geostatistics that has commonly been known as
multiple-point geostatistics because it uses more than two-point statistics (correlation), traditionally represented by
the variogram, to model spatial phenomena"–Provided by publisher.
   ISBN 978-1-118-66275-5 (hardback)
   1. Geology–Statistical methods. 2. Geological modeling. I. Caers, Jef, author. II. Title.
   QE33.2.S82M37 2015
   551.01'5195–dc23

                                                                                      2014035660

A catalogue record for this book is available from the British Library.

Wiley also publishes its books in a variety of electronic formats. Some content that appears in print may not be
available in electronic books.

Cover image: Courtesy of NASA Earth Observatory.

Set in 9.5/13pt Meridien by Aptara Inc., New Delhi, India

1   2015

Contents

Preface

Arguably, one of the important challenges in modeling, whether statistical or physical, is the presence and availability of "big data" and the advancement of "big simulations." With an increased focus on the Earth's resources, energy, and the environment comes an increased need for understanding, modeling, and simulating the processes that take place on our planet. This need is driven by a quest to forecast. Forecasting is required for decision making and for addressing engineering-type questions. How much will temperature increase? How much original oil is in place? What will be the volume and shape of the injected CO_2 plume? Where should one place a well for aquifer storage and recovery? The problems are complex; the questions and their answers are often simple.

In addressing such complex problems, uncertainty becomes an integral component. The general lack of understanding of the processes taking place and the lack of data to constrain the physical parameters of such processes make forecasting an exercise in quantifying uncertainty. As a result, forecasting methods often have two components in modeling: a stochastic and a physical component. Physical models produce deterministic outcomes or forecasts; hence, they lack the ability to produce realistic models of uncertainty in such forecasts. On the other hand, stochastic processes can only mimic physics, and although they produce models of uncertainty, these often present poor physical realism or, worse, are physically implausible. The challenge in many forecasting problems is to find the right middle ground for the intended purpose: produce physically realistic models that include the critical elements of uncertainty and are therefore able to answer the simple questions posed.

To some extent, geostatistical methods can historically be framed within this context of forecasting and within the quest for realism and truth. In the past, applications were mostly in the area of subsurface geology, in particular mineral resources, and then later oil and gas resources (as well as groundwater and hydrogeology). Perhaps a key recognition early on was that an assumption of independently and identically distributed (IID) samples taken from a spatially distributed phenomenon, such as an ore body, is a geologically ("physically") unrealistic assumption. Mineral grades show a clear spatial structure that is the direct result of the physical genesis of such deposits. The goal of geostatistics then (and still) was not to model the genesis of that deposit by means of a physical process, but to produce estimates based on a model of spatial continuity that is as realistic as possible. The predominant model was the semivariogram, which is a statistical model, not a physical one, yet captures some elements of physical variability. Management of mineral resources constitutes a data-rich environment. Although the semivariogram is a rather limited model for describing complex physical realities, the presence of a large amount of drill holes (actual observations of physical reality) made this model of spatial continuity a plausible and successful one in the early stages of applications of geostatistics. The second major application, at least historically, is the modeling of subsurface reservoirs, where direct observations (wells) are sparse and the purpose is to forecast flow in porous media, which in itself requires physical models. In this way, two physical realities are present: the physics of deposition of clastics (sedimentation) or carbonates (growth), and the physics of fluid flow in porous media. Realism is sought in both cases. Many publications showed that geological models of the subsurface that were built based on multi-Gaussian processes (and the semivariogram as a basic parameter) lack geological realism in order to produce realistic forecasts of flow. Although any such evaluation is dependent on the nature of the flow problem considered, it appears to be the case in the large majority of practical flow-forecasting problems. A second problem in data-poor environments concerns the inference of semivariogram parameters. With data based on only a few wells, at best, one can infer some vertical semivariogram properties, but modelers were left to guess most other modeling parameters.

As a consequence, at least in reservoir modeling, Boolean (or object-based) models became fashionable because of their geological realism and flow-forecasting ability. Such models were calibrated from a richness of information available in analog outcrop models. The 1990s saw an expansion of geostatistical techniques in the traditional fields as well as application in several nongeological areas, in particular the environmental sciences. Considering the International Geostatistics Congress proceedings as a particular sample, one finds in the 1988 Avignon Congress only ~10% of applications outside traditional fields, whereas in the 2000 Cape Town Congress, environmental applications alone cover about ~20% of the papers. The 1990s therefore saw a shift in geostatistics that was twofold. Firstly, the early applications and theory that developed around semivariograms, various flavors of kriging, and multi-Gaussian simulation, including hard and soft data, were rapidly maturing. Secondly, the International Geostatistics Congress, which is held every 4 years and had long been the single platform for dissemination of novel research, saw its unique role wane because of the advent of more application-focused conferences (e.g. Petroleum Geostatistics, geoENV, and Spatial Statistics). In terms of research, and particularly in terms of the development of new methods, a drive toward non-Gaussian model development can be observed, perhaps now scattered over various areas of science and presented in various disjunctive conferences and journals. Some of the non-Gaussian methods still rely on semivariograms (or covariance functions in the statistical literature), such as the pluri-Gaussian methods or Karhune–Loeve expansions, whereas others rely on developments in the field of image analysis.

The Markov Random field (MRF), although its theory was originally developed in the 1980s, saw a proliferation of applications in both spatial and space–time modeling. The development of methods remained classical, however: data were used to fit parametric models, whether semivariograms, MRF parameters, or using traditional statistical methodologies (e.g., maximum likelihood and least squares); models were then used for estimation, or for simulation by sampling posterior distributions. Development of theoretical models is clearly based on probability theory and its extension such as Bayesian methods.

Multiple-point geostatistics, abbreviated throughout the literature as MPS, was primordially born out of a need to address the issue of lack of physical realism as well as the lack of control in the simulated fields in traditional modeling. As Matheron stated in his seminal contribution, parameters of traditional statistical models need not have a physical equivalent. Although for a theoretical probabilistic model there may be a "true" parameter, such as the Poisson intensity θ, there exists no physical property in the real world known as θ. One only has a set of true point locations within a domain when studying point processes. The data are the only physical reality. The goal of MPS is to mimic physical reality, and the vehicle to achieve this is the training image. Perhaps the name "multiple-point" suggests that this is a field of study that focuses on higher-order statistics only, but this is only partially true. The second component, namely the source of such statistics (an order of 2 or higher), is the use of a representation of the physical reality: the training image. We believe that the most important contribution in this new field, and this first book, lies in the use of training images to inform and hence include physical reality in stochastic modeling. This is a completely new contribution; it is, without exaggeration, a paradigm shift. Most of the methods covered do not follow the traditional paradigm of first parametric (or even nonparametric) modeling from data, then estimating or sampling from the given parametric model, building on probability theory only. We propose methods that skip this intermediate step (of parameterized or nonparametric models) and directly lift what is desired, whether it is the estimate or the sample or realization from the training images. The methods we propose are therefore no longer solely steeped in statistical science or probability theory (as is most of geostatistics); we borrow from computer science as well and create hybridization between these fields. For that reason, some would no longer term this "geostatistics". Labels are but labels; what matters is the content behind them.

This book is therefore a book about spatial and spatiotemporal modeling in the physical sciences (sedimentology, mineralogy, climate, environment, etc.). We do not claim any applications (yet) in other areas where spatial statistics are used (e.g., health or finance), although such applications are likely to occur in the future. This book is therefore all about practice and solving real problems, not to create more theory. The primary goal of engineering is to address engineering questions; it is not just the creation of stochastic models. However, within stochastic modeling itself, the goal is not the posterior probability distribution function (pdf) or the model parameters; rather, it is the estimates of that reality or the simulation of that reality. All other intermediate steps, whether inferring parameters or assessing convergence of samples, are but intermediate steps to the creation of a physical reality.

This book is constructed in three major parts. In Part I, we provide by means of a virtual case, a motivation and illustration of what MPS is, the major conceptual elements of MPS, what it aims to achieve, and how training images are generated and used. Part I therefore also serves as a platform to review some assumptions that are fundamental to spatial and spatio-temporal modeling. The aim is to illustrate that a simple problem of spatial estimation and simulation can be solved with and without random function theory.

In Part II, we cover quite exhaustively the various technical details of the methodologies and algorithms currently developed in this field. Starting from basic building blocks in statistical science and computer science, the glue of algorithmic development, we provide an overview of most existing algorithms. We treat important concepts in modeling such as nonstationary and multivariate modeling, the evaluation of consistency between data and model, the construction of training images, and how such training images can be used to formulate and solve spatial inverse problems.

In Part III, we provide the application of these methods to three major application areas: reservoir modeling, mineral resources modeling, and climate science. The last part serves as an illustration of the methodology development in Part I; it should not be seen as an exhaustive list of applications but, rather, as a template for future development.

Accompanying this book is a website with a collection of training images and example test cases: http://www.trainingimages.org. In the book, we provide a reference list per chapter. For a complete and updated reference list, please visit the website http://www.trainingimages.org. PowerPoint slides of all figures in the book can also be accessed and downloaded at www.wiley.com/go/caers/multiplepointgeostatistics.

Acknowledgments

This book would not have come to existence without contributions from a number of students and colleagues who helped us with their challenging comments, provided outcomes of their own research, and reviewed parts of this book. We are particularly grateful to Celine Scheidt, Lewis Li, Pejman Tahmasebi, Kashif Mahmud, Sanjeev Jha, Siyao Xu, Satomi Suzuki, Sarah Alsaif, and Cheolkyun Jeong for contributing to some of the most innovative aspects of the research results presented in this book, and they provided us with excellent figures. Thanks are also due to Lewis Li for coding and analyzing the "universal kriging with training image" method in Part I.

We also want to thank Thomas Romary for discussions relating to Part I and for reviewing this part, to Thomas Hansen for reviewing the chapter on inverse modeling, to Odd Kolbjørnsen for reviewing the chapter on MRF, and to Philippe Renard and Julian Straubhaar for reviewing the chapters on nonstationarity and training image construction. Among the people who contributed to this book, we especially thank Alexandre Boucher, Cristian Peréz, and Julián Ortiz for coauthoring the chapters on the mining case study. Alex also provided inspiration and SGEMS code regarding the case study in Part I.

We also want to thank the anonymous reviewers of the initial proposal for this book, for unanimously endorsing the book project and making valuable suggestions. We acknowledge funding from the Stanford Center for Reservoir Forecasting and the National Centre for Groundwater Research and Training. Importantly, the support of the University of New South Wales (Australia) and of ETH Zürich are acknowledged, which both employed Gregoire Mariethoz during the time he was writing this book. Without the support of these institutions, this book would not exist.

Finally, we would like to thank Fiona Seymour and her team at Wiley, who made the publication of this book a smooth one.

PART I
Concepts

CHAPTER 1

Hiking in the Sierra Nevada

1.1 An imaginary outdoor adventure company: Buena Sierra

As is the case for any applied science, no geostatistical application is without context. This context matters; it determines modeling choices, parameter choices, and the level of detail required in such modeling. In this short first chapter, we introduce an imagined context that has elements common to many applications of geostatistics: sparse local data, indirect (secondary) or trend information, a transfer function or decision variable, as well as a specific study target. The idea of doing so is to remain general by employing a synthetic example whose elements can be linked or translated into one's own area of application.

Consider an imaginary hiking company, Buena Sierra, a start-up company interested in organizing hiking adventures in the Sierra Nevada Mountains in the area shown in Figure I.1.1(left). The company drops customers over a range of locations to hike over a famous but challenging mountain range and meets them at the other end of that range for pickup. Customers require sufficient supplies in what is considered a strenuous trip over rocky terrain, with high elevation changes on possibly hot summer days. Imagine, however, that this area lies in the vicinity of a military base; hence, no detailed topographic or digital elevation model from satellite observation is available at this point. Instead, the company must rely on sparse point information obtained from weather stations in the area, dotted over the landscape; see Figure I.1.1(right). We consider that the exact elevation of these weather stations has been determined. The company now needs to plan for the adventure trip. This would require determining the quantity of supplies needed for each customer, which would require knowing the length of the path and the cumulative elevation gain because both correlate well with effort. The hike will generally move from west to east. The starting location can be any location on the west side from grid cell (100,1) to grid cell (180,1) (see Figure I.1.2).

To make predictions about path length and cumulative elevation gain, a small routing computer program is written; although it simplifies real hiking, the program is considered adequate for this situation. More advanced routing could be applied, but this will not change the intended message of this imaginary example.

Multiple-point Geostatistics: Stochastic Modeling with Training Images, First Edition. Gregoire Mariethoz and Jef Caers.
© 2015 John Wiley & Sons, Ltd. Published 2015 by John Wiley & Sons, Ltd.
Companion website: www.wiley.com/go/caers/multiplepointgeostatistics

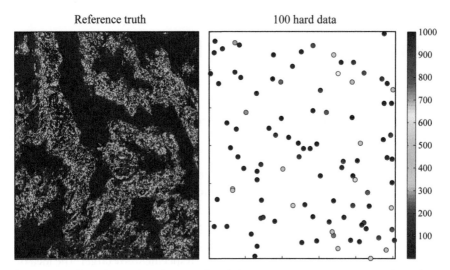

Figure I.1.1 (left) Walker Lake exhaustive digital elevation map (size: 260×300 pixels) grid; and (right) 100 extracted sample data. The colorbar represents elevation in units of ft.

The program requires as input a digital elevation map (DEM) of the area gridded on a certain grid. The program has as input a certain point on the west side, then walks by scanning for the direction that has the smallest elevation change. The program simulates two types of hikers: the minimal-effort (lazy) hiker and the maximal-effort (achiever) hiker. In both cases, the program assumes the hiker thinks only locally, namely, follows a path that is based on where they

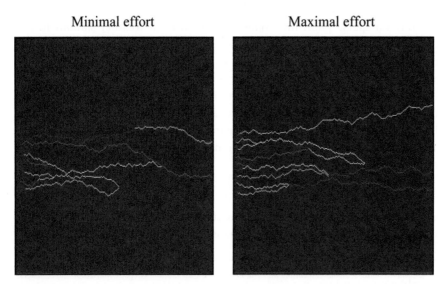

Figure I.1.2 Visualization of the 80 paths taken by hikers of two types: (left) minimal effort; and (right) maximal effort. The color indicates how frequently that portion of the path is taken, with redder color denoting higher frequency.

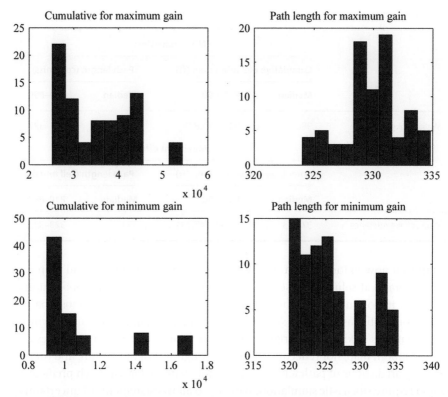

Figure I.1.3 Histograms of the cumulative elevation gain and path length for the minimal- and maximal-effort hiker. Cumulative elevation gain in units of ft, path length in units of grid cells.

are and what lies just ahead. The minimal hiker takes a path of local least resistance (steepest downhill or least uphill). The achiever hiker takes a path of maximal ascent (or minimal descent). Note that the computer program represents a deterministic transfer function: given a single DEM map, a single starting point, and a specific hiker type, it outputs a single deterministic hiking route. If the actual reference, Walker Lake, is used as input, then given starting locations from grid cell (100,1) to (180,1) on the west side, a total of 80 outcomes are generated. These 80 outcomes can be shown as a histogram; see Figure I.1.3. The resulting path statistics for both minimal effort and maximal effort are shown in Table I.1.1, which are summarized with quantiles (the eighth lowest, or P10; the 40th lowest, or P50; and the 72nd lowest, or P90).

1.2   **What lies ahead**

The problem evidently is that no DEM is available. How, then, would one proceed with forecasting path length and cumulative elevation change, and thereby make recommendations for Buena Sierra? We start in this Part I from very basic

Table I.1.1 Summary statistics

	Minimum effort			
	Cumulative elevation gain (ft)		Path length (cell units)	
	Median	P10–P90	Median	P10–P90
Walker Lake reference	9862	9434–14,875	324	311–327
	Maximum effort			
	Cumulative elevation gain (ft)		Path length (cell units)	
	Median	P10–P90	Median	P10–P90
Walker Lake reference	37,783	35,335–47,731	331	323–333

notions on how to formulate a theory for these kinds of problems, and then we present practical solutions based on that theory. This provides an opportunity to review important notions and assumptions that are common to most spatial prediction problems. The first such theory formulates spatial estimation, which in geostatistics is known as kriging. It is well known that kriging provides an overly smooth map, not reflecting the actual roughness of the terrain, and therefore any predictions of path length or elevation gain are biased. Such prediction would require stochastic simulation, which also allows statements of uncertainty about the calculated route statistics. Nevertheless, we will start with developing kriging because we will show how the traditional kriging (Chapter I.2) can be formulated without relying on the notions of expectation, probability, or random function theory, as long as a training image is available (Chapter I.3). The solution obtained is strikingly similar to traditional kriging, yet at no instance will we rely on random function theory.

Next, we will review stochastic simulation, which traditionally has relied on the same variogram and random function notions as kriging. In particular, we will review Gaussian theory and some popular methods that have been derived from this theory (Chapter I.4). Next, we will show, in a similar vein as for kriging, that the random function theory is not needed to perform stochastic simulation (Chapter I.5). We will present three alternative algorithms as an introduction to the many algorithms presented in Part II. These methods are compared in their ability to solve the practical problem discussed here (Chapter I.6).

CHAPTER 2

Spatial estimation based on random function theory

2.1 Assumptions of stationarity

In this chapter, we mostly review spatial estimation, a general term for estimating or guessing the outcome at unmeasured geographic locations from locations where measurements ("hard data") are available. As is the case for many statistical methods of estimation, the specification of a criterion of "best" is required. There will be only one guess or one estimate that can be given, once such a criterion has been specified. The variable being considered in the example case is the digital elevation map (DEM). One cannot directly estimate the path statistics.

Consider first a nonspatial problem, such as estimating the weight of a specific chair in a classroom. To represent this problem, we introduce the following notation. The true weight of that chair is unknown, denoted as Z, a random variable representing an unknown truth. A particular outcome, for example $z = 7\,\mathrm{kg}$, is written with a small letter. Suppose that all other chairs in that room are similar to the chair in question and we know the weight of those chairs, denoted as $\{z_1, z_2, \dots, z_n\}$. Based on these data, we make a histogram of the set of chairs. In doing so, an assumption is made: pooling all the weight data into a single plot, such as a histogram, entails that they are "similar" or "comparable". In probability theory, this is often referred to as "the population": a set of outcomes whose values can be grouped. They can be grouped for various reasons: similar origin, similar manufacturer, similar species, similar geological layer, similar location, and so on. However, such pooling requires a decision of what this reference population is. If one would pool tables into the set of chairs, then such pooling will lead to possibly very different results later on, and possibly very erroneous results. In many geostatistics books, this pooling and the accompanying assumption have been termed an "assumption of stationarity".

Only once a decision of stationarity has been made can estimation based on data proceed. Any guess – or, in statistical terms, any specific estimator – will be some unique function of the data, returning a single value:

$$z^* = g(z_1, z_2, \dots, z_n) \tag{I.2.1}$$

Multiple-point Geostatistics: Stochastic Modeling with Training Images, First Edition. Gregoire Mariethoz and Jef Caers.
© 2015 John Wiley & Sons, Ltd. Published 2015 by John Wiley & Sons, Ltd.
Companion website: www.wiley.com/go/caers/multiplepointgeostatistics

Many functions g could be considered, hence we need to specify or state some desirable properties for it. A property often stated as desirable is that of unbiasedness: namely, if a guess is made and denoted as Z^* (that guess is not yet known; it is therefore a random variable by itself), then unbiasedness can be stated based on the notion of expectation:

$$\text{Unbiasedness condition: } E[Z - Z^*] = 0 \qquad (\text{I.2.2})$$

Although this condition is common in many probability theory books, it is nontrivial for most first readers. The question is often: what is this an expectation of? What are we "averaging" over? To make such averaging feasible, one would need repeated situations, yet there are no such repeated situations: there is only one single specific chair with an actual weight that is estimated, and hence there will be only one difference between the true weight and our guess. Hence, why this "expectation"?

The unbiasedness therefore invokes a second assumption of stationary: the particular guessing procedure, if applied to (infinitely large) *similar* situations, will have the property of being, on average, equal to the truth. Suppose now that a reality exists where we would have many rooms, each such room containing a set of chairs and a specific chair for which we want to estimate the weight. Then, we need to assume that the situations presenting themselves in all these rooms form yet another population: the population of rooms. Making such a population requires, in a similar vein, an assumption of stationarity. The difficulty is that these alternative-world rooms never exist or are never truly considered; they are imaginary theoretical constructs.

A second condition often posed relates to our attitude toward making a mistake or the consequences of making errors. In the context of the chair, the particular person involved could reason as follows: overestimating the weight of the chair may be of less concern, but underestimating the weight may lead to injury upon attempting to lift it (supposing the estimating person has back problems). Clearly, making errors, whether positive or negative, may have different consequences. In the case of the chair, different attitudes may be taken. A thresholding function could be defined, where underestimating has a given consequence (a fixed hospital bill) over a certain weight value, or may gradually increase due to the increasing severity of the injury. In general, there is a function L, termed the loss function, quantifying our attitude to mistakes or to consequences of the error $z - z^*$. This leads to a second property that could be deemed desirable for any estimate or guess: minimize an expected loss, or,

$$E[L(Z - Z^*)] \text{ is minimal} \qquad (\text{I.2.3})$$

We return to the question: what are we averaging over? Averaging requires repetition of similar situations. Again, we need to consider imaginary parallel

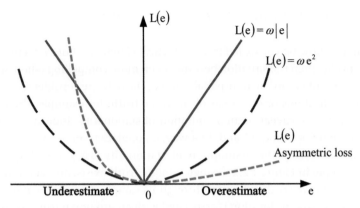

Figure I.2.1 Example loss functions; the most common choice is the parabola (least squares).

universes where the same situation occurs; and now, as a third assumption of stationarity, we assume that our attitude of loss will be similar in all those situations.

The most common loss function is to assume a parabola, as in Figure I.2.1, which is not necessarily applicable to the situation of the chair. The reason for assuming this simple squared function is not necessarily practical or aligned with reality, but rather is a mathematical convenience and driven by the elegance of the resulting solution, namely,

$$\min_{Z^*} E[(Z - Z^*)^2] \Leftrightarrow Z^* = E[Z] \tag{I.2.4}$$

In other words, the expected value minimizes a squared loss function. This basic result is the foundation of least squares theory. A simple arithmetic average is then an unbiased guess of the population mean (expected value), and, as stated in many statistical books, this occurs under the condition of independently identically distributed (IID) data. The latter entails the first hypothesis of stationarity (identically) and includes a kind of sampling that is not biased toward certain values (independently). However, as stated above, two additional assumptions of stationarity are required to get to this result.

In summary, the following stationarity hypotheses are needed to make any estimation procedure feasible, whether nonspatial or spatial:

1 Pooling of data: the creation of populations
2 Properties of estimators are defined through expectation, referring to repeated estimations in similar circumstances.
3 Loss incurred due to errors in the estimation (difference with the truth) refers to repeated situations with a similar attitude of loss.

None of these hypotheses can be tested objectively with any data; they are fundamental to the construction of the theory.

2.2   Assumption of stationarity in spatial problems

If this sounds a little bit construed from a practical viewpoint, but perfectly sound mathematically, then the situation becomes even more compelling when dealing with a spatial context. In the nonspatial context, the data are considered multiple alternative realizations or outcomes of the same truth: for example, the weight of a chair, with the caveat of an assumption of stationarity. Indeed, otherwise each chair would be "unique" and, in a way, a population on its own.

Consider analyzing the assumption of stationarity using the simple constructed example in Figure I.2.2: a single unique truth exists, and at a few locations, this unknown truth is known through its sample data. In notation, each unmeasured geographic location is associated with an unknown truth, which we denote as $Z(\mathbf{x})$, $\mathbf{x} = (x, y, z)$, or, if space–time is considered, $\mathbf{x} = (x, y, z, t)$. $Z(\mathbf{x})$ is considered to be a random function (as opposed to a random variable before). The term "function" refers to the fact that the outcome associated with each $\mathbf{x}$ is unique, and it also suggests a systematic variation (noting that "pure random" is a specific form of such systematic variation). If a grid is specified, then we need to deal only with a finite set of such $Z(\mathbf{x})$: $\{Z(\mathbf{x}), \mathbf{x} \in \text{Grid}\}$. At a finite set of locations, samples are recorded: $\{z(\mathbf{x}_\alpha), \alpha = 1, \ldots, n\}$.

Clearly, no repeated data exist on each $Z(\mathbf{x})$, as is the case for the nonspatial case. The only information available is measurements taken at a limited set of locations. The assumption of stationarity that is needed to make any estimation possible now requires including a geographical element, namely, an area over which pooling of data is allowed. For example, if we make a histogram of sample data over the area of study, then clearly we have made a geographical assumption of stationarity: the data at a location in that area can be pooled into a single plot, and that plot is meaningful.

We now return to the problem of estimation, in this case spatial estimation. We would like to determine at each uninformed location a "best guess" of that unknown value given some data. The problem is now spatial due to the indexing

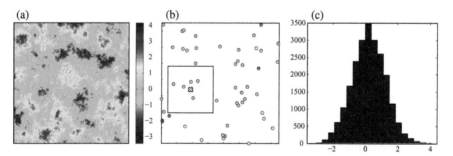

Figure I.2.2 (a) A single unique truth; (b) some sample data taken from it; and (c) its histogram. The goal is to estimate the value at the unsampled location marked with X.

with $\mathbf{x}$, hence the estimator becomes $Z^*(\mathbf{x})$. If unbiasedness is a desirable property, then

$$\text{unbiasedness condition: } E\left[Z(\mathbf{x}) - Z^*(\mathbf{x})\right] = 0 \tag{I.2.5}$$

What are we averaging over? One can imagine two types of averaging: in the first, we could average spatially, meaning over all possible $\mathbf{x}$. A second type of averaging is over all possible similar situations that could possibly occur in the universe. Again, to make this happen, we could invoke an alternate reality with many parallel universes where a similar situation as in Figure I.2.2 occurs; and, on average, over all the alternate realities, our guess would be equal to the truth.

2.3   The kriging solution

Now that some basic notions common to spatial estimation have been established, as well as basic assumptions needed to use and apply probability theory (expectation) to formulate such problems, we establish the most commonly used spatial estimation method in geostatistics: kriging. This section is mostly a simple review of basic equations of ordinary kriging, but perhaps with a more explicit statement and discussion of their underlying assumptions. Later, we will rewrite these equations without any use of expectation or random function theory.

2.3.1   Unbiasedness condition

We consider the situation in Figure I.2.2 where an estimate at only one location is required. First, we specify the function in Equation (I.2.1) as a simple linear sum:

$$z^*(\mathbf{x}) = \sum_{\alpha=1}^{n} \lambda_\alpha z(\mathbf{x}_\alpha) \tag{I.2.6}$$

or, written in terms of random variables:

$$Z^*(\mathbf{x}) = \sum_{\alpha=1}^{n} \lambda_\alpha Z(\mathbf{x}_\alpha) \tag{I.2.7}$$

Plugging this estimator into the unbiasedness condition

$$E[Z(\mathbf{x}) - Z^*(\mathbf{x})] = 0 \tag{I.2.8}$$

leads to

$$\sum_{\alpha=1}^{n} \lambda_\alpha E[Z(\mathbf{x}_\alpha)] = E[Z(\mathbf{x})] \tag{I.2.9}$$

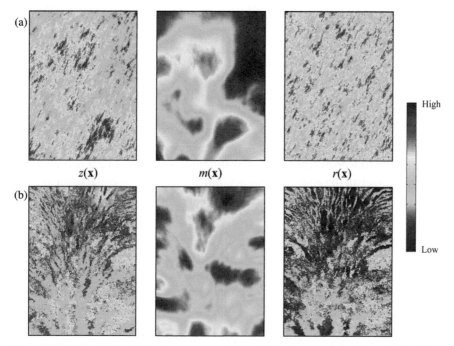

Figure I.2.3 (a) Rock density in a homogeneous layer of a carbonate reservoir; and (b) rock density in a heterogeneous deltaic reservoir.

Should the expected value be stationary, meaning constant over the domain, then the unbiasedness condition becomes

$$\sum_{\alpha=1}^{n} \lambda_\alpha = 1 \qquad (I.2.10)$$

The assumption of a stationary expected value is rarely useful in practical modeling. It assumes that a phenomenon under study can be decomposed as in Figure I.2.3a, or, in mathematical terms:

$$Z(\mathbf{x}) = M + R(\mathbf{x}) \qquad (I.2.11)$$

where M is some unknown but spatially constant expected value (hence, a random variable itself) and $R(\mathbf{u})$ is often termed the residual, which is spatially varying. In a Bayesian context, this expected value could have a prior itself (Omre, 1987), but most times we assume this expected value to be a constant, then

$$Z(\mathbf{x}) = m + R(\mathbf{x}) \qquad (I.2.12)$$

Very few phenomena vary spatially in the same way as in Figure I.2.2, and certainly not the DEM under study, with its systematic variation of mountain and lake beds. In this context, one can propose an extension as follows:

$$Z(\mathbf{x}) = m(\mathbf{x}) + R(\mathbf{x}) \quad \text{with} \quad E[R(\mathbf{x})] = 0 \quad \forall \mathbf{x} \tag{I.2.13}$$

The phenomenon is now decomposed into two parts: (1) a slowly varying expected value, often termed "trend"; and (2) a second part, R, that varies faster than the first part and whose expected value equals zero – this is often termed the "residual". We first discuss to what extent this decomposition is useful or even realistic. The decomposition would only be useful if the estimation or modeling of the two components m and R is easier than the direct estimation or modeling of Z. If this is not the case, then the decomposition is made purely for mathematical reasons. Consider two phenomena shown in Figure I.2.3. The question is whether it is useful to write each image as

$$z(\mathbf{x}) = m(\mathbf{x}) + r(\mathbf{x}) \quad \forall \mathbf{x} \in A \tag{I.2.14}$$

In Figure I.2.3(a), we can easily make such decomposition meaningful: a slowly varying and highly varying decomposition is achieved. One can imagine that modeling each component is easier than directly modeling the z-phenomenon. This is no longer the case in Figure I.2.3(b). The trend in this image lies on certain channel properties (width and orientation), not on the image z itself. The decomposition does not achieve an easier modeling task: the residual r looks like z. From a purely mathematical point of view, all phenomena can be written as a sum of two other phenomena; the more fundamental question lies in whether this is meaningful, makes further modeling easier and, perhaps more importantly, leads to better predictions in the given modeling context.

2.3.2 Minimizing squared loss

Consider now a case where the decomposition in Equation (I.2.13) is meaningful for the phenomenon being studied. Next, we need a specification of loss, as discussed in this chapter. Consider the following specification of loss:

$$Var[Z(\mathbf{x}) - Z^*(\mathbf{x})] \text{ is minimal} \tag{I.2.15}$$

which simplifies in combination with the unbiasedness condition to

$$E[(Z(\mathbf{x}) - Z^*(\mathbf{x}))^2] \text{ is minimal} \tag{I.2.16}$$

which, using Equation (I.2.13) and Equation (I.2.7), can be rewritten as

$$E\left[\left(E[Z(\mathbf{x})] + R(\mathbf{x}) - \sum_{\alpha=1}^{n} \lambda_\alpha (E[Z(\mathbf{x}_\alpha)] + R(\mathbf{x}_\alpha))\right)^2\right] \text{ is minimal} \tag{I.2.17}$$

We can write the difference between truth and estimator as follows:

$$Z^*(\mathbf{x}) - Z(\mathbf{x}) = \sum_{\alpha=1}^{n} \lambda_\alpha Z(\mathbf{x}_\alpha) - Z(\mathbf{x}) = \sum_{\alpha=0}^{n} \lambda_\alpha Z(\mathbf{x}_\alpha) = \sum_{\alpha=0}^{n} \lambda_\alpha (E[Z(\mathbf{x}_\alpha)] + R(\mathbf{x}_\alpha))$$

$$(I.2.18)$$

with

$$\lambda_0 = -1; \ \mathbf{x} = \mathbf{x}_0 \tag{I.2.19}$$

The unbiasedness condition can be written as follows:

$$\sum_{\alpha=0}^{n} \lambda_\alpha E[Z(\mathbf{x}_\alpha)] = 0 \tag{I.2.20}$$

Some simple algebra then leads to

$$E[(Z(\mathbf{x}) - Z^*(\mathbf{x}))^2] \text{ is minimal} \Leftrightarrow$$

$$E[(R(\mathbf{x}))^2] + 2\sum_{\alpha=1}^{n} \lambda_\alpha E[R(\mathbf{x})R(\mathbf{x}_\alpha)] + \sum_{\alpha=1}^{n}\sum_{\beta=1}^{n} \lambda_\alpha \lambda_\beta E[R(\mathbf{x}_\alpha)R(\mathbf{x}_\beta)] \text{ is minimal}$$

$$(I.2.21)$$

One notices how the expected value of Z has disappeared from the loss specification, and only residual expectations remain. This is possible only because we assume

$$E[Z(\mathbf{x})] = m(\mathbf{x}) \tag{I.2.22}$$

In other words, the expected value is not randomized (assumed to be a random variable) itself. The combination of an unbiasedness condition and a loss specification has resulted in the following minimization problem with linear constraints:

$$\begin{cases} E[(R(\mathbf{x}))^2] + \sum_{\alpha=1}^{n} \lambda_\alpha E[R(\mathbf{x})R(\mathbf{x}_\alpha)] + \sum_{\alpha=1}^{n}\sum_{\beta=1}^{n} \lambda_\alpha \lambda_\beta E[R(\mathbf{x}_\alpha)R(\mathbf{x}_\beta)] \text{ is minimal} \\ \sum_{\alpha=0}^{n} \lambda_\alpha E[Z(\mathbf{x}_\alpha)] = 0 \end{cases} \tag{I.2.23}$$

Using the Lagrange formalism, the following augmented function is being minimized:

$$\begin{cases} S(\lambda_\alpha, \alpha = 1, \dots, n; \mu) = s(\lambda_\alpha, \alpha = 1, \dots, n) + \mu\left(\sum_{\alpha=0}^{n} \lambda_\alpha E[Z(\mathbf{x}_\alpha)]\right) \\ \text{with } s(\lambda_\alpha, \alpha = 1, \dots, n; \mu) = E[(R(\mathbf{x}))^2] + 2\sum_{\alpha=1}^{n} \lambda_\alpha E[R(\mathbf{x})R(\mathbf{x}_\alpha)] \\ + \sum_{\alpha=1}^{n}\sum_{\beta=1}^{n} \lambda_\alpha \lambda_\beta E[R(\mathbf{x}_\alpha)R(\mathbf{x}_\beta)] \\ \lambda_0 = -1; \ \mathbf{x}_0 = \mathbf{x} \end{cases} \tag{I.2.24}$$

Calculating and equating derivatives to zero result in the following systems of linear equations with linear constraints:

$$
\begin{cases}
\sum_{\beta=1}^{n} \lambda_\beta E[R(\mathbf{x}_\alpha)R(\mathbf{x}_\beta)] + \mu = E[R(\mathbf{x})R(\mathbf{x}_\alpha)] \quad \alpha = 1, \dots, n \\
\sum_{\beta=1}^{n} \lambda_\beta E[Z(\mathbf{x}_\beta)] = E[Z(\mathbf{x})]
\end{cases}
\tag{I.2.25}
$$

The linear system of size $(n+1) \times (n+1)$ can be solved once the following terms are specified:

$E[R(\mathbf{x}_\alpha)R(\mathbf{x}_\beta)]$: the covariance of the residuals between different data locations

$E[R(\mathbf{x})R(\mathbf{x}_\alpha)]$: the covariance of the residual between data location
   and the location to be estimated

$E[Z(\mathbf{x}_\alpha)]$: the expected value at the data locations

$E[Z(\mathbf{x})]$: the expected value at the location to be estimated

Several roadblocks are still in place to obtain any kind of numerical values for these terms:
- There are no repeated data to estimate $E[Z(\mathbf{x}_\alpha)]$ or $E[Z(\mathbf{x})]$ without making additional assumptions.
- Even if we had such repeated data, it is on Z, not on R, hence we cannot estimate the above covariances of R.

It is clear that additional simplifications and assumptions need to be invoked before any numerical calculations can be carried out. First, one could assume the expected value to be the same at all geographical equations. This would take care of the bottom two terms in the above list and also simplify the linear constraint to

$$
\sum_{\beta=1}^{n} \lambda_\beta = 1
\tag{I.2.26}
$$

Given a stationary expected value, one can in addition assume a stationary residual. If we introduce the notation of covariance for R as follows:

$$
Cov[R(\mathbf{x}'), R(\mathbf{x})] = E[R(\mathbf{x}')R(\mathbf{x})]
\tag{I.2.27}
$$

Then, under an assumption of stationary expected value and stationary covariance, the notation can be simplified as follows:

$$
Cov[R(\mathbf{x}'), R(\mathbf{x})] = Cov_R(\mathbf{x}' - \mathbf{x})
\tag{I.2.28}
$$

In other words, the covariance is only a function of the distance between geo-graphical locations, not the exact place where geographically one is located. As a consequence, the linear system now becomes

$$
\begin{cases}
\displaystyle\sum_{\beta=1}^{n} \lambda_\beta Cov_R(\mathbf{x}_\alpha - \mathbf{x}_\beta) + \mu = Cov_R(\mathbf{x} - \mathbf{x}_\alpha) \quad \alpha = 1, \ldots, n \\
\displaystyle\sum_{\beta=1}^{n} \lambda_\beta = 1
\end{cases}
\tag{I.2.29}
$$

an $(n + 1) \times (n + 1)$ system of equations that is traditionally known as the ordi-nary kriging system. Once the system is solved – namely, numerical values for λ and μ are obtained – then the minimum of Equation (I.2.16) can be algebraically expressed as

$$
\text{var}_{\min}(\mathbf{x}) = Var(Z) - \sum_{\alpha=1}^{n} \lambda_\alpha Cov_R(\mathbf{x}_\alpha - \mathbf{x}) - \mu
\tag{I.2.30}
$$

which is commonly known as the ordinary kriging variance. Next, we deal with obtaining numerical values for the covariance terms to solve Equation (I.2.29).

2.4 Estimating covariances

Solving the system of linear equations to estimate the unsampled value at the location highlighted in Figure I.2.2 requires specifying covariance values in Equa-tion (I.2.29). In traditional approaches, this calculation is only possible through the assumption in Equation (I.2.28): "covariance is only function of the distance between geographical locations". This assumption allows pooling data pairs with similar distance (exact distance replicates rarely exist with irregular data) into a single scatterplot from which the covariance value can be calculated. This exer-cise can be repeated for various distances. The very existence of this single scat-terplot is an explicit statement or expression of stationarity: data from differ-ent locations are pooled into one single plot. These covariance values are then grouped based on distances calculated along the same (or similar for irregular data) directions. The above linear system calls for covariances on R, not on Z. For this simple case, this poses no problem, as the mean appears fairly constant over the domain. More difficult situations are discussed in this section.

In geostatistics, semivariograms are commonly estimated. Without any assumption of stationarity, these semivariograms are defined as follows:

$$
2\gamma[R(\mathbf{x}'), R(\mathbf{x})] = E[(R(\mathbf{x}') - R(\mathbf{x}))^2]
\tag{I.2.31}
$$

Similar to the covariance, under the assumption of stationarity, the semivariogram becomes a function of distance in location only:

$$\gamma[R(\mathbf{x}'), R(\mathbf{x})] = \gamma_R(\mathbf{x}' - \mathbf{x}) \tag{I.2.32}$$

Under the same assumption of stationarity, the covariance and semivariogram can be related as follows:

$$Cov_R(\mathbf{x}' - \mathbf{x}) = Var(R) - \gamma_R(\mathbf{x}' - \mathbf{x}) \tag{I.2.33}$$

Figure I.2.4 shows the omnidirectional variogram of Z for the data in Figure I.2.2. In general, the semivariogram is often preferred over the covariance because it does not require knowledge of the mean. In practice, the mean is not known (and not necessarily constant; this is discussed later in this chapter), which introduces a bias when the mean is estimated from data (Chilès and Delfiner, 1999, p. 33).

In many practical cases, one notices that the covariances and semivariograms of the sample data look "noisy"; they do not exhibit the expected crisp nugget, range, or sill parameters that geostatisticians interpret and model using parametric models. Take, for example, the Walker Lake data set in Figure I.2.5 and the semivariogram calculated on the exhaustive and sample data. Some of the experimental semivariograms appear as almost a pure nugget effect for certain directions (e.g., the east-west [EW] direction). This problem is common in many practical applications of geostatistics. Lack of clear structure should evidently not be confused with a lack of spatial correlation (compare the experimental semivariogram of the sample data with the exhaustive image in Figure I.2.5).). This problem of obtaining meaningful semivariograms has been one of the motivating factors of looking beyond the data alone to calculate spatial statistics such as covariance or semivariograms.

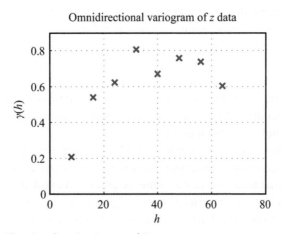

Figure I.2.4 Omnidirectional semivariogram of Z.

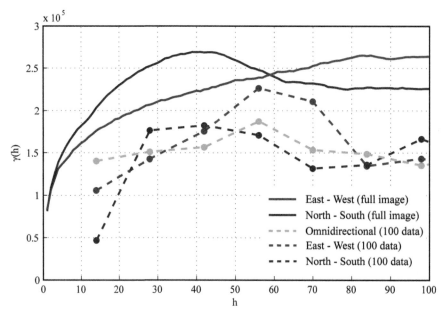

Figure I.2.5 Semivariogram of the exhaustive Walker Lake data set versus the sample variogram.

Why is obtaining meaningful semivariograms often problematic?

- The data may be too few to allow for a meaningful calculation of experimental statistics.
- The sample data may have a skewed histogram, affecting the robustness of the traditional covariance and semivariogram estimators. In the Walker Lake case, one could take a uniform score transform of the data and calculate semivariograms, then borrow ranges and anisotropy from these variograms.
- The assumption of stationarity may not be valid, meaning data pairs are pooled into the scatterplot that do not belong to the same plot.
  - Any attempt to calculate nonstationary spatial statistics, for example by limiting the pooling to a local geographical region around the point being estimated, will only lead to further decimation of the already sparse data.
  - Relying on the decomposition (Equation (I.2.13)) and subtracting some trend from the data does not necessarily solve the problem, either. As discussed, Equation (I.2.13) is just one very simple way of expressing a trend, and it may not apply to all phenomena.
- The spatial phenomenon may not show much spatial correlation of order 2. The semivariogram and covariance capture spatial variability by only considering two spatial locations at one time ($\mathbf{x}'$ and $\mathbf{x}$). What if the phenomenon exhibits a much stronger correlation at a higher order, but not a lower order?

In practice, modelers often borrow covariance or semivariogram statistics from elsewhere, whether it is another geographical area or based on their own domain knowledge. Even if the actual statistics may just look like pure nugget effect, they may rely on expertise or prior knowledge of the phenomenon to attribute covariance properties, such as range, that are not reflected by the actual experimental values. Some practitioners may go even further and consider other data. These could be data from the same geographical area or from another area deemed analog to the area or phenomenon under study. For example, in subsurface modeling, modelers often calculate vertical spatial statistics such as the vertical semivariogram range from vertical wells and borrow horizontal ranges from geophysical data such as seismic data (Mukerji et al., 1997).

2.5 Semivariogram modeling

The experimental semivariogram needs to be modeled because kriging requires evaluation of semivariograms for distances that may not be present in the sample data. Additionally, semivariograms calculated from sample data are subject to small sample fluctuations that may require smoothing. In the case of the experimental semivariograms of Figure I.2.4, a spherical model with a range equal to 25 and no nugget can easily be justified.

The model semivariogram is different from the exhaustive semivariogram calculated on the single unique truth. Additional theoretical constructs are needed to explain and motivate the existence of this difference, which in geostatistics is known as the ergodic property. In nonspatial univariate problems, sample statistics such as the arithmetic average are known to converge to the population mean, when the sample size tends toward infinity. In spatial modeling, this does not occur, because the domain is finite, or at least within that finite domain one considers a grid with a finite amount of locations. One can now take two viewpoints and create a theoretical construct of an infinite domain (Matheron, 1978; Chilès and Delfiner, 1999) or assume the finite domain can be refined at infinitely small resolution (Matheron, 1978; Stein, 1999; Rivoirard and Romary, 2011). If one assumes the single unique truth to be a realization of a unspecified stationary random function $Z(\mathbf{x})$, then the assumption of ergodicity implies that statistics calculated on any realizations (including the unique truth) of that stationary random function converge to the statistics of $Z(\mathbf{x})$ as the spatial area tends toward infinity (if one takes the infinite domain ergodic approach). Intuitively stated, the theory calls for imagining that a stationary stochastic process exists (a random function) over an infinite domain and that such a process created our unique single truth that we only observe in a finite part of the domain. If that is the case, and if that process is stationary, then one can imagine that even a single finite domain realization informs the spatial statistics of the infinite domain process. It would also mean that any sampling of that finite domain informs these statistics. Additionally, if that finite domain increases, then any semivariogram calculated from it (such as the semivariogram of the exhaustive image in Figure I.2.4) will converge to the model semivariogram (ergodicity in the semivariogram or covariance). It should be stressed that this is a purely theoretical construct to motivate the use, in actual practice nonetheless, of a smooth semivariogram model; there is no practical reality or verifiability (from any data) in stating such an ergodic property, and hence the use of a smooth semivariogram model.

2.6   Using a limited neighborhood

One can now perform ordinary kriging over a regular grid instead of a single location. In doing so, one can solve a global kriging problem involving all sample data values. For the simple case of Figure I.2.2, this can be easily done since only a 50 × 50 covariance matrix needs to be inverted and because the statistics in the domain can be modeled as stationary; see Figure I.2.6(a). In most practical cases, some form of nonstationary kriging is required (see Chapter I.3) or the amount of sample data becomes prohibitively large. In such cases, a neighborhood

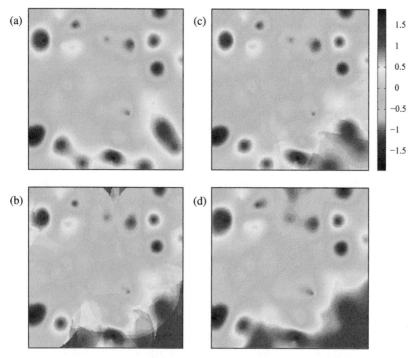

Figure I.2.6 (a) Global kriging using all 50 sample data at all estimated locations; (b) a local moving neighborhood; (c) using a minimum of 12 samples; and (d) using penalty. Parts (a–c) are calculated using SGEMS (Remy et al., 2008), and (d) is calculated using the R-package RGeostats.

around each location being estimated is constructed, and only data within that neighborhood are used. The method of selecting neighboring data and ignoring other data is justified because these other data would have received a zero or negligible kriging weight (the screening effect; Chilès and Delfiner, 1999). The use of a moving neighborhood is justified by the assumption of local stationarity in the mean, and it is common to most of Matheron's developments in geostatistics. The unknown mean is considered constant in a neighborhood around the target location but slowly varying over the domain (Equation (I.2.13)). We now consider the application of such moving neighborhood ordinary kriging to Figure I.2.2, on a 150×150 grid with a simple circular neighborhood of size 50 and a maximum amount of samples set to 15. Clearly, artifacts in the shape of "circus tents" occur, induced by using a moving neighborhood. This is usually the case when the outermost data of the moving neighborhood are suddenly removed from that neighborhood, yet at the same time these data may have a nonnegligible weight associated with them. Several solutions have been proposed to deal with this problem:

- Use more sophisticated neighborhood searches, such as octant search (see Chapter II.2).
- Use, instead of a fixed search radius, a minimum amount of samples. Figure I.2.6(c) shows that in this case, setting this minimum to 12 results in most artifacts to disappear.
- Assign an increasing penalty to data that are increasingly further away (Rivoirard and Romary, 2011). One can add an error variance term:

$$penalty = \sum_{\alpha=1}^{n} \lambda_\alpha^2 \varepsilon(\mathbf{x}_\alpha) \qquad (I.2.34)$$

to the least squares problem in Equation (I.2.24). This penalty is zero at the location being estimated and increases as samples are more distant, hence such samples get progressively less weight. The resulting kriging system then has a penalty added to the diagonal of the covariance matrix:

$$\begin{cases} \sum_{\beta=1}^{n} \lambda_\beta Cov_R(\mathbf{x}_\alpha - \mathbf{x}_\beta) + \mu + \sum_{\alpha=1}^{n} \lambda_\alpha \varepsilon(\mathbf{x}_\alpha) = Cov_R(\mathbf{x} - \mathbf{x}_\alpha) \quad \alpha = 1, \ldots, n \\ \sum_{\beta=1}^{n} \lambda_\beta = 1 \end{cases} \qquad (I.2.35)$$

Figure I.2.6(d) shows that applying this to the example removes most artifacts.

- A finite domain kriging (Matheron, 1978; Deutsch, 1993, 1994) consists of solving several kriging systems of increasing size, starting from using only the closest data points to using all data points in the neighborhood, then simply averaging the multiple weights obtained. This approach requires more CPU time due to the multiple krigings.

Even if this is not a book on linear geostatistics, the discussion on using limited neighborhoods is of relevance to the training-image-based methods presented later in the book. Most spatial statistical approaches make use of the limited neighborhood concept. Markov random fields make use of a limited neighborhood in the model formulation of the random function (spatial law; see Chapter II.4). In other situations, a limited neighborhood is needed to make methodologies practical for solving large problems, such as in sequential simulation. Grids with 10^9 cells are no longer uncommon in practice, and such large-scale problems are specifically targeted in this book. Often associated with the use of a limited neighborhood is the presence of artifacts, as is the case in Figure I.2.6(b). In other methods, particularly in stochastic simulation, such artifacts may manifest themselves in various forms other than the above example, such as in reduced spatial uncertainty, poor data conditioning, or poor reproduction of statistics. Similar to the above example, specific "tricks" need to be implemented to mitigate them.

2.7 Universal kriging

In theory, ordinary kriging (OK) was developed for situations where the slowly varying mean is unknown. When in practice OK is combined with a limited neighborhood, various viewpoints exist on the resulting kriging map and the use of the estimates. OK can be seen as simple kriging (SK, or kriging with a known and stationary mean) where the mean is estimated by kriging (Goovaerts, 1997; Chilès and Delfiner, 1999). If the phenomenon has a clear trend, then, when using a limited neighborhood, that estimated mean will vary. Some authors (Journel and Rossi, 1989; Goovaerts, 1997) argue that this takes care of most cases involving a trend, such as all cases involving interpolation within the finite domain. It is argued that more involved techniques are only needed when extrapolation is made. Alternatively, one can argue for a more explicit model for the trend, which, in principle, need not rely on the limited neighborhood concept. In universal kriging (UK; Matheron, 1978), later also termed "kriging with a trend function" (KT; Goovaerts, 1997), a trend model is proposed as

$$E[Z(\mathbf{x})] = \sum_{k=0}^{K} a_k f_k(\mathbf{x}) \tag{I.2.36}$$

with typically $a_0 = 1$. This leads to the following system of equations:

$$\begin{cases} \sum_{\beta=1}^{n} \lambda_\beta Cov_R(\mathbf{x}_\alpha - \mathbf{x}_\beta) + \sum_{k=0}^{K} \mu_k f_k(\mathbf{x}_\alpha) = Cov_R(\mathbf{x} - \mathbf{x}_\alpha) \quad \alpha = 1, \dots, n \\ \sum_{\beta=1}^{n} \lambda_\beta f_k(\mathbf{x}_\alpha) = f_k(\mathbf{x}) \qquad k = 0, \dots, K \end{cases} \tag{I.2.37}$$

2.8   Semivariogram modeling for universal kriging

For the system in Equation (I.2.37) to be solved, the residual covariance has to be known. In practice, the estimation of the covariance or semivariogram parameters causes several problems, regardless of the inference method chosen (e.g., the moments method or maximum-likelihood method under a Gaussian hypothesis). The main problem is that the true residuals are never observed because the trend or drift is unknown. Consider the Walker Lake case of Figure I.1.1. The assumption of a constant mean becomes questionable; hence, the covariance of Z is no longer useful to determine the covariance of R. In such cases, various approaches are considered. One could consider directions along which the trend does not exist. This can be done in relatively simple trend cases: for example, if the trend is north-south, one considers the EW direction. This approach can only work for rather simple trend cases. Other approaches (Delfiner, 1976) rely on filtering the trend from linear combinations of the sample data, but such approaches require sample data to be plenty and on a (pseudo-) regular grid.

In more general cases, the estimated residuals have to be computed from an estimated trend. It has also been shown (see Matheron, 1969; Armstrong, 1984) that the empirical semivariogram of the residuals always presents a negative bias (for a detailed study of the bias, see Cressie, 1993; Beckers and Boogaets, 1998). Alternatively, using the data, one can estimate the trend by regression using ordinary least squares (OLS), then subtract that estimated trend from the data to model the covariance of R. However, the application of OLS requires uncorrelated samples, and hence estimates under correlated R are biased. As a result, the covariance of R calculated this way may be biased as well (Cressie, 1993; Chilès and Delfiner, 1999). In fact, even if the true covariance would be known and the drift was estimated by generalized least squares (GLS), the empirical variogram of the residuals would still be biased. This is directly resulting from the fact that the covariance of R and the trend functions f are not orthogonal, which is a requirement (Matheron, 1971). This bias is unfortunately unavoidable and a direct consequence of the way the random function model is defined as theoretically ideal, but it is practically unattainable for most real cases.

Therefore, in practice, ordinary least squares (OLS) are generally used to estimate the residuals as a first guess. This OLS estimate of the trend is suboptimal but unbiased, and, as a consequence, the empirical variogram computed from the OLS-based estimated residuals will also carry a potentially large bias, in particular with small sample data sets. Semivariogram estimation from the residuals should ideally be conducted in an iterative manner as follows (e.g., as proposed in Gelfand et al., 2010):

- Estimate the trend using OLS; then, compute the residuals $r(\mathbf{x}_\alpha)$ at data locations.
- Model the residual semivariogram based on these data.

- Estimate the trend using GLS using the previously estimated semivariogram as a model for covariance of the residuals; compute the residuals $r(\mathbf{x}_\alpha)$ at data locations.
- Model the residual semivariogram based on these data.
- Solve universal kriging (I.2.37).

2.9 Simple trend example case

We illustrate the result of the above procedure on a simple example in Figure I.2.7. The truth is generated using the following random function model:

$$Z(\mathbf{x}) = \frac{y}{200} + \sigma^2 R(\mathbf{x}) \qquad (\text{I.2.38})$$

where R is a stationary Gaussian field with zero mean, unit variance, an exponential variogram with a range equal to 25 and $\sigma^2 = 0.05$. We assume the exact trend to be linear but with unknown parameters; hence only trend terms x and y are included in Equation (I.2.36). The OLS estimate based on 50-sample data (Figure I.2.7(right)) is shown in (Figure I.2.8(left)), and it has both an x and a y component. The universal kriging estimate with the trend estimate that is obtained using OLS is shown in Figure I.2.7(right), displaying a clear trend in the estimates.

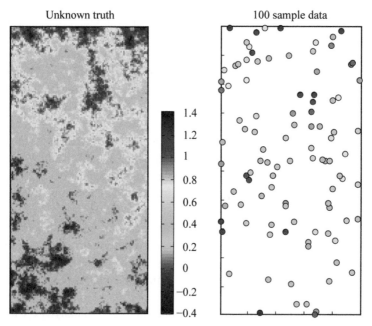

Figure I.2.7 Simple trend case study: (left) the unknown truth (one realization, see Equation (I.2.38)); and (right) the sample data.

| OLS estimate of trend | With OLS | With two-step procedure |

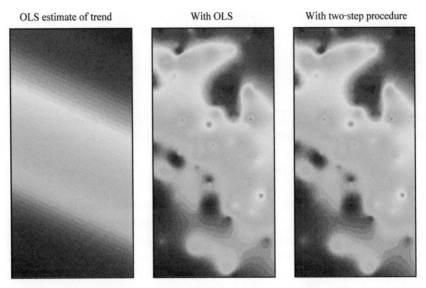

Figure I.2.8 (left) OLS estimate of trend; (middle) a universal kriging estimate based on the OLS estimate of trend; and (right) a universal kriging estimate based on the two-step procedure (using R-package RGeostats).

2.10 Nonstationary covariances

The ideal situation of Figure I.2.2(a) or the example in Figure I.2.7, with decomposition into trend and residual does not always occur. In the statistical literature, alternative approaches have been formulated relying on the direct specification of nonstationary semivariogram models (see, e.g., Switzer, 1989; Sampson and Guttorp, 1992; Higdon et al., 1999; Higdon, 2002; Mateu et al., 2013). These approaches rely on the random function model approach, often leading to certain (limited) classes of admissible covariance models. The parameter estimation of such models remains difficult for real data sets and often requires a substantial amount of data. For example, Sampson and Guttorp (1992) rely on a deformation of the underlying space where the random function is modeled as

$$Z(\mathbf{x}) = Y(f(\mathbf{x})) \tag{I.2.39}$$

where f is a bijective function modeling the transformation of space, and hence Z is considered stationary up to a deformation of space. This practice is quite common, for example in reservoir modeling (Mallet, 2002; Caers, 2005), where a coordinate transformation is needed to restore the system to its original depositional coordinates. Such deformation takes care of issues related to modeling appropriate distances; it does not necessarily turn the random function into a stationary one. The above formulation requires the estimation of both f and Y, which is not possible without multiple realizations of the random function Z. Other approaches rely on a Karhunen–Loève expansion (Loève, 1955) of the random field as follows:

$$Z(\mathbf{x}) = \sum_{i=1}^{\infty} \xi_i \sqrt{\lambda_i} \phi_i(\mathbf{x}) \qquad (I.2.40)$$

where, with eigenvalues $\sqrt{\lambda_i}$ and eigenvectors $\phi_i(\mathbf{x})$, ξ_i are independent Gaussian deviates. This decomposition then results in the following covariance:

$$C(\mathbf{x}) = \sum_{i=1}^{\infty} \lambda_i \phi_i(x) \phi_i(y) \phi_i(z) \qquad (I.2.41)$$

The series can be truncated, and a set of basic functions chosen. The problem now becomes one of choosing these basic functions (wavelets, splines or polynomials). In Chapter II.9, we discuss the choice of such basic functions in the context of both Gaussian and non-Gaussian random functions. Regardless of the elegance of such a theory, the problem is that, in practice, these methods are really only usable by the expert mathematician, where a detailed insight in the various model choices, the model parameters, as well as their often computationally demanding estimation through likelihood models is needed.

2.11 Assessment

The issues raised in this chapter on universal kriging discouraged early practitioners; for example, Armstrong (1984) stated, "Although the theory of universal kriging is mathematically correct, the difficulty in estimating the variogram and the drift at the same time make it unworkable in practice, except in those rare cases where the variogram is known a priori". More than 40 years after its introduction by Matheron (1969), UK remains practically challenged: the general model formulation, combining a deterministic drift and a correlated residual, hides technical difficulties in its implementation. The numerous different existing estimation methods all have their drawbacks: some are technically difficult to implement, and some suffer from bias. Most of the theory works fine as long as experimental variograms allow meaningful interpretation of range, sill, and anisotropy, or if a variogram is known a priori. Such a dichotomy between theory and practice has two main reasons: (1) there are too few samples, and (2) the data are on the z-variable, not the residual or mean. Hence, although the decomposition (Equation (I.2.13)) is perfectly fine from a mathematical viewpoint, it may not be practical, even if the decomposition is in fact meaningful (Figure I.2.3(b)). The two-step procedure of first getting a rough estimate of the nonstationary mean using OLS, then using the residual variogram to get a better estimate using GLS provides a way out, but it also shows that the elegant theory may need to rely on a more "messy" practice. Perhaps the most important point concerns the "separation of scales" in Equation (I.2.13) that needs to exist between the trend and the residual. The two components should be as close as possible to orthogonality for the inference to be reliable. It is in the philosophy of the UK model that the drift functions capture the large-scale fluctuations while the residual structure governs the smaller ones. The difficulty lies in the separation between these two components, which will in the absence of large amounts of data need to rely on subjective modeling choices.

As a result, applying these procedures may require considerable expertise: the modeling of trend and variograms is not always intuitive to the nonstatistician user. It is often difficult to relate trend and variogram parameters to actual physical quantities of the phenomenon being modeled. To achieve success, one has to be an expert in both statistical modeling and the area of application being considered. In this book, we provide an alternative view and an alternative set of tools where spatial continuity information is not communicated in abstract mathematical models but through explicit analog images. Such "training images" serve as a more intuitive vehicle, a fully explicit depiction of the "physical world" where the debate is now more meaningful and relevant to application experts. In Chapter I.3, we show how universal kriging with semivariograms can be turned into a universal kriging with training images.

References

Armstrong, M. 1984. Common problems seen in variograms. *Journal of the International Association for Mathematical Geology*, 16, 305–313.

Beckers, F. & Boogaets, P. 1998. Nonstationarity of the mean and unbiased variogram estimation: Extension of the weighted least-squares method. *Mathematical Geology*, 30, 223–240.

Caers, J. 2005. *Petroleum Geostatistics*, Society of Petroleum Engineers, Richardson, TX.

Chilès, J.-P. & Delfiner, P. 1999. *Geostatistics – Modeling Spatial Uncertainty*, John Wiley & Sons, New York.

Cressie, N. 1993. *Statistics for Spatial Data*, John Wiley & Sons, New York.

Delfiner, P. 1976. Linear estimation of nonstationary spatial phenomena. *In:* Guarascia, M., David, M. & Huijbregts, C. (eds.) *Advanced Geostatistics in the Mining Industry*. Reidel, Dordrecht.

Deutsch, C. V. 1993. Kriging in a finite domain. *Mathematical Geology*, 25, 41–52.

Deutsch, C. V. 1994. Kriging with strings of data. *Mathematical Geology*, 26, 623–638.

Gelfand, A., Guttorp, P., Diggle, P., & Fuentes, M. 2010. *Handbook of Spatial Statistics*. Taylor & Francis, Boca Raton, FL.

Goovaerts, P. 1997. *Geostatistics for Natural Resources Evaluation*, Oxford University Press, Oxford.

Higdon, D. 2002. Space and space-time modeling using process convolutions. In: *Quantitative Methods for Current Environmental Issues*, 37–56. Springer, London.

Higdon, D., Swall, J. & Kern, J. 1999. Non-stationary spatial modeling. *Bayesian Statistics*, 6, 761–768.

Journel, A. & Rossi, M. E. 1989. When do we need a trend model in kriging? *Mathematical Geology*, 21, 715–739.

Loève, M. 1955. *Probability Theory*, Van Nostrand, New York.

Mallet, J. L. 2002. *Geomodeling*, Oxford University Press, New York.

Mateu, J., Fernández-Avilés, G. & Montero, J. M. 2013. On a class of non-stationary, compactly supported spatial covariance functions. *Stochastic Environmental Research and Risk Assessment*, 27, 297–309.

Matheron, G. 1969. *Le krigeage universel*, Cahiers du Centre de Morphologie Mathématique, Fontainebleau, France.

Matheron, G. 1971. *The Theory of Regionalized Variables and Its Application*, Ecoles des Mines de Paris, Fontainebleau, France.

Matheron, G. 1978. *Estimer et choisir: essai sur la pratique des probabilités*. Ecoles des Mines de Paris, Fontainebleau, France.

Mukerji, T., Mavko, G. & Rio, P. 1997. Scales of reservoir heterogeneities and impact of seismic resolution on geostatistical integration. *Mathematical Geology*, 29, 933–950.

Omre, H. 1987. Bayesian kriging: Merging observations and qualified guesses in kriging. *Mathematical Geology*, 19, 25–39.

Rivoirard, J. & Romary, T. 2011. Continuity for kriging with moving neighborhood. *Mathematical Geosciences*, 43, 469–481.

Sampson, P. & Guttorp, P. 1992. Nonparametric estimation of nonstationary spatial covariance structure. *Journal of the American Statistical Association*, 87, 108–119.

Stein, M. 1999. *Interpolation of Spatial Data: Some Theory for Kriging*, Springer, Berlin.

Switzer, P. 1989. Non-stationary spatial covariances estimated from monitoring data. *In:* Armstrong, M. (ed.) *Geostatistics*, Kluwer Academic, Dordrecht, 127–138.

CHAPTER 3

Universal kriging with training images

3.1 Choosing for random function theory or not?

Most traditional geostatistics relies on random function theory to formulate problems of spatial estimation. This theory allows formulating the problem of spatial estimation, such as the system of ordinary kriging equations. Actual practice requires repeated data to solve the problem numerically, to provide actual results on which decisions can be made. Fundamental to the theory in Chapter I.2 is the notion of expectation, whether it is an expected value, a covariance, or a semivariogram. Assumptions of stationarity are required to turn theoretical models into usable, practical numerical equations. In fact, this assumption is true for all of statistical science dealing with the notion of expectation. Averaging cannot proceed without repeated data, some form of repetition or replication.

Anyone with some experience in teaching or using geostatistics in daily practice understands the often wide gap between elegant theory and ad hoc decisions, such as borrowing variogram model parameters from "elsewhere" that are needed to make geostatistics practical. The example experimental semivariogram of Figure I.2.5 is one such illustration. Dealing with trends in nonstationary kriging methods is another case. Often, considerable "finesse" is required to tune kriging systems by carefully selecting how many and which data points are used to avoid artifacts. The finite domain kriging in Figure I.2.6 is an example of such tuning. One viewpoint is that, regardless of this ad hoc practice, theory is needed to allow formulating and studying the problem in a consistent framework, without the tediousness of dealing with actual data and the need to make specific data-related approximations. This is a common approach in many areas of science, and we do not doubt its great value.

There could possibly be equal value in turning the problem into a more "engineering" approach from the get-go as follows. With theory come assumptions, as clearly demonstrated in the derivation of the ordinary kriging equations, and such assumptions can often not be verified by any data; they are assumptions fundamental to the theory only. In reality, there is no random function, just a single unique truth and some data on it; and, perhaps in addition, some analog

Multiple-point Geostatistics: Stochastic Modeling with Training Images, First Edition. Gregoire Mariethoz and Jef Caers.
© 2015 John Wiley & Sons, Ltd. Published 2015 by John Wiley & Sons, Ltd.
Companion website: www.wiley.com/go/caers/multiplepointgeostatistics

ideas about that truth or some analog information. It is therefore equally valuable to attempt to write equations without theoretical considerations a priori (e.g., the notion of expectation, stationarity, or ergodicity) and directly rely on the available information. Perhaps there is interest in devising methods with explicitly verifiable conditions, limiting any unverifiable assumption as much as possible. In fact, this idea is not new, and it was first proposed by Matheron (1978) as "schéma glissant" and later further suggested as "deterministic geostatistics" by Journel (1996). Here we take that work a step further: we provide actual implementations of these ideas and discuss them within the context of using a training image instead of a semivariogram model, or even without relying on any notion of probability theory, such as expectations or random variables.

3.2 Formulation of universal kriging with training images

3.2.1 Zero error-sum condition

Consider the same estimation problem as in Figure I.2.2(b) involving $n = 5$ data points around a target location $\mathbf{x}$. The data values, with their configuration around the location to be estimated, are termed the "data event" dev,

$$\mathbf{dev}(\mathbf{x}) = \{z(\mathbf{x}_1) = z(\mathbf{x} + \mathbf{h}_1); z(\mathbf{x}_2) = z(\mathbf{x} + \mathbf{h}_2); \ldots ; z(\mathbf{x}_n) = z(\mathbf{x} + \mathbf{h}_n)\}$$

(I.3.1)

the data configuration, or template (see Section II.2.3), is denoted as

$$\mathbf{dcf} = \{\mathbf{x} + \mathbf{h}_1; \mathbf{x} + \mathbf{h}_2; \ldots ; \mathbf{x} + \mathbf{h}_n\}$$

(I.3.2)

For this specific case, we have

$$\{\mathbf{dcf}; \mathbf{dev}(\mathbf{x})\} = \begin{bmatrix} 39 & 78 & -0.741 \\ 17 & 75 & -0.436 \\ 30 & 72 & 0.090 \\ 48 & 79 & -0.072 \\ 42 & 58 & -0.239 \end{bmatrix}$$

(I.3.3)

Imagine these to be the only data over the area available, meaning that no meaningful experimental covariance can be calculated or modeled from these data. Consider instead an analog image available, shown in Figure I.3.1, which provides an analog to the field of study: for short, a training image (*TI*). This TI is deemed representative, by a domain expert, for the spatial variation of the study problem. The qualifier "deemed representative" will be revisited at several points in this book; for now, we accept its perhaps vague definition.

One way of solving the estimation problem is to calculate the experimental variogram of the training image, model it, and use it in ordinary kriging or universal kriging; however, the same problems of trend versus residual would need

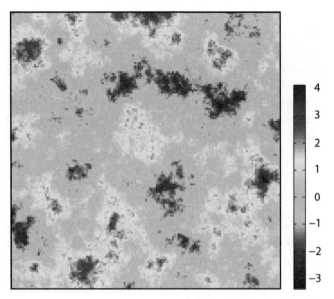

Figure I.3.1 An analog data set considered relevant for the domain being modeled. The size of this image (250×250) need not be the same as the model grid size.

to be addressed. Can this problem therefore be formulated and solved without relying on any kriging equation and, hence, rely on the notion of expectation? How would one formulate conditions on the estimator that can be verified and debated based on actual explicit information, such as Figure I.3.1, rather than relying on theoretical constructs?

In the same way as is done for linear estimation, we develop an "estimator by analogy". This new estimator, denoted $z^\#$, looks exactly the same as the random function–based estimator z^*:

$$z^\#(\mathbf{x}) = \sum_{\alpha=1}^{n} \lambda_\alpha z(\mathbf{x}_\alpha) \qquad (\text{I.3.4})$$

The choice of the dash (#) symbol is perhaps not as arbitrary as it looks, because we will develop methods based on "counting" and "frequencies", rather than random variables and probabilities. As a consequence, we will not write any capital letter $Z^\#$ or $Z(\mathbf{x}_\alpha)$, because we do not rely on the notion of random variables.

We can no longer rely on theoretical constructs such as unbiasedness, which would call for the definition of expectation. Instead, we introduce the following condition analogous to unbiasedness:

Zero error sum condition: "given an exhaustively sampled training image, similar to the study area, if estimation were to be performed over that image, with the same configuration of locations as the given data event, then the sum of the differences between the estimators and the actual values of the training image equals zero".

In other words, the condition calls for the explicit specification of another area, the *training image* that is exhaustively sampled and *deemed representative* of the area being studied. This explicit specification is avoided in the traditional unbiasedness condition and simply relegated to the $E[]$ (expectation) operator, without any explicit statement of what one is actually averaging over (see the above discussion). Mathematically, the zero error-sum condition is written as follows:

$$\sum_{\mathbf{x} \in TI_{dcf}} \left(z_{TI}^{\#}(\mathbf{x}) - z_{TI}(\mathbf{x}) \right) = 0 \tag{I.3.5}$$

where $z_{TI}(\mathbf{x})$ is a value in the training image at location $\mathbf{x}$. Clearly, the summing cannot be specified over the entire training image, only over the training image eroded (in the mathematical morphological sense) with the data configuration **dcf**, meaning that only those pixels of the TI are retained that can contain the entire data event. TI_{dcf} is this eroded area.

The zero error-sum condition is further developed as follows:

$$\sum_{\mathbf{x} \in TI_{dcf}} \left(z_{TI}^{\#}(\mathbf{x}) - z_{TI}(\mathbf{x}) \right) = \sum_{\mathbf{x} \in TI_{dcf}} \left(\sum_{\alpha=1}^{n} \lambda_{\alpha} z_{TI}(\mathbf{x} + \mathbf{h}_{\alpha}) - z_{TI}(\mathbf{x}) \right) = 0$$

$$\Rightarrow \sum_{\alpha=1}^{n} \lambda_{\alpha} \sum_{\mathbf{x} \in TI_{dcf}} (z_{TI}(\mathbf{x} + \mathbf{h}_{\alpha})) = \sum_{\mathbf{x} \in TI_{dcf}} z_{TI}(\mathbf{x})$$

$$\Rightarrow \sum_{\alpha=1}^{n} \lambda_{\alpha} \left(\frac{1}{\# \left| TI_{dcf} \right|} \sum_{\mathbf{x} \in TI_{dcf}} (z_{TI}(\mathbf{x} + \mathbf{h}_{\alpha})) \right) = \frac{1}{\# \left| TI_{dcf} \right|} \sum_{\mathbf{x} \in TI_{dcf}} z_{TI}(\mathbf{x}) \tag{I.3.6}$$

$\# \left| TI_{dcf} \right|$ is simply the number of grid cells or locations $\mathbf{x}$ in the eroded *TI*. It appears now that two averages are present:

$$av_{TI_{dcf}} = \frac{1}{\# \left| TI_{dcf} \right|} \sum_{\mathbf{x} \in TI_{dcf}} z_{TI}(\mathbf{x}): \text{ the average of the } z\text{-values in the } TI \text{ eroded}$$

by **dcf**.

$$av_{TI_{dcf}}(\mathbf{h}_{\alpha}) = \left(\frac{1}{\# \left| TI_{dcf} \right|} \sum_{\mathbf{x} \in TI_{dcf}} (z_{TI}(\mathbf{x} + \mathbf{h}_{\alpha})) \right): \text{ the average of the } z\text{-values in the}$$

TI, first eroded by the **dcf**, then translated by $-\mathbf{h}_{\alpha}$.

The zero error-sum condition therefore becomes

$$\sum_{\alpha=1}^{n} \lambda_{\alpha} av_{TI_{dcf}}(\mathbf{h}_{\alpha}) = av_{TI_{dcf}} \tag{I.3.7}$$

The comparison between the zero error-sum condition and the unbiasedness condition is compelling. The unbiasedness condition relies on expectation, yet it is not explicitly stated how such expected value would actually be calculated or what one is averaging over. The zero error-sum relies on averaging over an explicitly stated analog image. Note that due to the translation operation $-\mathbf{h}_\alpha$, any differences in these averages will reflect a "trend" in the training image. However, we need to be careful in using the term "trend" and state explicitly that such a trend is on the spatial average only, not on some other property of the training image. Specifically for the simple case, we calculate six averages:

$$av_{TI_{def}}\left(\mathbf{h}_1 : \mathbf{h}_n\right) = [0.079 \quad 0.062 \quad 0.082 \quad 0.081 \quad 0.117]$$

$$av_{TI_{def}} = 0.098$$

3.2.2 Minimum sum of square error condition

Similar to the development of kriging, we specify a condition of loss:

Minimum sum of square error condition: "given an exhaustively sampled training image, similar to the study area, if estimation were to be performed over that image, with the same configuration of locations as the given data event, then the sum of the squared differences between the (linear) estimators and the actual values of that training image is the smallest possible".

Mathematically, this translates into

$$\min \sum_{\mathbf{x}\in TI_{def}} \left(z_{TI}^{\#}(\mathbf{x}) - z_{TI}(\mathbf{x})\right)^2 \tag{I.3.8}$$

This sum of squared error can be further developed as follows:

$$
\begin{aligned}
sse &= \sum_{\mathbf{x}\in TI_{def}} \left(z_{TI}^{\#}(\mathbf{x}) - z_{TI}(\mathbf{x})\right)^2 \\
&= \sum_{\mathbf{x}\in TI_{def}} \left(\sum_{\alpha=1}^{n} \lambda_\alpha z_{TI}(\mathbf{x}+\mathbf{h}_\alpha) - z_{TI}(\mathbf{x})\right)^2 \\
&= \sum_{\mathbf{x}\in TI_{def}} \left(\sum_{\alpha=1}^{n}\sum_{\beta=1}^{n} \lambda_\alpha \lambda_\beta z_{TI}(\mathbf{x}+\mathbf{h}_\alpha) z_{TI}(\mathbf{x}+\mathbf{h}_\beta) - 2\sum_{\alpha=1}^{n} \lambda_\alpha z_{TI}(\mathbf{x}+\mathbf{h}_\alpha) z_{TI}(\mathbf{x}) + z_{TI}^2(\mathbf{x})\right) \\
&= \sum_{\alpha=1}^{n}\sum_{\beta=1}^{n} \lambda_\alpha \lambda_\beta \sum_{\mathbf{x}\in TI_{def}} z_{TI}(\mathbf{x}+\mathbf{h}_\alpha) z_{TI}(\mathbf{x}+\mathbf{h}_\beta) - 2\sum_{\alpha=1}^{n} \lambda_\alpha \sum_{\mathbf{x}\in TI_{def}} z_{TI}(\mathbf{x}+\mathbf{h}_\alpha) z_{TI}(\mathbf{x}) \\
&\quad + \sum_{\mathbf{x}\in TI_{def}} z_{TI}^2(\mathbf{x}) \tag{I.3.9}
\end{aligned}
$$

We introduce now a notation for the various sums of products (sop):

$$sop_{TI}(\mathbf{h}_\alpha, \mathbf{h}_\beta) = \frac{1}{\#|TI_{def}|} \sum_{\mathbf{x}\in TI_{def}} z_{TI}(\mathbf{x}+\mathbf{h}_\alpha) z_{TI}(\mathbf{x}+\mathbf{h}_\beta) \tag{I.3.10}$$

Hence, the *sse* can be written as

$$\frac{sse}{\#\left|TI_{dcf}\right|} = \sum_{\alpha=1}^{n}\sum_{\beta=1}^{n} \lambda_\alpha \lambda_\beta sop_{TI}(\mathbf{h}_\alpha, \mathbf{h}_\beta) - 2\sum_{\alpha=1}^{n}\lambda_\alpha sop_{TI}(\mathbf{h}_\alpha, \mathbf{0}) + sop_{TI}(\mathbf{0}, \mathbf{0}) \quad \text{(I.3.11)}$$

The notation *sop* is used to emphasize that the notation of covariance (requiring expectation) is not used or needed here. Minimization of the *sse* under the linear constraint proceeds in the exact same way as for kriging and results in the following normal system of equations:

$$\begin{cases} \displaystyle\sum_{\beta=1}^{n} \lambda_\beta sop_{TI}\left(\mathbf{h}_\alpha, \mathbf{h}_\beta\right) + \mu\, av_{TI_{dcf}}\left(\mathbf{h}_\alpha\right) = sop_{TI}\left(\mathbf{h}_\alpha, \mathbf{0}\right) \; \alpha = 1 \ldots n \\ \displaystyle\sum_{\alpha=1}^{n} \lambda_\alpha av_{TI_{dcf}}\left(\mathbf{h}_\alpha\right) = av_{TI_{dcf}} \end{cases} \quad \text{(I.3.12)}$$

The minimum achieved by solving this system is, then, the error variance under this procedure, and it is expressed as

$$sse_{min} = sop_{TI}\left(\mathbf{0}, \mathbf{0}\right) - \sum_{\alpha=1}^{n}\lambda_\alpha sop_{TI}\left(\mathbf{h}_\alpha, \mathbf{0}\right) - \mu\, av_{TI_{dcf}} \quad \text{(I.3.13)}$$

The sum of products for the five data cases is as follows:

$$sop(\mathbf{h}_\alpha, \mathbf{h}_\beta) = \begin{bmatrix} 0.963 & 0.155 & 0.406 & 0.485 & 0.099 \\ 0.155 & 1.003 & 0.366 & 0.098 & 0.065 \\ 0.406 & 0.366 & 0.983 & 0.150 & 0.198 \\ 0.485 & 0.098 & 0.150 & 0.960 & 0.022 \\ 0.099 & 0.065 & 0.198 & 0.022 & 0.962 \end{bmatrix}; sop(\mathbf{h}_\alpha, \mathbf{0}) = \begin{bmatrix} 0.079 \\ 0.062 \\ 0.082 \\ 0.081 \\ 0.117 \end{bmatrix}$$

Hence, the system to be inverted looks as follows:

$$\begin{bmatrix} 0.963 & 0.155 & 0.406 & 0.485 & 0.099 & 0.079 \\ 0.155 & 1.003 & 0.366 & 0.098 & 0.065 & 0.062 \\ 0.406 & 0.366 & 0.983 & 0.150 & 0.198 & 0.082 \\ 0.485 & 0.098 & 0.150 & 0.960 & 0.022 & 0.081 \\ 0.099 & 0.065 & 0.198 & 0.022 & 0.962 & 0.117 \\ 0.079 & 0.062 & 0.082 & 0.081 & 0.117 & 0.000 \end{bmatrix}\begin{bmatrix} \lambda_1 \\ \lambda_2 \\ \lambda_3 \\ \lambda_4 \\ \lambda_5 \\ \mu \end{bmatrix} = \begin{bmatrix} 0.504 \\ 0.145 \\ 0.397 \\ 0.380 \\ 0.339 \\ 0.098 \end{bmatrix}$$

Solving the linear system provides

$$\begin{bmatrix} \lambda_1 \\ \lambda_2 \\ \lambda_3 \\ \lambda_4 \\ \lambda_5 \\ \mu \end{bmatrix} = \begin{bmatrix} 0.322481 \\ 0.002136 \\ 0.203923 \\ 0.226853 \\ 0.321342 \\ -0.40362 \end{bmatrix}$$

Resulting in the estimate and sse_{min}:

$$z^{\#} = -0.314 \text{ and } sse_{min} = 0.551$$

3.3 Positive definiteness of the sop matrix

Solution of the system in Equation (I.3.12) requires that the matrix of *sop* values is positive definite. It is logical and intuitive that any experimental statistics properly eroded over an existing (real) image will provide such a system. In a similar sense, the experimental covariance carries the same property. A formal mathematical proof can also be given, noting that by definition,

$$\frac{1}{\#\left|TI_{def}\right|} \sum_{\mathbf{x} \in TI_{def}} \left(z_{TI}^{\#}(\mathbf{x})\right)^2 \geq 0$$

As a consequence, one finds

$$\frac{1}{\#\left|TI_{def}\right|} \sum_{\mathbf{x} \in TI_{def}} \sum_{\alpha=1}^{n} \sum_{\beta=1}^{n} \lambda_\alpha \lambda_\beta z_{TI}\left(\mathbf{x}+\mathbf{h}_\alpha\right) z_{TI}\left(\mathbf{x}+\mathbf{h}_\beta\right) \geq 0$$

or

$$\sum_{\alpha=1}^{n} \sum_{\beta=1}^{n} \lambda_\alpha \lambda_\beta sop_{TI}\left(\mathbf{h}_\alpha, \mathbf{h}_\beta\right) \geq 0 \quad \forall \lambda_\alpha, \lambda_\beta$$

Hence, the matrix of the sum of products is positive definite and has only one solution.

3.4 Simple kriging with training images

In linear geostatistics, simple kriging is a method whereby in addition to stationarity on the semivariogram, one assumes the mean to be stationary and known over the domain. In such cases, the mean becomes an additional "datum" that can help provide estimates with lesser estimation variance:

$$z^*(\mathbf{x}) = \lambda_0 m + \sum_{\alpha=1}^{n} \lambda_\alpha z(\mathbf{x}_\alpha) \text{ with } E\left[Z(\mathbf{x})\right] = m \quad \forall \mathbf{x} \qquad (I.3.14)$$

The unbiasedness condition then results into

$$\lambda_0 = 1 - \sum_{\alpha=1}^{n} \lambda_\alpha \qquad (I.3.15)$$

This then provides the following estimator:

$$z^*(\mathbf{x}) = m + \sum_{\alpha=1}^{n} \lambda_\alpha \left(z(\mathbf{x}_\alpha) - m\right) \qquad (I.3.16)$$

Simple kriging has its equivalent in kriging with training images. Consider now that, instead of an expected value, a properly eroded average over the training image is used in the estimator:

$$z^\#(\mathbf{x}) = av_{TI_{dcf}} + \sum_{\alpha=1}^{n} \lambda_\alpha \left(z(\mathbf{x}+\mathbf{h}_\alpha) - av_{TI_{dcf}}(\mathbf{h}_\alpha) \right) \tag{I.3.17}$$

Note that we use the eroded average, not the full training image average. The zero sum condition now becomes

$$\sum_{\mathbf{x} \in TI_{dcf}} \left(z^\#_{TI}(\mathbf{x}) - z_{TI}(\mathbf{x}) \right) = \sum_{\mathbf{x} \in TI_{dcf}} \left(av_{TI_{dcf}} + \sum_{\alpha=1}^{n} \lambda_\alpha \left(z(\mathbf{x}+\mathbf{h}_\alpha) - av_{TI_{dcf}}(\mathbf{h}_\alpha) \right) - z_{TI}(\mathbf{x}) \right)$$

$$= \# \left| TI_{dcf} \right| av_{TI_{dcf}} + \sum_{\alpha=1}^{n} \lambda_\alpha \sum_{\mathbf{x} \in TI_{dcf}} z_{TI}(\mathbf{x}+\mathbf{h}_\alpha) - \# \left| TI_{dcf} \right| \sum_{\alpha=1}^{n} \lambda_\alpha av_{TI_{dcf}}(\mathbf{h}_\alpha) - \sum_{\mathbf{x} \in TI_{dcf}} z_{TI}(\mathbf{x})$$

$$= av_{TI_{dcf}} + \sum_{\alpha=1}^{n} \lambda_\alpha av_{TI_{dcf}}(\mathbf{h}_\alpha) - \sum_{\alpha=1}^{n} \lambda_\alpha av_{TI_{dcf}}(\mathbf{h}_\alpha) - av_{TI_{dcf}} = 0 \tag{I.3.18}$$

Hence the estimator, Equation (I.3.17), verifies the zero sum condition.

3.5   Creating a map of estimates

Consider now the problem of creating maps of estimates using universal kriging with a training image. We use a constant search radius of 50; otherwise, there is no semivariogram estimation or modeling. In Figure I.3.2(a), we observe the same artifacts as before, and they are perhaps even more prevalent than in ordinary kriging (Figure I.2.6(b)). The reason for this is that the sums of product matrices, Equation (I.3.10), are constructed anew each time (due to the

(a)   (b)

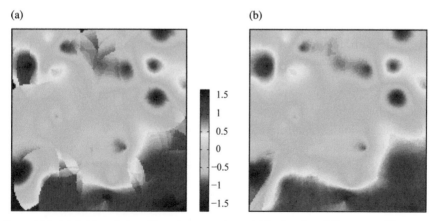

Figure I.3.2 (a) Artifacts induced by using a moving neighborhood; and (b) corrected by using finite domain kriging and by requiring a minimum of 12 neighboring points.

changing data configuration) and thereby eroding the training image over a different area. In traditional kriging, the variogram model used is fixed, and it is applied to all locations. Hence, one can expect a larger fluctuation in these sums of product matrices as compared to the kriging covariance matrices. To remove these artifacts, we apply the finite domain kriging approach and we use a flexible neighborhood by requiring a minimum of 12 kriging points; see Figure I.3.2(b).

3.6 Effect of the size of the training image

The training image in Figure I.3.1 is of a different size than the domain of study. We investigate how the size of the image affects the universal kriging with training image results. Consider therefore the use of a training image of the same size as the domain of study (150×150), obtained simply by cropping the training image in Figure I.3.1. The size of the training image does not have much effect on the estimates; see Figure I.3.3. The main impact lies in the minimum sum of squared error maps obtained through evaluating expression (Equation (I.3.13)) (the equivalent of a kriging variance). Figure I.3.4 compares the kriging variance obtained using a single global neighborhood with the minimum errors calculated using Equation (I.3.13). As the size of the image increases, the minimum *sse* becomes close to the global kriging results. This result is in line with the ergodic property on which traditional kriging relies: as the domain size increases, the experimental variogram of the training images converges to the "model" variogram of the random function (infinite domain ergodicity). In this case, the convergence appears to go fairly rapidly, far less than "infinite" size. Which minimum error variance is now the correct one? The ergodic property is a choice of the model, as discussed before, hence the reference "kriging

(a)   (b)

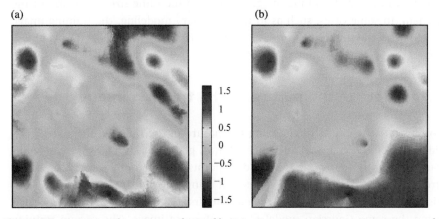

Figure I.3.3 Comparing the estimates obtained by (a) using a 150×150 size training image and (b) using a 250×250 size training image.

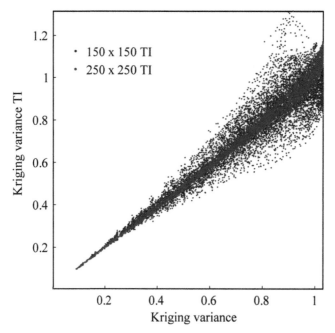

Figure I.3.4 Minimum error maps (kriging variances) compared to the kriging variance obtained using global ordinary kriging.

variance" in Figure I.3.4 need not be seen as the "correct estimation variance"; it could instead be seen as some limit case. The size of the training image is therefore equivalently regarded as a model choice. The dependency of the universal kriging with training image variance on the chosen size of the training image is, however, an important one and will be revisited at several instances in this book.

As a final note, one can, instead of choosing the size of the training image, choose a number of training images each with the same size as the size of the domain. In some cases, such as in nonstationary modeling, the training image has to be of the same size as the domain in order for the analogy to make any sense (see Section I.3.8 and Chapter II.5).

3.7   Effect of the nature of the training image

To illustrate the effects of changing the nature of the training image, we consider the reference shown in Figure I.3.5(a), generated using a spherical semivariogram with a range of 35. 100 data points are randomly sampled; see Figure I.3.5(b). To demonstrate the effects of using a dissimilar training image, two distinctly different TIs are constructed. The first (Figure I.3.6(a)) was generated using the same Gaussian model of the reference image. The second training image (Figure I.3.6(b)) was generated such that it contains distinct disks of higher

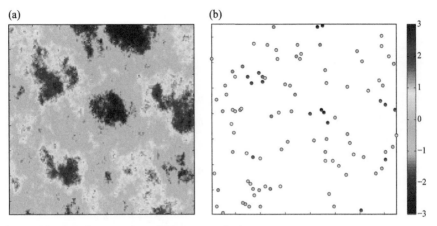

Figure I.3.5 (a) Reference case and (b) 100-sample data.

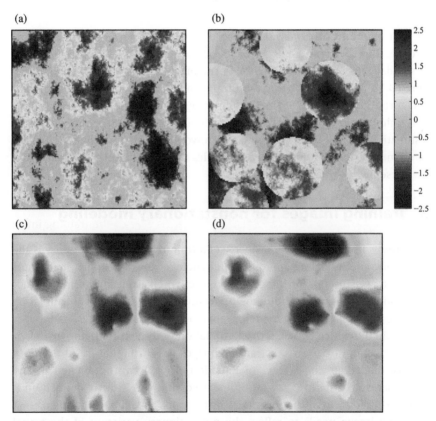

Figure I.3.6 (a–b) Two visually dissimilar training images each with their estimated map from 100 sample data (c–d).

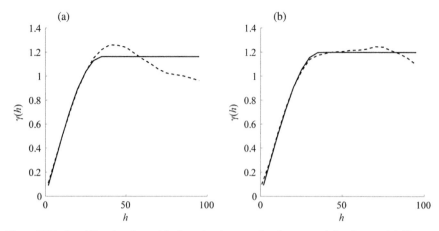

Figure I.3.7 Omnidirectional empirical semivariograms for the two training images (a) Figure I.3.6(a) and (b) Figure I.3.6(b) along with fitted semivariograms using package RGeoS. Both have nugget 0, a range of approximately 35, and sill of 0.985.

value than the rest of the map (like crop circles). However, despite their visual difference, the two training images have some similarities. Their histograms are both standard normal; their experimental semivariograms are also very similar (Figure I.3.7). Consequently, the resulting estimates are basically the same (compare Figure I.3.6(c) and I.3.6(d)). This illustrates clearly that in universal kriging with training images, only two-point statistics (sums of products) are borrowed for estimation. Higher-order statistics, which would be different for both images, have no impact. In multiple-point geostatistics (MPS), evidently, the idea is also to use these higher-order statistics.

3.8 Training images for nonstationary modeling

Stationarity is not necessarily a property of a training image. Stationarity is the property of some modeling approach (and choice), not of a single or multiple deterministic sets of data or images. Therefore, when using the term "nonstationary training images", it should be understood that this is short formulation for the use of a nonstationary modeling approach with such images.

The main difference between the formulation of universal kriging by means of covariances and by means of training images lies in the nature of the nonstationary modeling. The system of Equation (I.3.12) can be applied to any image, whatever its nature: the set of linear equations is directly defined on z, and no decomposition of the form (I.2.14) is made. This, however, means that the training image needs to reflect the nonstationary variation of the actual true field. The modeling questions are now relegated to the construction of such training images, instead of the residual or nonstationary covariances. An entire chapter of this book (Chapter II.5) is devoted to creating and using such

(a)   (b)

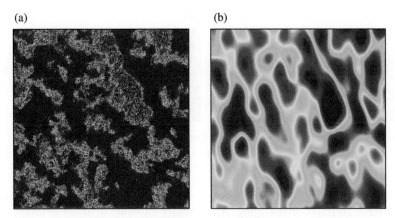

Figure I.3.8 (a) An exhaustive DEM deemed representative for the Walker Lake area. The size of the training image is 400×400. (b) the result of a simple smoother applied to (a).

"nonstationary training images" in stochastic modeling. In general, stochastic modeling with training images heavily borrows methods from pattern recognition, image analysis and processing, and aspects of computer graphics, as will become clear in Part II.

Consider the Walker Lake study in Figure I.1.1. The true field has a complexity that is not easily modeled with parametric functions. Instead, one may have the DEM of other areas that are exhaustively known, such as in Figure I.3.8(a). When comparing these two images (Figure I.1.1(left) and Figure I.3.8(a)), one can take two perspectives: local and global comparison. A global comparison entails that statistics calculated over the entire image compare well with statistics over the other image, whether this is a mean, covariance, or higher-order moment, or perhaps based on a wavelet, curvelet, or spectral-based comparison. A local comparison involves comparing images within co-located areas. The latter means that images need to have the same size. In Chapter II.8, we will present various quantitative global and local comparisons between exhaustive images and between images and data sets. The latter will be used to validate the training image with field data.

In order to apply universal kriging with training images in the nonstationary case, the training images need to be local representations. The training image of Figure I.3.8(a) provides only a global representation. To differentiate these two types of training images, we denote a global representation as $z_{TI,global}$, while what is desired for universal kriging with a training image is $z_{TI,local}$. The solution proposed is to transform such global training images into local ones using some form of image transformation. Several such methods will be presented in Part II of this book and are developed specifically for geostatistical applications.

To illustrate one such method, we first revisit the trend example of Figure I.2.7. Consider that we have at our disposal a training image (Figure I.3.9(a)), and with that image is provided its trend, Figure I.3.9(b). We term

(a)   (b)

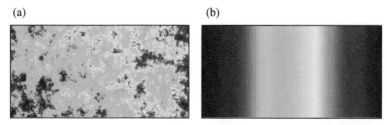

Figure I.3.9 (a) A training image for the simple trend case; and (b) its auxiliary variable.

this trend the "auxiliary variable", because its use will be quite different from the use of trends in traditional kriging. An auxiliary variable, denoted as aux_{TI}, can be seen as an "explanatory variable", informing or "explaining" some basic variation of the training image $z_{TI,global}$. The actual unit of this variable does not matter. and it is typically scaled to be between 0 and 1. This variable may be provided by the user, or it may be extracted from the training image using image analysis methods. There may be several such variables, each informing a particular aspect of the image (e.g. a local mean, a local orientation and a local scaling). Clearly, Figure I.3.9(a) is not a local training image: the domain size is 200×100, whereas the domain of study is 100×200. The quest is for a general solution to create $z_{TI,local}$.

This very question is also treated in the computer graphics literature (Hertzmann et al., 2001) where the goal is to create "image analogies"; in short, the idea is to generate an object or image B′ that satisfies the following logical relation:

$$A : A' :: B : B'$$

where B stands to B′ in the same vein as A stands to A′ (see Section II.6.4 for details). This simply means that the relationship between two objects, A and A′, is the same as the relation between two other objects, B and B′. The user provides the algorithm with three inputs: the training image A $(z_{TI,global})$, the auxiliary variable of the training image A′ (aux_{TI}), and the auxiliary variable of the domain B′ (aux). The resulting stochastic transformation then provides one or more $z_{TI,local}$.

In the simple trend example, we use as the auxiliary variable of the domain the ordinary least squares (OLS) estimate of Figure I.2.8(left). The question therefore is to create an image that looks like Figure I.3.9(c) but has the trend of Figure I.2.8(left) without relying on any generating algorithm that created these images. Several algorithms are presented in Part II.6 and Part II.7 that achieve this transformation. Figure I.3.10 shows three realizations that are generated using the direct sampling algorithm (Mariethoz et al., 2010). Any of these images can now serve as either a single or as a set of training images for universal kriging with training images.

(a)   (b)

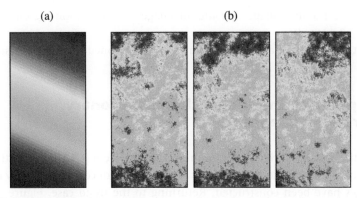

Figure I.3.10 Three nonstationary training images (b) for the simple trend case using the OLS estimate (a) as an auxiliary variable.

In cases such as Walker Lake (Figure I.3.8 (a)), the auxiliary variable belonging to the training image is not provided and hence needs to be generated. Since what we need to provide in kriging is some form of smoothly varying property (a trend), we achieve this by means of a simple local smoother. Next, we need to generate an auxiliary variable for the domain. This can be done by transforming the Walker Lake sample data into auxiliary data at the sample locations. We use a simple histogram transformation to achieve this task. Then a smooth interpolator is applied on these data to create a smooth auxiliary variable for the domain. The following set of steps summarize the procedure for creating the Walker Lake nonstationary training images:

1 Smooth the Walker Lake image training image (Figure I.3.8(a)) using a local smoother to obtain its auxiliary variable (Figure I.3.8(b)).
2 Transform the 100 Walker Lake data (Figure I.1.1(right)) into the same histogram as the auxiliary variable of Figure I.3.8(b) to obtain Walker Lake auxiliary sample data.
3 Perform a biharmonic spline interpolation to create the auxiliary variable of the domain; see Figure I.3.11(a).

(a)   (b)   (c)   (d)

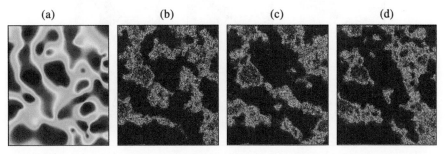

Figure I.3.11 Biharmonic spline (a) to create the domain auxiliary variable for the Walker Lake with three nonstationary training images (b–d) generated with direct sampling.

4 Create the training image(s) using an image transformation algorithm; see Figure I.3.11(b–d).

Note that the final training images are not conditioned to any of the sample data.

3.9 Spatial estimation with nonstationary training images

Because the universal kriging with training images system (I.3.12) is directly formulated on the z-variable, the system is readily solved once nonstationary training images have been constructed. Returning to the trend case, Figure I.3.12 compares the result with both universal kriging with a global neighborhood and ordinary kriging with a local neighborhood. The R-package RGeos was used for universal kriging, the semivariogram fitting presented in Figure I.3.13. Visually, the results are quite similar. To study the properties of the training image approach a bit more closely, a set of 100 random samples, each containing n sample locations, are extracted from the unknown truth; see Figure I.2.7(left). n is varied as follows: $n = 50, 100, 150,$ and 200. Both universal kriging with covariances and universal kriging with training images are repeated on each of the 100-sample sets, and then a relative mean squared error is calculated as follows:

$$\text{ReMSE} = \frac{1}{N\sigma^2} \sum_{i=1}^{N} \left(z^*(\mathbf{x}_i) - z(\mathbf{x}_i) \right)^2 \qquad (\text{I.3.19})$$

which represents a measure of difference between estimates and truth. N is the total number of grid cells being etimated, and σ^2 the variance. This ReMSE is averaged over all sample sets. ReMSE is close to unity when few samples are available and then tends to zero when more samples are used. Table I.3.1 shows that for this case, there is little difference between using just universal kriging

(a) (b) (c)

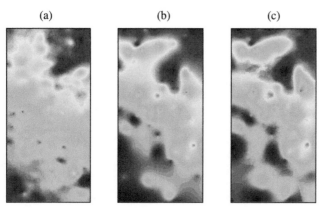

Figure I.3.12 Comparing the estimates obtained by (a) spatial estimation with training images; (b) universal kriging; and (c) ordinary kriging with local moving neighborhood.

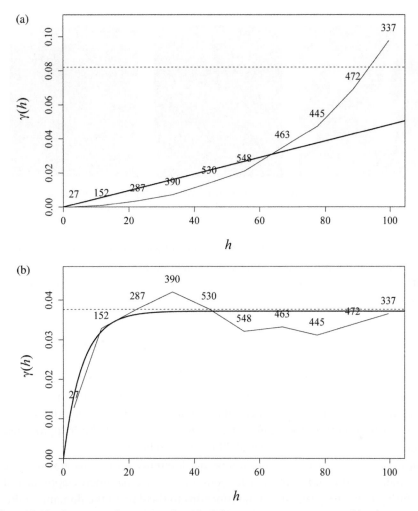

Figure I.3.13 The empirical raw (a) and residual (b) semivariograms computed for the nonstationary case. The automatic semivariogram fitting function in RGeostat was used. The number of pairs used for the calculation are plotted next to the experimental semivariogram.

Table I.3.1 Relative mean squared error ($100 \times$ variance over several sample sizes) for universal kriging (with OLS estimate of trend and with the two step GLS procedure) and using Universal Kriging with Training Images (UK-TI)

No. of samples	25	50	100	200
UK with OLS	0.88	0.77	0.65	0.64
UK with GLS	0.89	0.78	0.66	0.54
UK-TI	0.95	0.83	0.72	0.60

Note: In UK-TI, the finite correction was applied.

(a)   (b)   (c)

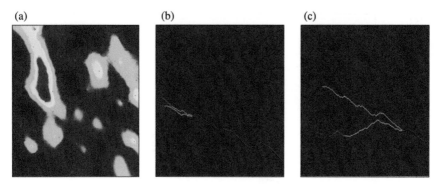

Figure I.3.14 (a) Estimate for Walker Lake using universal kriging with training image in Figure I.3.11(b); (b) path for minimal effort and (c) path for maximal effort.

with OLS estimates of the trend, using universal kriging with the two-step procedure, and using universal kriging with a single training image.

Figure I.3.14 shows the estimates for Walker Lake produced using the training image of Figure I.3.11(c).

3.10 Summary of methodological differences

The examples studied show a clear methodological difference between the two approaches: UK-v, a universal kriging relying on the variogram and random function theory, and, UK-TI, a method formulated outside any such theory and relying instead on the specification of an image. In comparing both approaches, a number of areas of methodological difference can be identified.

Modeling effort: Perhaps the most important difference is in terms of where the modeling effort takes place. UK-v often calls for a parametric approach of fitting semivariogram and trend functions to the data of the domain. Training image methods rely on the availability of images that are reflective of spatial variability of the domain but are not calculated from data, as is the case for variograms. The Walker Lake TI was simply deemed similar to variation of the variable of interest in the modeling domain. As a result, UK-TI separates conditioning from spatial model specification. Because training images are borrowed from elsewhere, a potentially important issue, in cases of more dense data, is the consistency between the training image and the actual data. In the cases presented here, this issue is not of great concern because the amount of sample data is small. The obvious question, relevant also for later developments in this book, lies in the possible inconsistency between data and training image. The specific validation in the Walker Lake case is very simple: verify whether both histogram and variogram are consistent between sample data and training image. Any deviation in histogram is easily taken care of through histogram identification (anamorphosis). The limited sample data and hence resulting noisy variogram leave considerable flexibility in choosing a training image. However, when the data amount increases, this possible inconsistency becomes important and needs to be addressed head-on. Chapter II.8 is devoted to this issue.

Nonstationary modeling: UK-v relies on a decomposition into trend and residual in the modeling domain (see the above discussion), whereas UK-TI relies on a decomposition in the training image domain. In UK-TI, if a trend is present, then the training image needs to contain this trend, and modeling of any trend takes place in that domain. The single "universality condition" applies regardless of the nature of the training image. If any trend is present in the training image, then this trend will be reflected through the eroded averages. The topic of dealing with nonstationary training images is therefore particularly relevant to make the UK-TI method applicable in real cases. In actual applications, the explicit construction of such nonstationary training images is useful as a communication tool with the domain expert. The nonstationary training image represents an explicit full 3D disclosure of the variation in the spatial (and temporal) domain that is in some applications more accessible (to the nonstatistician domain expert) than parameters of a residual covariance. The issue of nonstationarity is therefore of critical importance for methods that employ training images as well. We devote Chapter II.5 to this topic.

Neighborhood: UK-v can employ both a local and global neighborhood. For practical reasons, due to the often limited size of the TI, UK-TI can be used with local neighborhoods only. This is the direct consequence of using a random function model versus using a finite training image. The random function concept, in terms of macro-ergodicity (Matheron, 1978), relies on the notion that the model domain itself is part of an infinitely large area; as a result, the global neighborhood is in fact a local neighborhood. UK-TI does not rely on the property of ergodicity of random function theory; hence, one is required to specify models using a local neighborhood.

References

Hertzmann, A., Jacobs, C. E., Oliver, N., Curless, B. & Salesin, D. H. Image analogies. In *Proceedings of the ACM SIGGRAPH Conference on Computer Graphics* (SIGGRAPH 2001), Los Angeles, CA, 12–17 August 2001, 2001. 327–340.

Journel, A. G. 1996. Deterministic geostatistics: A new visit. *Geostatistics Wollongong*, 85.

Mariethoz, G., Renard, P. & Straubhaar, J. 2010. The direct sampling method to perform multiple-point geostatistical simulations. *Water Resources Research*, 46.

Matheron, G. 1978. *Estimer et choisir: essai sur la pratique des probabilités.*

CHAPTER 4

Stochastic simulations based on random function theory

4.1 The goal of stochastic simulations

If we retain the kriging map of Figure I.3.14(a) and apply the routing program (see Figure I.3.14(b–c)) to estimate the length of the path and cumulative elevation gain, we notice that the estimates provided by kriging (Table I.4.1) are far from the true values. Secondly, there is no way to establish any kind of uncertainty on these estimates. The fact that the single smooth kriging solution cannot be used to predict the adventure route statistics is a well-known issue. It is clear that the kriging map is not a reflection of spatial variability in terms of both the univariate distribution (the histogram will have too low variance) and the semivariogram (which typically would have a smooth behavior at the origin). Both will clearly affect the forecast of path statistics from such maps.

Particular to those fields where geostatistical models are used as input to transfer function, such as dynamic simulators, this smoothing property of the kriging solution is well studied. For such applications, stochastic simulation is used for prediction purposes. The goal is now to generate models that, taken as a whole, are representations of a truth, not a set of estimates at individual locations. The idea is to impart spatial characteristics to these models that they hopefully share with the actual field of study. For example, the mean, variance, and semivariogram of a stochastically simulated model should be similar to those of the true unknown field. But, because the true field is not known exhaustively, we can only estimate statistics from any available data. If such data are too sparse, or perhaps sampled in a biased fashion, then one often needs to rely on analog information or data sets that are deemed reflective of the spatial variability of the true field.

The statistical viewpoint taken in solving this spatial stochastic simulation problem is similar to the way in which univariate problems are treated. Instead of dealing with one variable, one now deals with a random function (also termed a "random field") discretized on a grid (in the statistical literature termed a "lattice"). One uses the data to estimate the parameters θ of the multivariate distribution underlying such a random function under study, after making the necessary

Multiple-point Geostatistics: Stochastic Modeling with Training Images, First Edition. Gregoire Mariethoz and Jef Caers.
© 2015 John Wiley & Sons, Ltd. Published 2015 by John Wiley & Sons, Ltd.
Companion website: www.wiley.com/go/caers/multiplepointgeostatistics

Table I.4.1 Comparison of path statistics between reference and UK-TI estimate

	Minimum effort			
	Cumulative elevation gain (ft)		Path length (cell units)	
	Median	P10–P90	Median	P10–P90
Walker Lake reference	9862	9434–14,875	324	311–327
UK-TI	926	911–1037	337	334–340
	Maximum effort			
	Cumulative elevation gain (ft)		Path length (cell units)	
	Median	P10–P90	Median	P10–P90
Walker Lake reference	37,783	35,335–47,731	331	323–333
UK-TI	3445	3312–3570	340	336

ergodic assumptions. The random function is specified using a parametric model, for example as a joint multivariate cumulative distribution function:

$$F(z_1, z_2, \dots, z_n; \boldsymbol{\theta}) = F(Z(\mathbf{x}_1) \leq z_1, Z(\mathbf{x}_2) \leq z_2, \dots Z(\mathbf{x}_N) \leq z_N; \boldsymbol{\theta}) \quad (\text{I.4.1})$$

The ergodic property allows for inference of the parameters $\boldsymbol{\theta}$ from a limited sample data set. Then, using Monte Carlo simulation, samples are generated from the model Equation (I.4.1) with parameter estimates $\hat{\boldsymbol{\theta}}$. The samples are termed "realizations of the random function". Unlike the univariate case, there is no statistical test, based on limited sample data, that can be used to verify if the chosen multivariate distribution (I.4.1) is appropriate or not. Additionally, few multivariate distributions are analytically known (or practical, for that matter). Perhaps the most general is the Markov random field model, which will be discussed in Chapter II.4. Probably the most popular, for reasons of mathematical tractability, is the multivariate Gaussian distribution, which we will discuss in Section I.4.2.

4.2   Stochastic simulation: Gaussian theory

Methods for Gaussian simulation rely on the specification of the spatial law using a multivariate Gaussian distribution. Consider the random vector $\mathbf{Y}$ as the collection of random variables on a grid:

$$\mathbf{Y} = \left(Y(\mathbf{x}_1), Y(\mathbf{x}_2), \dots, Y(\mathbf{x}_N)\right)^T \quad (\text{I.4.2})$$

with outcome (realization)

$$\mathbf{y} = \left(y(\mathbf{x}_1), y(\mathbf{x}_2), \dots, y(\mathbf{x}_N)\right)^T \quad (\text{I.4.3})$$

Consider the expected value

$$m = E\left[Y(\mathbf{x})\right] \quad \forall \mathbf{x} \tag{I.4.4}$$

and the $N \times N$ covariance matrix C consisting of entries

$$C_{ij} = E\left[\left(Y(\mathbf{x}_i) - m\right)\left(Y(\mathbf{x}_j) - m\right)\right] \tag{I.4.5}$$

then the unconditional multivariate Gaussian density for **Y** is written as

$$f_{\mathbf{Y}}(\mathbf{y}) = \frac{1}{\sqrt{2\pi^N \det(C)}} \exp\left(-\frac{1}{2}(\mathbf{y} - \mathbf{m})^T C^{-1} (\mathbf{y} - \mathbf{m})\right) \tag{I.4.6}$$

The multivariate Gaussian has many convenient properties:
- The marginal of Y_i is univariate Gaussian.
- Any univariate or multivariate conditional is Gaussian.
- Any subset of Y has a multivariate Gaussian distribution.
- Any linear combination of Y_i is univariate Gaussian.

In statistical terminology, this distribution has two "parameters", the mean and covariance. The traditional statistical approach is to estimate these parameters from data, then sample from the distribution. The same issues now arise as in the case of kriging, namely, that there is only one single truth (and not multiple samples) and that such truth is only partially observed through sample data. Assumptions of stationarity are required for inference on these parameters to be possible.

In practical cases, one is interested in the conditional multivariate density. Consider that locations on the grid are informed with sample data:

$$\mathbf{Y}_\alpha = \left\{y(\mathbf{x}_\alpha), \alpha = 1, \ldots, n\right\} \tag{I.4.7}$$

Similar to kriging, there are now two approaches to deal with conditioning: a global and a local approach. The global approach requires specification of the conditional multivariate density over the entire grid, then a globally specified multi-Gaussian distribution is sampled. However, the problem of conditioning in the multivariate Gaussian case can be decoupled from the simulation part using what is termed "conditioning with kriging". To achieve this, one relies on the property that the simple kriging estimator identifies the conditional expectation of the conditional Gaussian density. This means that simple kriging performed with a global neighborhood has the following property:

$$y^*(\mathbf{x}) = m + \sum_{\alpha=1}^{n} \lambda_\alpha (y(\mathbf{x}_\alpha) - m)$$
$$= E\left[Y(\mathbf{x}) \mid \left\{y(\mathbf{x}_\alpha), \alpha = 1, \ldots, n\right\}\right] \tag{I.4.8}$$

with the weights λ_α solution of the simple kriging system of equations. In addition, it can be shown that the kriging estimates $\mathbf{Y}^*$ and the vector $\mathbf{Y}^* - \mathbf{Y}$ are independent. This means that one can decompose the problem of conditioning into two parts: a kriging part ($\mathbf{Y}^*$) and an unconditional simulation part ($\mathbf{Y}$), resulting in conditioning through kriging as follows:

1 Calculate the kriging estimates $\mathbf{y}^*$.
2 Generate an unconditional Gaussian realization $\mathbf{y}_{uc}$.
3 Extract the simulated values at the data locations, and perform kriging of the differences between simulated values and data values to get $\mathbf{y}_{uc}^*$.
4 Return as a conditional simulation $\mathbf{y}^* + \mathbf{y}_{uc} - \mathbf{y}_{uc}^*$.

For step 2, many algorithms are available (for an overview, see Chilès and Delfiner, 1999; Lantuejoul, 2002).

The alternative is to take a local conditional approach, which relies on the following properties:

1 Every local univariate conditional distribution is also Gaussian.
2 The local conditional mean and conditional variance of that univariate conditional Gaussian are identified by means of local simple kriging.

The problem then becomes one of patching together local solutions into a realization over the entire grid. This strategy, which is quite common in covariance-based geostatistics, will be the norm in multiple-point geostatistics (MPS). One way to achieve this "patching" is through sequential conditioning, which applies to any multivariate density, not just the multivariate Gaussian. In the unconditional case, the sequential conditioning is as follows:

$$f_{\mathbf{Y}}(\mathbf{Y}) = f\left[y(\mathbf{x}_{(1)})\right] \times f\left[y(\mathbf{x}_{(2)})|y(\mathbf{x}_{(1)})\right] \times f\left[y(\mathbf{x}_{(3)})|y(\mathbf{x}_{(2)}), y(\mathbf{x}_{(1)})\right] \times \cdots$$
$$\cdots \times f\left[y(\mathbf{x}_{(N)})|y(\mathbf{x}_{(N-1)}) \ldots y(\mathbf{x}_{(1)})\right] \tag{I.4.9}$$

where the notation (i) refers to a random permutation of the grid locations (the questions relative to the simulation path will be discussed further in Section II.2.6). In the presence of conditioning data, the same sequential conditioning can be achieved, except that each univariate distribution is now additionally conditioned to the sample data (Equation (I.4.7)).

4.3   The sequential Gaussian simulation algorithm

Although Gaussian theory is well known, practical methods for sampling (simulating) realizations often deviate from ideal theory. The reason is modeling flexibility and attempting to achieve reasonable CPU time for large problems. For these two reasons, conditional sequential Gaussian simulation is one of the most-used methods in geostatistics for generating conditional Gaussian realizations, hence we take here some time to discuss the various practical implementation issues. The use of a limited neighborhood in combination with the sequential sampling has the advantage of flexibility in conditioning to a variety of data, but it also may cause certain artifacts:

• *Flexibility*: the simple kriging, required by theory, is often replaced with other forms of kriging (e.g., co-kriging and trend kriging) to account for other local

(a)   (b)   (c)

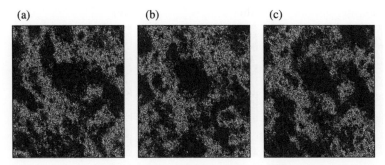

Figure I.4.1 (a–c) Three sequential Gaussian simulations of the Walker Lake case.

data within the neighborhood, and hence it conditions the simulated realization to such data. In addition, one can locally modify certain semivariogram parameters, such as the direction of major anisotropy or the ratio of major to minor ranges, to induce nonstationary variation into the generated realization.
- *Artifacts*: similar to local kriging, a neighborhood is defined around the location to be simulated. However, unlike kriging, artifacts are often not clearly visible because the simulated models need not exhibit the smooth character of kriged maps. The use of a limited neighborhood may, however, affect reproduction of semivariogram ranges, result in artifacts near conditioning data, or cause a reduction in variability between the simulated realizations.

In Figure I.4.1, we apply the sequential Gaussian simulation algorithm to the Walker Lake case. Universal kriging with external drift with a local neighborhood was used (a single trend function, namely, the left image of Figure I.3.11). The realizations are additionally constrained to the 100-sample data.

4.4   Properties of multi-Gaussian realizations

The multi-Gaussian law provides a theoretically sound and convenient framework that can address a large variety of problems, in particular those involving large amounts of (direct or indirect) conditioning data. Consider, for example, the Walker Lake example with two cases of hard-data information available: one case with 5% of the image informed by point data, and a second case with only 0.1%; see Figure I.4.2. In the former case, the variability in the resulting realization is largely driven by the data, whereas in the latter case it is driven by the spatial characteristics of the multivariate Gaussian distribution. It is clear that a "destructuration" of sorts is taking place, where realizations become spatially "disorganized".

This property of the multi-Gaussian distribution, also termed the "maximum entropy property" (Deutsch and Journel, 1992; Journel and Deutsch, 1993), is well documented (Goovaerts, 1997), both theoretically and regarding

(a)   (b)   (c)

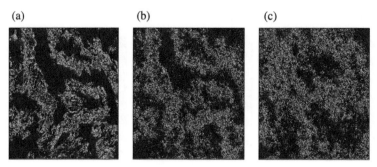

Figure I.4.2 (a) Exhaustive Walker Lake image; (b) a Gaussian sample constrained to 5% of randomly sampled data from (a); and (c) the same Gaussian law, but now only constrained to 0.1% of the exhaustive image.

its impact in practical applications (for subsurface flow examples, see Gómez-Hernández and Wen, 1994, 1998; Zinn and Harvey, 2003). One important property is that spatial correlations of extreme high and low values tend rapidly toward zero. As a result (without any conditioning data), the realization (see Figure I.4.2(c)) consists of "islands" of high values, rather than connected mountains as would be desired in this application.

Practical procedures (Deutsch and Journel, 1992) exist to check whether the multivariate Gaussian distribution is appropriate for the given data. A first check can be made regarding the univariate marginal distribution. However, any deviation from a Gaussian marginal can be addressed by a simple transformation and anamorphosis. Such univariate monotonic transformation does not change the maximum entropy characteristics in terms of destructuration of extremes, which is a higher-order characteristic.

Other methods often are limited to the bivariate marginal of the multi-Gaussian distribution and test whether indicator variograms of the data align with what is expected from theory. However, such methods can really only work when data are plenty, in which case the multi-Gaussian model may work well for the intended application whether it is consistent or not with such data (see Figure I.4.2(b)). In addition, such tests are limited to second-order moments and do not test higher-order moments or statistics.

Next, we will describe some alternatives to the multi-Gaussian model that still rely on covariances to define the spatial model.

4.5   Beyond Gaussian or beyond covariance?

Given the properties of the multi-Gaussian distribution and its inherent limitations in modeling a wide variety of natural phenomena, alternatives have been developed in both the statistical and geostatistical literature, such as Boolean models (Lantuejoul, 2002), latent Gaussian models (Rue et al., 2009), and skew-normal random fields (Genton, 2004), to name just a few. It is not the goal of this book to provide an exhaustive overview of these methods, but rather to present some of the basic philosophies in developing such non-Gaussian models to shed light on the rationale behind MPS. What is important to recognize is that although these approaches are non-Gaussian, in many cases they still rely on the covariance as a fundamental parameter of the model. In that sense, these techniques rely on the same philosophy of first estimating a lower-order parametric model, then using that model for estimation and simulation, while higher-order statistics are left to the intrinsic properties of the model. Consider two such cases: a direct sequential simulation model and a pluri-Gaussian model.

Direct sequential simulation (DSSIM; Soares, 2001) relies on an extension of the sequential Gaussian simulation, where instead of a local Gaussian conditional distribution, one uses any type of distribution. This allows simulating non-Gaussian variables directly without the need for a transformation of the variable. The practical advantage is that it allows conditioning to a larger variety of data, such as block average or multiscale data (e.g., Journel, 1999). Because the direct simulation relies on the semivariogram as a model of spatial continuity, the features in the simulated realization share their amorphous character with the multi-Gaussian simulation (see, e.g., Figure I.4.3), although with specific choices in the type of local conditional distributions, the connectivity may be increased (Caers, 2000).

Pluri-Gaussian models (see Armstrong et al., 2003 for an overview) aim at simulating categorical variables by combining multiple multi-Gaussian variables using multiple thresholds. The principle of the method is to first generate two continuous Gaussian fields using standard multi-Gaussian techniques (note that

(a)   (b)

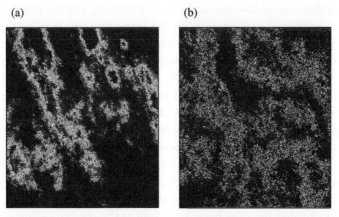

Figure I.4.3 (a) Realizations of Walker Lake generated with direct sequential simulation conditioned to 5% of conditioning data; and (b) sequential Gaussian simulation (Figure I.4.2(b)) for comparison. Although DSSIM realizations often show better connected high and low values, they display a similar amorphous character as multi-Gaussian models.

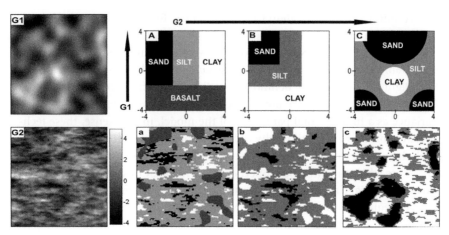

Figure I.4.4 Two Gaussian realizations (G1 and G2) and three rules diagrams (A–C), resulting in unconditional pluri-Gaussian realizations (a–c) (after (Mariethoz et al., 2009).

the method can be generalized to any number of initial fields). These fields are then truncated to produce categories, but the thresholding depends on the value of both Gaussian fields.

Figure I.4.4 shows an illustrative example of a pluri-Gaussian model applied to modeling the spatial distribution of lithofacies. First, two Gaussian random fields G1 and G2 are generated. These two fields are truncated to create the four lithofacies. The pluri-Gaussian approach consists of defining the relations between various categories in a rule diagram, in this case a lithotype rule diagram. Three examples of defining such rules through a diagram (A, B, and C) are shown in Figure I.4.4. In these diagrams, the two axes correspond to the values of the underlying multi-Gaussian fields (G1 and G2), and the grayscale levels correspond to different rock types. With the truncation rule A, G2 is truncated with two thresholds, defining the sand, silt, and clay facies. Another threshold is used with G1, delimiting the basalt facies. The result (Figure I.4.1a) is that whenever G1 has a low value, the basalt facies is present. At locations with a higher value of G1, sand, silt, or clay is present according to the value of G2. Because the Gaussian field G1 represents a smooth spatial variability (Gaussian semivariogram), the boundary between the basalt facies and the other facies has a smoother shape than the boundaries between the sands, silts, and clay. Lithotype rule B has three facies that are in a fixed order, and silt is a transition facies between sand and clay. In lithotype rule C, the rock types are defined by discontinuous zones in the lithotype rule, generating complex effects.

Modelers can alter the rules diagram as well as the semivariogram models to achieve desired spatial relationships and variability of the generated categories. Modeling of such desired relationships is therefore indirect through a set of parametric models (rules diagram and semivariogram models). Modelers will need, by means of trial and error, to obtain a desired image (Figure I.4.4(a–c)) by iterating on various combinations of parameters. It is also clear from Figure I.4.4 that the resulting models inherit the amorphous features typical of the multi-Gaussian distribution. Conditioning of pluri-Gaussian is indirect and needs to be accomplished using a Gibbs sampler (Armstrong et al., 2003).

References

Armstrong, M., Galli, A. G., Loc'h, G. L., Geoffroy, F. & Eschard, R. 2003. *Plurigaussian Simulations in Geosciences*, Berlin, Springer.

Caers, J. 2000. Adding local accuracy to direct sequential simulation. *Mathematical Geology*, 32, 815–850.

Chilès, J.-P. & Delfiner, P. 1999. *Geostatistics – Modeling Spatial Uncertainty*, New York, John Wiley & Sons.

Deutsch, C. & Journel, A. 1992. *GSLIB: Geostatistical Software Library*, New York, Oxford University Press.

Genton, M. G. (ed.) 2004. *Skew-Elliptical Distributions and Their Applications: A Journey Beyond Normality*, Chapman & Hall/CRC, Boca Raton, FL.

Gómez-hernández, J. J. & Wen, X. H. 1994. Probabilistic assessment of travel times in groundwater modeling. *Stochastic Hydrology and Hydraulics*, 8, 19–55.

Gómez-hernández, J. J. & Wen, X. H. 1998. To be or not to be multi-Gaussian? A reflection on stochastic hydrogeology. *Advances in Water Resources*, 21, 47–61.

Goovaerts, P. 1997. *Geostatistics for Natural Resources Evaluation*, Oxford University Press, Oxford.

Journel, A. 1999. Conditioning geostatistical operations to nonlinear volume averages. *Mathematical Geology*, 31, 931–953.

Journel, A. & Deutsch, C. 1993. Entropy and spatial disorder. *Mathematical Geology*, 25, 329–355.

Lantuejoul, C. 2002. *Geostatistical Simulation: Models and Algorithms*, Berlin, Springer.

Mariethoz, G., Renard, P., Cornaton, F. & Jaquet, O. 2009. Truncated plurigaussian simulations to characterize aquifer heterogeneity. *Ground Water*, 47, 13–24.

Rue, H., Martino, S. & Chopin, N. 2009. Approximate Bayesian inference for latent Gaussian models by using integrated nested Laplace approximations. *Journal of the Royal Statistical Society. Series B: Statistical Methodology*, 71, 319–392.

Soares, A. 2001. Direct sequential simulation and cosimulation. *Mathematical Geology*, 33, 911–926.

Zinn, B. & Harvey, C. 2003. When good statistical models of aquifer heterogeneity go bad: A comparison of flow, dispersion, and mass transfer in connected and multivariate Gaussian hydraulic conductivity fields. *Water Resources Research*, 39, WR001146.

CHAPTER 5

Stochastic simulation without random function theory

5.1 Direct sampling

5.1.1 Relying on information theory

As in the case of universal kriging with training images, one can develop stochastic simulation without relying on the random function approach. In other words, the realizations (and the unknown truth) are not considered as outcomes of an unknown stochastic process. We now need to recognize that in practice, what is often ultimately desired is not a multivariate distribution and its parameter estimates, but the realizations generated. Such realizations can be obtained in many ways: not necessarily as realizations of a stochastic process, but also as realizations of any algorithm that has a built-in stochastic component (a pseudo-random function generator that, perhaps ironically, constitutes a deterministic algorithm). A posteriori, one could then interpret these realizations as outcomes of an unspecified random function. However, there is little practical relevance in doing so. In a similar vein, universal kriging with training images can be interpreted as some form of random function–based kriging with the sum of products as experimental covariances. However, such an a posteriori interpretation does not add anything to the ultimate solution, which is the obtained map. The question, similar to the above kriging situation, is then: can we get these realizations directly from an image, rather than taking the route of models specification, parameter estimation, and samplers? The main methodological difference then consists in putting the modeling effort into designing such an explicit image, rather than first focusing on mathematical constructs and parameter inference.

To achieve this, we return to the local neighborhood concept, which was used in the UK-TI approach. Namely, we assume that what occurs at location $\mathbf{x}$ is mostly a function of what is near location $\mathbf{x}$: as long as enough information is near $\mathbf{x}$, one can determine the value at $\mathbf{x}$, or its uncertainty. This is particularly true when the field of study has a lot of data (i.e., the data drive the major trends) or when the variation of that variable over the domain is not changing too much (i.e., the absence of a trend). When variation is driven by trend

---

Multiple-point Geostatistics: Stochastic Modeling with Training Images, First Edition. Gregoire Mariethoz and Jef Caers.
© 2015 John Wiley & Sons, Ltd. Published 2015 by John Wiley & Sons, Ltd.
Companion website: www.wiley.com/go/caers/multiplepointgeostatistics

(long-range correlation) or when very little data are present, the nonstationary modeling component becomes important and hence will be explicitly specified.

As before, we denote by $\mathbf{dev}(\mathbf{x})$ the ensemble of the n closest sample values near a location of interest $\mathbf{x}$ in the modeling domain. The random function approach would call for the modeling of a conditional probability:

$$F(z) = P(Z(\mathbf{x}) \le z | \mathbf{dev}(\mathbf{x})) \qquad (I.5.1)$$

Theoretical models of these conditional probabilities can be derived after making multivariate Gaussian assumptions or assuming a Markov random field (MRF) model. Here, we completely abandon this approach and directly aim at obtaining a value for z. To that end, we focus on a concept of distance (or similarity) instead of probability. The training image is scanned, and, at each location $\mathbf{y}$ in the training image TI, the corresponding data event $\mathbf{dev}_{TI}(\mathbf{y})$ is retrieved. A distance is calculated as a measure of difference between the data event of the domain and the data event of the training image. We denote this distance as $d(\mathbf{dev}(\mathbf{x}), \mathbf{dev}_{TI}(\mathbf{y}))$. The scanning and distance calculations are performed using a random path. The idea is to find a location $\mathbf{y}'$ where the distance is below a specified threshold t. We rely on a simple principle formulated in information theory (Shannon, 1948): that the first location found this way is a representative "sample" of all locations in the training image that have a distance under the specified threshold.

When a suitable location is found, its value $z_{TI}(\mathbf{y})$ is assigned to the location $\mathbf{x}$ in the domain. The procedure we describe here is the same as the one described by Shannon for the generation of random sequences of English text. The following quote describes the method:

> One opens a book at random and selects a letter at random on the page. This letter is recorded. The book is then opened to another page and one reads until this letter is encountered. The succeeding letter is then recorded. Turning to another page this second letter is searched for and the succeeding letter recorded, etc. (Shannon, 1948)

We illustrate this with an actual text; see Figure I.5.1. The book in question (the classical novel of Gustave Flaubert, *Bouvard et Pécuchet*, published in 1881) contains 250 pages. To reconstruct a page that looks like any page in that book, we can start with selecting a random letter (consider, e.g., the letter "e"). Then, in order to sample the next symbol (one of the 256 ASCII characters) to be placed after that first letter, we find by random search the next symbol and write down that symbol after the letter "e". This procedure is continued, each time increasing the amount of "conditioning" letters. The page created in Figure I.5.1 uses up to 13 letters (13 neighbors). Statistically, this can be seen as a Markov chain with a neighborhood of order 13, but we never called upon this notion, nor did we estimate any model of a 13th-order Markov chain. The procedure described

Flaubert: Bouvard et Pecuchet (250p.)	Direct sample realization of Bouvard et Pecuchet
CHAPITRE I Comme il faisait une chaleur de 33 degrés, le boulevard Bourdon se trouvait absolument désert. Plus bas le canal Saint-Martin, fermé par les deux écluses étalait en ligne droite son eau couleur d'encre. Il y avait au milieu, un bateau plein de bois, et sur la berge deux rangs de barriques. Au delà du canal, entre les maisons que séparent des chantiers le grand ciel pur se découpait en plaques d'outremer, et sous la réverbération du soleil, les façades blanches, les toits d'ardoises, les quais de granit éblouissaient. Une rumeur confuse montait du loin dans l'atmosphère tiède ; et tout semblait engourdi par le désoeuvrement du dimanche et la tristesse des jours d'été. Deux hommes parurent. L'un venait de la Bastille, l'autre du Jardin des Plantes. Le plus grand, vêtu de toile, marchait le chapeau en arrière, le gilet déboutonné et sa cravate à la main. Le plus petit, dont le corps disparaissait dans une redingote marron, baissait la tête sous une casquette à visière pointue.	toutes les affections et indique les pièces de vers qu'il faut imaginer pour se reconnaître son inférieur. L'abbé ripostait par la lettre du roi Abgar, les Actes de Pilate et le témoignage de la conscience, à la tradition des peuples, au besoin d'un peu de champagne, dont les détritus amèneraient d'autres ouvrages, Montgaillard, Prudhomme, Gallois, Lacretelle, etc. ; et les contradictions de ces livres ne valant pas une observation pour Josué -- et quant aux Juges, l'auteur nous prévient qu'à l'époque dont il fait l'histoire, particulièrement la féculerie et un nouveau genre de fromages. Pécuchet se mit à réfléchir -- La fenêtre était ouverte, la nuit tranquille. Enfin, avec les plus grandes formaient au loin comme une falaise surplombant la campagne. Puis le notaire tenant à son étude, Foureau fut choisi -- un rustre, un crétin. Le docteur s'en indigna. Fruit sec des concours, il regrettait Mélie, et le pastel de la dame en robe Louis XV, le gênait avec son décolletage.

Figure I.5.1 (Left) A page from the book *Bouvard et Pécuchet*, by Gustave Flaubert, published in 1881 (English version published by H. S. Nichols in 1896). (Right) a direct sampling realization of one page.

never called upon the notion of a random variable, or random function, and it never specified or estimated a probability model explicitly.

In the book example, the distance threshold is equal to $t = 0$: we are seeking an exact match. In the spatial context, due to the limited size of the training image, this match will not necessarily be perfect, in particular for continuous variables. Nonzero thresholds therefore need to be specified. The particular value chosen for this threshold will affect the results considerably, and the sensitivity of these tuning parameters will be discussed in Parts II and III of this book.

5.1.2 Application of direct sampling to Walker Lake

The application of direct sampling is now immediate and follows the results of Figure I.3.11, except that now the simulations are also conditioned to the local sample data; see Figure I.5.2. This conditioning is achieved in the same way as for any sequential method, namely, the conditioning data are allocated to the grid and frozen during simulation. The conditional direct sampling algorithm is described in detail in Chapter II.3.

The above procedure generates realizations that contain only z-values sampled from the training image. Evidently, this will not generate sample values

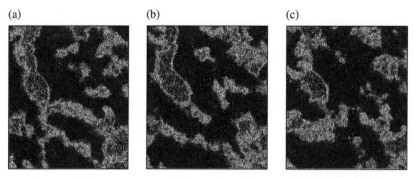

Figure I.5.2 (a–c) Three conditional realizations of Walker Lake digital elevation map using direct sampling.

above the maximum training image value or below its lowest value; no extrapolation of the histogram is performed. If such realizations are used in applications where connectivity of high values is important, then this is often not a problem. If applications call for such extrapolation, then one needs to explicitly generate these extreme values. This can be done by taking either the histogram of the training or the sample data (whichever is deemed relevant) and modeling it with a parametric distribution model. In many codes (e.g., GSLIB (Deutsch and Journel, 1992) and SGEMS (Remy et al., 2009)), this is done with simple interpolations and tail extrapolations based on the cumulative distribution. This modeled distribution can then be used to resample a new histogram. The values in the training image are then modified by a univariate histogram transform, resulting in extreme values being present as per the specified tail extrapolations.

5.2   The extended normal equation

5.2.1   Formulation

The direct sampling approach does not rely on any random function theory; it directly focuses on obtaining the sample. In this way, it has a lot in common with the universal kriging with training images approach, except that the goal is simulation; therefore, more information is borrowed from the training image than sums of products and eroded averages. In this section, we introduce another approach to stochastic simulation with training images, now using concepts of random function theory. These methods borrow some ideas related to random concepts theory, but they are not fully rigorous in following the complete theory. The main problem now lies in specifying the conditional probability, Equation (I.5.1), without explicitly specifying the full spatial law of Equation (I.4.1). The idea is to focus on the local problem, Equation (I.5.1), but to sequentially solve it throughout the entire domain. Most MPS methods operate this way: they solve

(a)   (b)   (c)

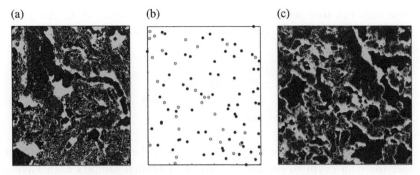

Figure I.5.3 (a) Walker Lake exhaustive data categorized; (b) corresponding 100 sample data; and (c) training image, categorized.

a local problem and then create a "global" realization by patching local problems together.

The method presented in this section is only applicable to discrete variables. We therefore consider a categorical variant of the Walker Lake case, where the topography is categorized into three classes: mountains (higher than 100 ft from reference level), foothills (between 10 and 100 ft from reference level), and lakebed (reference level from 0 to 10 ft); see Figure I.5.3. The focus now lies on specifying the conditional probability of a categorical variable $S(\mathbf{x})$:

$$P\left[S(\mathbf{x}) = s_k | \mathbf{dev}(\mathbf{x})\right] = f(\mathbf{x}, k, \mathbf{dev}(\mathbf{x})) \quad k = 1, \ldots, K \qquad (\text{I.5.2})$$

with the neighborhood of n data points more explicitly written as

$$\mathbf{dev}(\mathbf{x}) = \left\{s(\mathbf{x} + \mathbf{h}_1), \ldots, s(\mathbf{x} + \mathbf{h}_\alpha), \ldots, s(\mathbf{x} + \mathbf{h}_n)\right\} \qquad (\text{I.5.3})$$

If this function is available for any k and for any $\mathbf{dev}(\mathbf{x})$, then a realization can be generated using sequential simulation. In sequential Gaussian simulation, we know the shape of the local conditional distribution (Gaussian), and we calculate its mean and variance through kriging. In the traditional approach, such kriging consists of a system of equations where the relationships between the unknown and the data values as well as amongst the data values are considered only in pairs ("two-point") and measured using covariances. Now, we do not assume to know the shape of this local probability distribution function and attempt to reconstruct it based on the provided image. To achieve this, we establish a relationship between the data-event $\mathbf{dev}(\mathbf{x})$ in its entirety and the unknown. To establish this, we first specify a single binary indicator variable that states whether or not this particular data event, in its entirety, exists or not:

$$I_{\mathbf{dev}}(\mathbf{x}) = \begin{cases} 1 & if \quad S(\mathbf{x} + \mathbf{h}_\alpha) = s(\mathbf{x} + \mathbf{h}_\alpha) \quad \forall \alpha \\ 0 & else \end{cases} \qquad (\text{I.5.4})$$

with the subscript **dev** referring to the neighboring samples at **x**. The single unknown to be estimated is composed of the traditional indicator variables (Goovaerts, 1997) as

$$
I_k\left(\mathbf{x}\right) = \begin{cases} 1 & if \quad S(\mathbf{x}) = s_k \\ 0 & else \end{cases} \qquad k = 1, \dots, K \qquad (I.5.5)
$$

The $\forall \alpha$ in Equation I.5.4 is very important here, because we do no longer consider the data event to consist of a set of individual samples each having a "correlation" with the unknown and amongst themselves. Instead, the data event is a pattern, taken as a whole, for which we want to establish the uncertain relationship with the unknown, by means of a conditional probability. The consequence is that now kriging proceeds with one data point (the pattern or data event) and no longer with n data points. The simple kriging expression of this one-event problem is written as (Journel, 1993)

$$
P(I_k\left(\mathbf{x}\right) = 1 | I_{\mathbf{dev}}(\mathbf{x})) = E\left[I_k\left(\mathbf{x}\right)\right] + \lambda\left(1 - E\left[I_{\mathbf{dev}}(\mathbf{x})\right]\right) \qquad (I.5.6)
$$

which relies on the property that an expectation of a binary random variable is the probability of that binary variable being equal to 1. The "1" in Equation (I.5.6) is the single indicator value, indicating the presence of that single unique data event. Because only one "datum" is considered, the kriging system consists of one equation (the single normal equation; Guardiano and Srivastava, 1993):

$$
\lambda Var\left(I_{\mathbf{dev}}(\mathbf{x})\right) = Cov\left(I_k\left(\mathbf{x}\right), I_{\mathbf{dev}}(\mathbf{x})\right) \qquad (I.5.7)
$$

with

$$
\begin{aligned}
Var\left(I_{\mathbf{dev}}(\mathbf{x})\right) &= E\left[I_{\mathbf{dev}}(\mathbf{x})\right]\left(1 - E\left[I_{\mathbf{dev}}(\mathbf{x})\right]\right) \\
Cov\left(I_k\left(\mathbf{x}\right), I_{\mathbf{dev}}(\mathbf{x})\right) &= E\left[I_k\left(\mathbf{x}\right) I_{\mathbf{dev}}(\mathbf{x})\right] - E\left[I_k\left(\mathbf{x}\right)\right] E\left[I_{\mathbf{dev}}(\mathbf{x})\right] \\
E\left[I_k\left(\mathbf{x}\right) I_{\mathbf{dev}}(\mathbf{x})\right] &= P\left(I_k\left(\mathbf{x}\right) = 1, I_{\mathbf{dev}}(\mathbf{x}) = 1\right)
\end{aligned} \qquad (I.5.8)
$$

Plugging Equation (I.5.8) into Equation (I.5.7) results in

$$
P(I_k\left(\mathbf{x}\right) = 1 | I_{\mathbf{dev}}(\mathbf{x}) = 1) = \frac{P(I_k\left(\mathbf{x}\right) = 1, I_{\mathbf{dev}}(\mathbf{x}) = 1)}{P(I_{\mathbf{dev}}(\mathbf{x}) = 1)} \qquad (I.5.9)
$$

which identifies the definition of conditional probability.

Although the original formulation of the single normal equation relies on the use of expectation (Journel, 1993), it is in fact not needed to get a concurrent result. Consider simple kriging with training images presented in Section I.3.4. Instead of estimating a continuous z-value, we now estimate a set of indicators.

Consider for simplicity and without loss of generality the binary case. The averages over any eroded binary training image, namely,

$$av_{TI_{def}} = \frac{1}{\# \left| TI_{def} \right|} \sum_{\mathbf{x} \in TI_{def}} i_{TI}(\mathbf{x}) \overset{\Delta}{=} freq_{TI_{def}}(i) \qquad (I.5.10)$$

are the frequency that the binary variable i equals "1" over the eroded area in the training image TI. The simple kriging with training image estimator should now consider only one single datum: the datum $i_{\mathbf{dev}}(\mathbf{x}) = 1$. Consider the following eroded average:

$$av_{TI_{def}}(i_{\mathbf{dev}}) = \frac{1}{\# \left| TI_{def} \right|} \sum_{\mathbf{x} \in TI_{def}} i_{\mathbf{dev},TI}(\mathbf{x}) \overset{\Delta}{=} freq_{TI_{def}}(i_{\mathbf{dev}}) \qquad (I.5.11)$$

which is the frequency of $i_{\mathbf{N}}(\mathbf{x})$ equaling "1" in the training image. The simple kriging with training image estimator, Equation (I.3.17), now becomes

$$i^{\#}(\mathbf{x}) = freq_{TI_{def}}(i) + \lambda \left(i_{\mathbf{dev}}(\mathbf{x}) - freq_{TI_{def}}(i_{\mathbf{dev}}) \right) \qquad (I.5.12)$$

The sum or square error condition is then as follows:

$$\text{sse condition: } \min \sum_{\mathbf{x} \in TI_{def}} \left(i^{\#}_{TI}(\mathbf{x}) - i_{TI}(\mathbf{x}) \right)^2 \qquad (I.5.13)$$

This sum of squares can be worked out as follows:

$$\min \sum_{\mathbf{x} \in TI_{def}} \left(i^{\#}_{TI}(\mathbf{x}) - i_{TI}(\mathbf{x}) \right)^2 \Leftrightarrow$$

$$\sum_{\mathbf{x} \in TI_{def}} 2(freq_{TI_{def}}(i) + \lambda(i_{\mathbf{dev},TI}(\mathbf{x}) - freq_{TI_{def}}(i_N)) - i_{TI}(\mathbf{x}))(i_{\mathbf{dev},TI} - freq_{TI_{def}}(i_{\mathbf{dev}})) = 0$$

$$\Rightarrow \# |TI_{def}| \lambda freq_{TI_{def}}(i_{\mathbf{dev}})(1 - freq_{TI_{def}}(i_{\mathbf{dev}}))$$

$$- \sum_{\mathbf{x} \in TI_{def}} i_{\mathbf{dev},TI}(\mathbf{x}) i_{TI}(\mathbf{x}) + \# |TI_{def}| freq_{TI_{def}}(i) freq_{TI_{def}}(i_{\mathbf{dev}}) = 0$$

$$\Rightarrow \lambda = \frac{freq_{TI_{def}}(i, i_{\mathbf{dev}}) - freq_{TI_{def}}(i) freq_{TI_{def}}(i_{\mathbf{dev}})}{freq_{TI_{def}}(i_{\mathbf{dev}})(1 - freq_{TI_{def}}(i_{\mathbf{dev}}))} \qquad (I.5.14)$$

with

$$freq_{TI_{def}}(i, i_{\mathbf{dev}}) = \frac{1}{\# \left| TI_{def} \right|} \sum_{\mathbf{x} \in TI_{def}} i_{\mathbf{dev},TI}(\mathbf{x}) i_{TI}(\mathbf{x}) \qquad (I.5.15)$$

being the frequency that both $i_{\mathbf{N}}(\mathbf{x}) = 1$ and $i(\mathbf{x}) = 1$ in the training image. Plugging the simple kriging with training image solution for λ back into the estimator yields

$$i^{\#}(\mathbf{x}) = \frac{freq_{TI_{def}}(i, i_{\mathbf{dev}})}{freq_{TI_{def}}(i_{\mathbf{dev}})} \qquad (I.5.16)$$

which is simply the frequentist version of conditional probability. In other words, the original developments of Journel (1993) and Guardiano and Srivastava

(1993) do not need to call upon the notion of probability or expectation. After all, the way one estimates in practice the probability, Equation (I.5.9), is by means of the frequencies, Equation (I.5.16).

5.2.2   The RAM solution

Although the basic idea of this approach is quite straightforward, its practical implementation and application to possibly large three-dimensional (3D) data sets are not. The sequential simulation requires scanning the training image anew at each location to retrieve the statistics in Equation (I.5.16). For large training images and large simulation grids, this becomes impractical. In addition, the availability of sufficient replicates to calculate Equation (I.5.16) is not necessarily guaranteed. This is a simple consequence of the sequential nature of the algorithm, namely, the data event (values and configuration) changes each time, in fact becomes larger in size as the simulation proceeds. Due to the limited size of the training image, an exact replicate of a larger data event is unlikely to be present in the training image. Just limiting the size of the neighborhood (the solution in kriging or sequential Gaussian simulation), in terms of some area or volume around the location to be simulated does not solve the problem of lack of replicates either.

Strebelle (2002) proposed solutions to this CPU and lack-of-replicates problem. The former is addressed by turning a CPU problem into a RAM problem, by storing a large amount of frequencies into a dynamic data structure termed a "search tree" (Roberts, 1998) prior to simulation. This was feasible due to the advent of gigabyte-size RAM on personal computers around the turn of the twenty-first century. The latter issue is addressed by reducing progressively the size of the data event until "enough" replicates (10 to 20 replicates; Strebelle, 2002, p. 7) are found in the training image. This implies, however, that the training image has a repetitive character, and the global scanning and storing of such events in a single tree are intrinsic assumptions of stationarity of the gathered statistics over the training image domain. The details of the resulting SNESIM (single normal equation simulation) algorithm as well as its additional features (multigrid, servo-system, and subgrid implementation) are addressed in Chapters II.2 and II.3.

5.2.3   Single normal equations simulation for Walker Lake

We illustrate the application of the SNESIM algorithm to the Walker Lake categorical case. Three conditional simulations constrained to the 100-point data are shown in Figure I.5.4. Clearly, the simulated realizations have a repetitive character (stationarity). Despite the presence of a single large mountain feature in the training image itself, the resulting simulated realizations do not reflect any such structure and only consist of smaller mountain and lake areas.

To alleviate this issue, the SNESIM realization can additionally be constrained to "soft information", although this procedure is different than for sequential

(a)   (b)   (c)

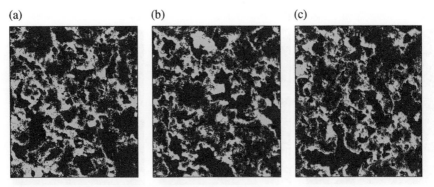

Figure I.5.4 (a–c) Three SNESIM realizations conditioned to the 100-point data.

Gaussian simulation (external drift) or direct sampling (auxiliary variable). Consider first the training image with its auxiliary variable (Figure I.3.8). This pair of images provides a statistical relationship between the presence of a certain category and the auxiliary variable. This relationship is used in conjunction with Bayes' rule to establish the conditional distribution:

$$P(S = s_k | aux) = \frac{f(aux | S = s_k) P(S = s_k)}{\sum_k f(aux | S = s_k) P(S = s_k)} \tag{I.5.17}$$

Both the density $f(aux | S = s_k)$ and the prior $P(S = s_k)$ can be easily estimated from the pair of training image and auxiliary variable. The function established, namely,

$$\phi_k(aux) = P(S = s_k | aux) \tag{I.5.18}$$

is then applied to the domain auxiliary variable to obtain three probability maps, one for each category; see Figure I.5.5.

(a)   (b)   (c)

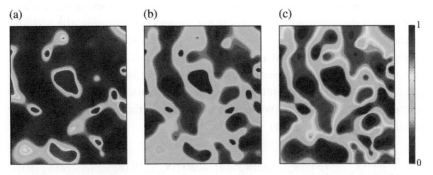

Figure I.5.5 Probability maps for each category, (a) lake, (b) foothill and (c) mountain, obtained from calibration on the training image and its auxiliary variable.

(a)   (b)   (c)

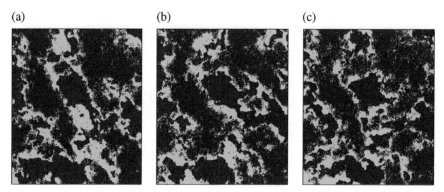

Figure I.5.6 (a–c) Three SNESIM realizations constrained to both point data and soft probabilities.

Realizations now need to be constrained to both the 100-point data and the probability maps provided in Figure I.5.5. In terms of sequential simulation, this calls for combining two probabilities at each location: a probability obtained from the training image (the search tree) and the probability calibrated from the soft information, as well as possibly any prior probabilities (proportion of categories). One such model used here is the tau model (see Section II.2.8.3.2), which we use with both tau parameters equal to unity. Three realizations generated in this fashion are shown in Figure I.5.6. The inclusion of soft probabilities has resulted in realizations that contain the single mountainous region in the left half of the image.

5.2.4 The problem of conditioning

Having introduced two methods (DS and SNESIM), we analyze the issues that can arise in data conditioning, in particular hard data conditioning. In both methods, conditioning is at least algorithmically trivial: hard data are frozen initially on the grid in their data locations; then, simulations are constructed around them. This way of conditioning is not necessarily without bias and some of these biases will be further documented in Chapter II.8. Consider, for example, the SNESIM algorithm that employs a random path for sequential simulation. Each simulated realization with SNESIM uses a different random path. Such a random path is nothing more than a permutation q of the grid locations $\mathbf{x}_1, \ldots, \mathbf{x}_n$ into $\mathbf{x}_{q(1)}, \ldots, \mathbf{x}_{q(n)}$. Due to the permuting path, the set of L unconditional realizations generated by SNESIM can be written (a posteriori to their simulation) as (see Toftaker and Tjelmeland, 2013)

$$p(\mathbf{z}) = \sum_{\ell=1}^{L} p\left[z(\mathbf{x}_{q_\ell(1)})\right] p\left[z(\mathbf{x}_{q_\ell(2)})|z(\mathbf{x}_{q_\ell(1)})\right] \ldots p\left[z(\mathbf{x}_{q_\ell(n)})|z(\mathbf{x}_{q_\ell(n-1)}) \ldots z(\mathbf{x}_{q_\ell(1)})\right].$$

(I.5.19)

An almost infinite set of realizations would sum over all possible permutations. Should a set of n_{hd} hard data be present, listed in vector **hd**, then the SNESIM-generated posterior (conditional) distribution is

$$p(\mathbf{z}|\mathbf{hd}) = \sum_{\ell=1}^{L} p\left[z(\mathbf{x}_{q_\ell(1)})|\mathbf{hd}\right] \times p\left[z(\mathbf{x}_{q_\ell(2)})|z(\mathbf{x}_{q_\ell(1)}), \mathbf{hd}\right]$$

$$\times p\left[z(\mathbf{x}_{q_\ell(n)})|z(\mathbf{x}_{q_\ell(n-n_{hd}-1)}) \dots z(\mathbf{x}_{q_\ell(1)}), \mathbf{hd}\right]$$

(I.5.20)

In other words, hard data are always placed at the beginning of the path – as if they were visited first, but with their conditional probability being a spike at the hard data value. Denoting any one of the permutations in the sum of Equation (I.5.19) as

$$p_\ell(\mathbf{z}) = p\left(z(\mathbf{x}_{q_\ell(1)})\right) p\left(z(\mathbf{x}_{q_\ell(2)})|z(\mathbf{x}_{q_\ell(1)})\right) \dots p\left(z(\mathbf{x}_{q_\ell(n)})|z(\mathbf{x}_{q_\ell(n-1)}) \dots z(\mathbf{x}_{q_\ell(1)})\right)$$

(I.5.21)

Then, Bayes' rule can be applied twice as follows:

$$p_\ell(\mathbf{z}|\mathbf{hd}) = K_\ell(\mathbf{hd})p_\ell(\mathbf{hd}|\mathbf{z})p_\ell(\mathbf{z})$$

$$p(\mathbf{z}|\mathbf{hd}) = K(\mathbf{hd})p(\mathbf{hd}|\mathbf{z})p(\mathbf{z})$$

(I.5.22)

The two normalization constants are different, but both are evidently functions of the hard data **hd**. After substitution, Equation (I.5.20) can be rewritten as

$$p(\mathbf{z}|\mathbf{hd}) = \sum_{\ell=1}^{L} w_\ell(\mathbf{hd})p_\ell(\mathbf{z}|\mathbf{hd})$$

(I.5.23)

while the actual conditional spatial law represented by SNESIM is

$$p(\mathbf{z}|\mathbf{hd}) = \sum_{\ell=1}^{L} p_\ell(\mathbf{z}|\mathbf{hd})$$

(I.5.24)

It is clear that conditioning in SNESIM does not follow Bayes' rule. The difference between Equations (I.5.23) and (I.5.24) is a direct consequence of the path implemented in SNESIM where hard data are always taken first in the permutations. Note that the weights in Equation (I.5.23) cannot be calculated because this would require summing over all possible outcomes **z**, hence an a posteriori correction is not possible. The approximation made here is for practical purposes often minor, certainly with respect to other uncertainties (e.g., the particular choice of the training image[s], uncertainty in the data itself, and relocation issues). Regardless, we therefore believe there is usefulness in formulating the algorithmic nature of MPS within stochastic laws such as MRF (Chapter II.4), then studying their properties within this context, explicitly formulating any approximations taken in MPS methods (e.g., as presented in Toftaker and Tjelmeland, 2013), and possibly suggesting corrections based on such studies.

5.3   Simulation by texture synthesis

5.3.1   Computer graphics

So far, we have illustrated how random function theory is not necessary to either perform estimation or generate realizations of natural phenomena, as long as a training image is present. The use of training images is not particular to the latest development in geostatistics; it has been used extensively in computer graphics, where "exemplars" are the equivalent of training images. Computer graphics (a term that predates geostatistics; see Carlson, 2003) is a field of computer science that develops technology for the digital synthesis and manipulation of visual content. Of particular interest are methods for texture synthesis where the aim is to algorithmically recreate large, possibly 3D images from small 2D exemplars (see Figure I.5.7). Such texture synthesis techniques were developed for animation movies and video games, where graphical realism and computationally efficient processing are key aspects (similar to the aims of MPS). In the texture synthesis technology, one can observe a similar transition from random function–based models (parametric MRF-based texture synthesis, e.g., Cross and Jain, 1983) to nonparametric methods (de Bonet, 1997) much like the MPS

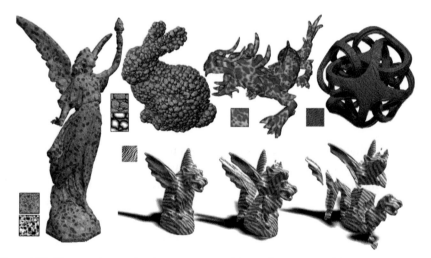

Figure I.5.7 The use of exemplars in computer graphics applications. Models are carved out from unconditional 3D blocks of exemplar-based textures, of size 1024^3 voxels. For computational efficiency, only the voxels that need to be displayed are simulated. 2D exemplars are used to generate 3D models. Exemplars are shown as small inserts (Dong et al., 2008).

methods presented in this book. This transition was made because statistical models, such as MRF, were viewed as too cumbersome and too demanding computationally (Wei et al., 2009).

One particular form of texture synthesis relies on creating a new texture by copying and stitching together textures at various offsets (Efros and Freeman, 2001; Liang et al., 2001; Kwatra et al., 2003; Lefebvre and Hoppe, 2005; Lasram and Lefebvre, 2012). A similar development in MPS was termed "stochastic simulation of patterns" (Arpat and Caers, 2007; Honarkhah and Caers, 2010; Parra and Ortiz, 2011; Tahmasebi et al., 2012; Rezaee et al., 2013). The idea is to use the unit of "pattern" (MPS) or "patch" (computer graphics), which is a building block, much like a puzzle piece for reconstructing a realization. Stochastic modeling then becomes a form of stochastic jigsaw puzzle. The training image is decomposed into pieces that are stitched together to generate a realization (see Figure I.5.8). Probabilistic modeling is then completely abandoned. Instead, the notion of distance between patches is introduced as a way to select patches from the training image such that they best fit together in the realization. At the same time, these patches are constrained to any point data, trend, or soft information.

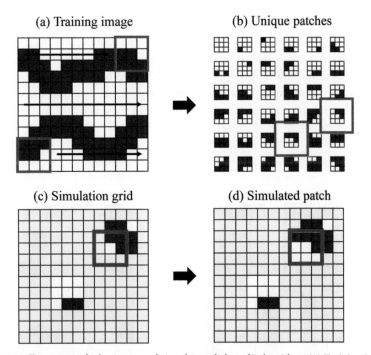

(a) Training image   (b) Unique patches

(c) Simulation grid   (d) Simulated patch

Figure I.5.8 Illustration of a basic unconditional "patch-based" algorithm. (a) Training image; (b) set of unique patches extracted from it; (c) stage of simulation where two non-overlapping patches have been simulated; and (d) selection (random) of one of the two patches that best fit (in terms of some distance) the local data event.

5.3.2 Image quilting

Efros and Freeman (2001) originally presented an unconditional image-quilting (IQ) algorithm to synthesize a texture by using a given example. The key idea is to generate similar textures by considering small pieces of existing textures and then sewing these pieces together in a coherent manner. IQ works by assembling such patches along a unilateral path or raster path (Figure I.5.10). However, patches are not only selected based on a distance criterion but also cut such that the overlap error is minimized. This is particularly useful in simulating continuous and smooth features without inducing a patchlike artifact (top-right plot in Figure I.5.10). The cutting is accomplished by means of Dijkstra's algorithm (Dijkstra, 1959).

In order to make this algorithm practical for typical geostatistical applications, the issue of conditioning to specific forms of data (e.g., point data) needs to be addressed. Indeed, computer graphics is often not concerned with "physical data" (i.e., data that have a volume and physical attributes associated with them). The use of patches and raster paths in computer graphics makes for fast CPU times (million-cell models in seconds), but additional algorithmic elements need to be introduced to make conditioning feasible and hence geostatistical applications realistic. For example, point conditioning can be addressed by only looking for patches that are consistent with any point data as well as using a template that looks ahead of the raster path for such point data (Parra and Ortiz, 2011). Conditioning to auxiliary variables is performed in the same fashion as is done in direct sampling. Chapter II.3 provides details on the implementation of image quilting and how such conditioning proceeds. Figure I.5.11 shows three realizations of the Walker Lake case using the image-quilting algorithm. The patch size used is 30×30 pixels, and the overlap area is constituted of six pixels.

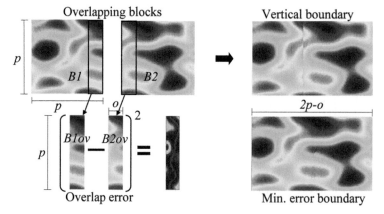

Figure I.5.9 Illustration of quilting of two patterns along a raster path.

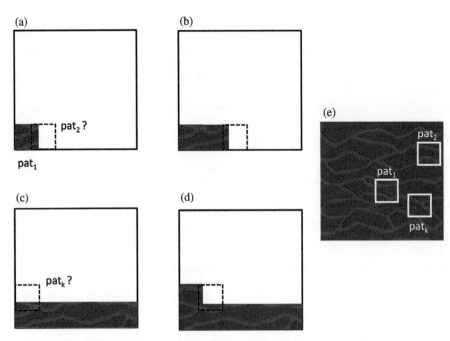

Figure I.5.10 Illustration of image quilting along a raster path. (a) Raster path starts at some corner; and (b) a next pattern is selected based on a small overlap and quilted with the previous pattern; see Figure I.5.8. (c–d) This procedure is continued along the next row. (e) Training image used for this example.

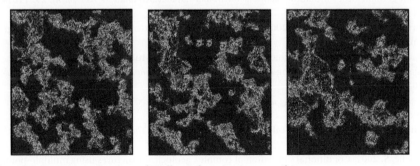

Figure I.5.11 Three realizations of Walker Lake using image quilting.

References

Arpat, B. & Caers, J. 2007. Conditional simulations with patterns. *Mathematical Geology*, 39, 177–203.

Carlson, W. 2003. *A Critical History of Computer Graphics and Animation*, The Ohio State University, Columbus.

Cross, G. R. & Jain, A. K. 1983. Markov Random field texture models. *IEEE Transactions on Pattern Analysis and Machine Intelligence*, PAMI-5, 25–39.

De Bonet, J. S. Multiresolution sampling procedure for analysis and synthesis of texture images. *Proceedings of the ACM SIGGRAPH Conference on Computer Graphics*, 1997. 361–368.

Deutsch, C. & Journel, A. 1992. *GSLIB: Geostatistical Software Library*, Oxford University Press, New York.

Dijkstra, E. 1959. A note on two problems in connexion with graphs. *Numerische Mathematik*, 1, 269–271.

Dong, Y., Lefebvre, S., Tong, X. & Drettakis, G. 2008. Lazy solid texture synthesis. *Computer Graphics Forum*, 27, 1165–1174.

Efros, A. A. & Freeman, W. T. Image quilting for texture synthesis and transfer *Proceedings of the ACM SIGGRAPH Conference on Computer Graphics*, 2001. 341–346.

Goovaerts, P. 1997. *Geostatistics for Natural Resources Evaluation*, Oxford University Press, Oxford.

Guardiano, F. & Srivastava, M. 1993. Multivariate geostatistics: Beyond bivariate moments. *In:* Soares, A. (ed.) *Geostatistics-Troia*. Dordrecht: Kluwer Academic.

Honarkhah, M. & Caers, J. 2010. Stochastic simulation of patterns using distance-based pattern modeling. *Mathematical Geosciences*, 42, 487–517.

Journel, A. 1993. Geostatistics: Roadblocks and challenges. *In:* Soares, A. (ed.) *Geostatistics Tróia '92.* Springer Netherlands.

Kwatra, N., Schödl, A., Essa, I., Turk, G. & Bobick, A. 2003. Graphcut textures: Image and video synthesis using graph cuts. *ACM Transactions on Graphics*, 22, 277–286.

Lasram, A. & Lefebvre, S. Parallel patch-based texture synthesis. *In:* C. Dachsbacher, J. M., and J. Pantaleoni, ed. *High Performance Graphics*, Paris, June 25–27, 2012, 2012.

Lefebvre, S. & Hoppe, H. 2005. Parallel controllable texture synthesis. *ACM Transactions on Graphics*, 24, 777–786.

Liang, L., Liu, C., Xu, Y. Q., Guo, B. & Shum, H. Y. 2001. Real-time texture synthesis by patch-based sampling. *ACM Transactions on Graphics*, 20, 127–150.

Parra, A. & Ortiz, J. M. 2011. Adapting a texture synthesis algorithm for conditional multiple point geostatistical simulation. *Stochastic Environmental Research and Risk Assessment*, 25, 1101–1111.

Remy, N., Boucher, A. & Wu, J. 2009. *Applied Geostatistics with SGeMS: A User's Guide*, Cambridge University Press, Cambridge.

Rezaee, H., Mariethoz, G., Koneshloo, M. & Asghari, O. 2013. Multiple-point geostatistical simulation using the bunch-pasting direct sampling method. *Computers and Geosciences*, 54, 293–308.

Roberts, E. S. 1998. *Programming abstractions in C: A second course in computer science*, Addison-Wesley, Reading, MA.

Shannon, C. 1948. A mathematical theory of communication. *The Bell System Technical Journal*, 379–423.

Strebelle, S. 2002. Conditional simulation of complex geological structures using multiple-point statistics. *Mathematical Geology*, 34, 1–22.

Tahmasebi, P., Hezarkhani, A. & Sahimi, M. 2012. Multiple-point geostatistical modeling based on the cross-correlation functions. *Computational Geosciences*, 16, 779–797.

Toftaker, H. & Tjelmeland, H. 2013. Construction of binary multi-grid Markov random field prior models from training images. *Mathematical Geosciences*, 45, 383–409.

Wei, L., Lefebvre, S., Kwatra, N. & Yurk, G. State of the art in example-based texture synthesis. Eurographics 2009, 30 March–3 April, Munich, Germany, 2009.

CHAPTER 6
Returning to the Sierra Nevada

We now return to the Walker Lake problem and the purpose of the stochastic modeling study. As stated in the introduction, geostatistical models are rarely an end goal. The various generated realizations need to be further processed for decision-making purposes. In the Walker Lake case, this concerns determining the length of the hiking path and the cumulative elevation gain.

This far, three sets of realizations have been generated:

- **MG-set** (multi-Gaussian set): a set of 50 realizations generated using sequential Gaussian simulation where the trend is modeled using the external drift method. All realizations are conditioned to the 50 sample data. The residual variogram was estimated from the 100 sample data.
- **DS-set** (direct sampling set): a set of 50 realizations generated using direct sampling. The trend model was derived from biharmonic spline smoothing and used as an auxiliary variable. All realizations are constrained to the 50 sample data.
- **IQ-set** (image-quilting set): a set of 50 realizations generated using image quilting. A similar setup in terms of trend and conditioning was used as in the direct sampling.

In generating these sets, we fixed some elements of the modeling exercise that would in most applications need randomization as well. For example, we enforced all realizations to have the same reference univariate distribution, at least within some minor (ergodic) fluctuation. This was achieved by using the true Walker Lake histogram as a reference. This ensures that all realizations reflect the same histogram, and hence comparisons between the results make abstraction of the uncertainty in the histogram uncertainty. In the multi-Gaussian model, we used only one (fixed) set of parameters for the semivariogram, and hence uncertainty in the semivariogram parameters was not considered.

Figure I.6.1 shows a comparison of the experimental semivariogram calculated from a representative realization from each set. It is clear that no major differences exist between these semivariograms. This also means that both sets are statistically similar in terms of first and second moment (histogram and semivariogram) but differ clearly in terms of higher-order (or multipoint) statistics.

Multiple-point Geostatistics: Stochastic Modeling with Training Images, First Edition. Gregoire Mariethoz and Jef Caers.
© 2015 John Wiley & Sons, Ltd. Published 2015 by John Wiley & Sons, Ltd.
Companion website: www.wiley.com/go/caers/multiplepointgeostatistics

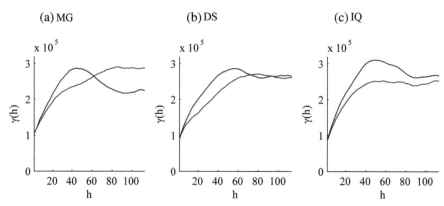

Figure I.6.1 Comparison of the experimental semivariogram of a representative realization from the three sets of realizations considered, along different directions. Blue: east-west semivariograms. Red: north-south semivariograms.

Note also that each set of realizations is constrained to the same trend data and the same 100-point data.

Applying the hiking algorithm for a given mode (minimal or maximal effort) results in 80 paths per realization from each set. This means that for each set, 4000 paths are generated. 50 histograms of cumulative elevation gain (CEG) and path length (PL) can then be generated for each set. We focus on a few statistics of this histogram, namely, a low value (P10, or eighth lowest), P50 (a middle value), and P90 (representing a high, namely, the eighth highest). The reference values, extracted from Walker Lake, were reported in Table I.1.1. The path statistics extracted from the realization are a representation of the uncertainty in the quantile parameters of the true distribution. Figures I.6.2, I.6.3, I.6.4, I.6.5, I.6.6, and I.6.7 summarize the results for both the minimal-effort and maximal-effort hiker for each set of realizations and make a comparison with the Walker Lake reference.

In the minimal-effort case, all three models (MG, DS, and IQ) capture the reference truth reasonably well, whereas in the maximal-effort case, the MG clearly underestimates the cumulative elevation gain statistics. To analyze why this occurs, we plot the elevation along a path from a single MG realization and a single DS realization; see Figure I.6.8. The clear difference in character between these two functions is due to the fact that the path is nonlinear and, of course, due to the difference in higher-order statistics between the MG and DS models. We would like to draw attention to two stretches along this path, which runs from west to east. The first stretch is the large-scale mountainous area (a climb), and the second is the flat or lake area. In terms of the high values (the extremes), we notice the destructuring effect of the MG model: few peaks of high values are met along the route, and these peaks tend to be isolated. In contrast, the path taken on a DS realization appears to have many peaks when traversing the mountain. In the flat area, the DS model is truly flat, while the MG model exhibits several peaks. This is easily attributed to the homogeneity of variance in the MG model.

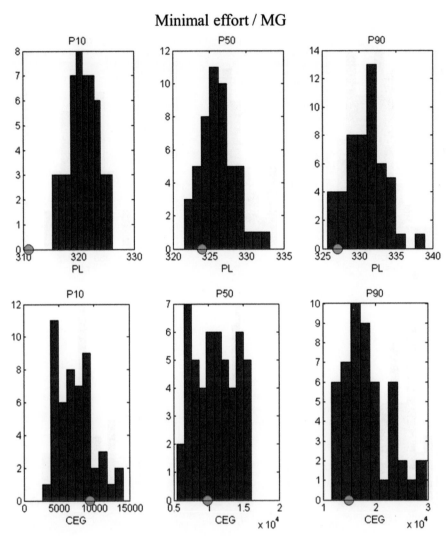

Figure I.6.2 Path statistics (PL = path length, CEG = cumulative elevation gain) calculated for minimal effort for the multi-Gaussian (MG) set.

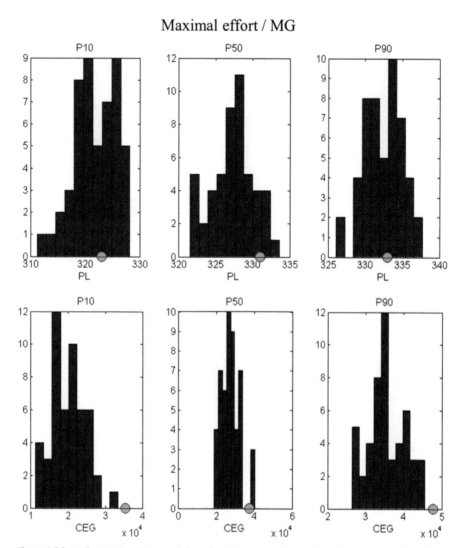

Figure I.6.3 Path statistics (PL = path length, CEG = cumulative elevation gain) calculated for maximal effort for the multi-Gaussian (MG) set.

Minimal effort / DS

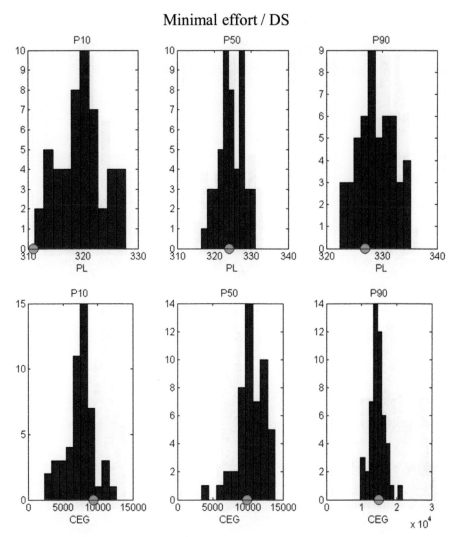

Figure I.6.4 Path statistics (PL = path length, CEG = cumulative elevation gain) calculated for minimal effort for the direct-sampling (DS) set.

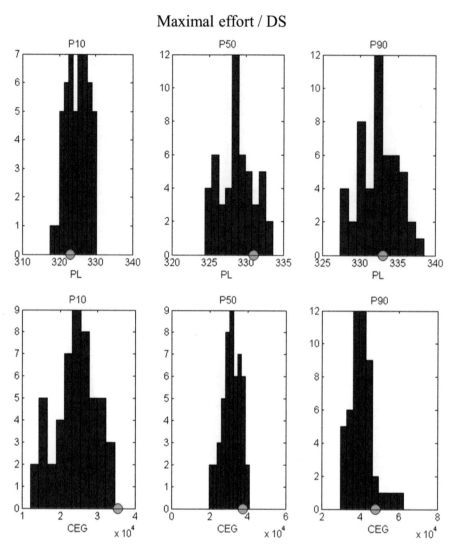

Figure I.6.5 Path statistics (PL = path length, CEG = cumulative elevation gain) calculated for maximal effort for the direct-sampling (DS) set.

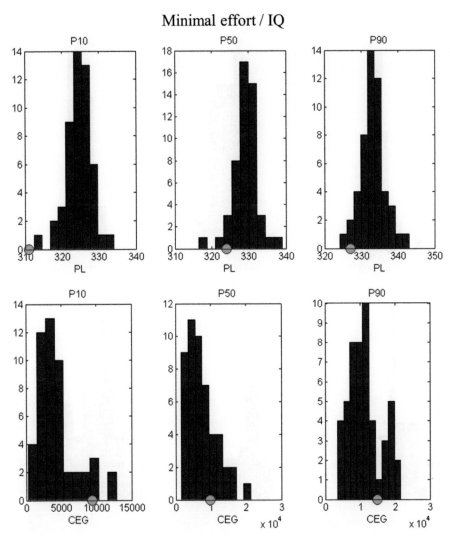

Figure I.6.6 Path statistics (PL = path length, CEG = cumulative elevation gain) calculated for minimal effort for the image-quilting (IQ) set.

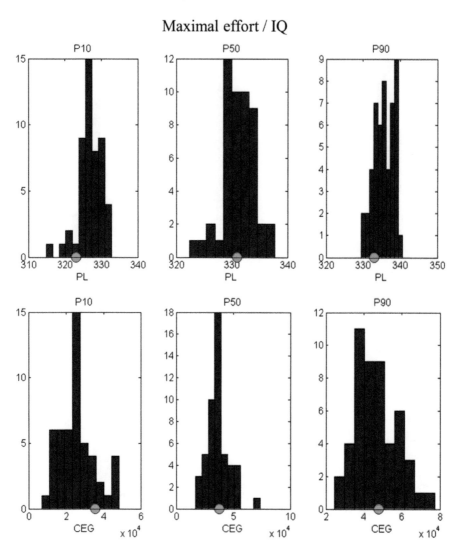

Figure I.6.7 Path statistics (PL = path length, CEG = cumulative elevation gain) calculated for maximal effort for the image-quilting (IQ) set.

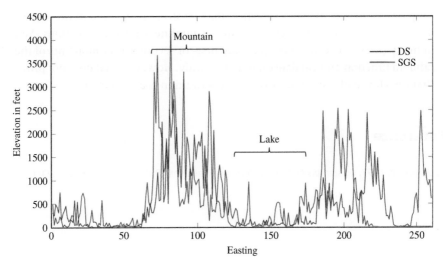

Figure I.6.8 Elevation along a path taken by a hiker walking on an MG model and a hiker walking on a DS model, for a maximal-effort hiker.

The purpose of this study is not to show that the MG is a poor model or that working with semivariograms should be abandoned. On the contrary, there are many ways to solve this problem without relying on training images. For example, one could use a hierarchical approach whereby first mountains versus valleys are modeled, and then the detailed elevation change within each such a category is modeled with a different multi-Gaussian distribution. Either indicator methods (Goovaerts, 1997) or pluri-Gaussian methods can be used for such purposes. However, the difficulty in applying such methods is the lack of data that are often needed to infer model parameters. If a binary hierarchical model is used, then three semivariogram models need to be estimated: one for the binary variable and two for the elevation change within each category. In many cases, insufficient data are present to do such modeling, and statistics need to be borrowed from elsewhere.

Instead, this study shows that

- Geostatistical models are rarely an end goal, and the transfer of uncertainty to nonlinear responses is not easy to understand a priori. Note that in reality, no "reference" exists, and hence it would be impossible to design an objective set of criteria that evaluate which modeling approach is best suited for a particular context. A case in point is that the MG model works well in one case (minimal effort) but not in another (maximal effort).
- What matters to most geostatistical applications are the generated realizations and not necessarily the intermediate multivariate distributions and their parameter estimates. The focus on such cases should be on the texture, patterns, or spatial continuity (to use a few terms with similar meaning) generated and how it impacts the response or transfer functions.

- Responses (such as the path statistics) may be strongly dependent on higher-order statistics. This is particularly true for nonlinear transfer functions. Yet, in many cases, one can at best calculate first- or second-order moments of the random function and validate on that basis only. However, that does not prove that the choice of the full multivariate distribution is a correct one.

Reference

Goovaerts, P. 1997. *Geostatistics for Natural Resources Evaluation*, Oxford University Press, Oxford.

PART II

Methods

PART II

CHAPTER 1

Introduction

In Part I, we used a virtual case study to motivate both theoretically as well as by example the validity of using analog information contained in an image for geostatistical modeling. In other words: the raison d'être for this book. In this second part, we focus on the various technical contributions to achieve this practically: the how-to guide. Similar to most geostatistical developments, theory is just a first step toward implementation. Theoretical equations will need to be transformed into usable computer code, simply because there is no application without such code. Next to concepts of statistics and probability, multiple-point geostatistics (MPS) heavily relies on modern computer science concepts, a feat that makes it distinguishable from most applied geostatistics. Our technical description in this Part II will therefore contain a mix of statistical as well as computer science concepts.

One may wonder at first why so many different algorithms have been developed within MPS. Just as there is no single search algorithm in computer science, there is not, and should not, be a single algorithm in MPS. Each algorithm has its own characteristics, advantages, and disadvantages. The key is to understand which algorithm applies where, based on the nature of the data set, the training image, the computational needs, and other practical considerations. The case studies we present in Part III offer a way to understand these factors and hopefully can be used as templates for other types of applications in the future.

Using training images is a radical change compared to most previous approaches. In several applications, the data available are not sufficient to directly infer the most important spatial features. However, additional knowledge of the physical phenomena is often available – such as informal concepts or guesses, prior databases, physical models, and historical data sets – that should be integrated into the models. The premise of using training images is that the information from this additional knowledge can be conveniently formulated as an "example" (i.e., the training image). Information on a given spatial process can, however, be defined with a variety of formalisms that can be more or less explicit. Certain constraints are appropriately handled by the classical statistical and geostatistical frameworks. The analysis of data might suggest, with a great level of confidence, that the field considered has a given probability distribution. Such

Multiple-point Geostatistics: Stochastic Modeling with Training Images, First Edition. Gregoire Mariethoz and Jef Caers.
© 2015 John Wiley & Sons, Ltd. Published 2015 by John Wiley & Sons, Ltd.
Companion website: www.wiley.com/go/caers/multiplepointgeostatistics

a piece of information can be expressed in probabilistic terms and satisfactorily modeled by the adjustment of a probability density function.

Multiple-point-based geostatistics are designed to handle more constraining information, which carries a high level of knowledge and is often not conveyed in the classical statistical formalism. Strictly adhering to the classical framework would theoretically be possible by using higher-order statistical moments, but these would be difficult to track analytically, and rigorous inference may often be impossible. Therefore, although the classical statistical and geostatistical tools are applicable when dealing with training images, they mostly allow for integrating less constraining information, such as proportions or spatial covariance. Essentially, obtaining random fields with the same histogram and variogram as the data, or as a given training image, is satisfactory under such constraints.

Let us consider a type of knowledge, loosely described as "the high values are connected and spatially organized in channels". In Earth science, the need for such spatial criteria is at least as common as imposing a marginal probability distribution. However, constraining models to this sort of information is very difficult with the usual statistical tools. Sinuosity is very poorly characterized by simple lag relationships that define spatial dependence as a function of spatial distance. Hence, variograms are not appropriate in this case. There is a need for generic ways to include such information in spatial models. Considering patterns, defined as spatial arrangements of values, allows such generality. Channelized structures are an example that is specific to 3D geological modeling, but similarly complex patterns can occur in most geoscience applications, for example:

- Different spatial patterns for high and low values in continuous remote-sensing data;
- The contact between different land use types;
- Nonlinear spatial relationships between ore grades and geological types in mining applications;
- Complex relationships between variables measured in both time and space and at different resolutions (e.g., rainfall time series).

On the other hand, in many cases a training image can convey so much spatial information that it is too restrictive. Imposing the training image in its entirety as a constraint would mean that all properties of the training image, at all possible scales, need to be reproduced in the realizations. Such a strong constraint leaves only one possible outcome: the simulation has to be identical to the training image, which is evidently not desirable. At the other end of the information spectrum, it is possible to generate realizations that only reproduce the histogram by randomly drawing values from the training image. This results in no constraints on the spatial arrangement of the values. The aim of a multiple-point simulation algorithm lies in between these two extremes: we want to impose constraints that correspond to a given level of knowledge on the processes modeled, but at the same time we need to release some features of the training image in order to introduce variability. MPS algorithms offer a variety of ways to control which features of the training image should be regarded as information and which ones should be discarded.

Often, the desired outcome is to obtain similar small-scale features as the training image, but at a larger scale these features should be significantly randomized. Strategies aimed at filtering out the large-scale patterns, such as templates, multiple grids, and random simulation paths, allow accomplishing such an outcome. The pattern selection can be based on scale, but also on orientation, on magnitude, or on a localized covariate. Certain features may not be desired and need to be filtered out, for example because they are too infrequent in the training image or too location specific. In this regard, the different simulation algorithms provide a range of options that result in completely different features of the training image to be retained. Hence, the simulation methods not only differ in the way they implement storage, filtering, and assemblage of the patterns; they also reflect different modeling assumptions and choices that are made regarding the relevant degree of information to retain from a given training image.

Training images are often nonlocal (although there are significant exceptions when we consider nonstationary modeling), meaning that they are not referenced in a real-world coordinate system, and the information they contain is of a textural nature. Other types of information can be local: these come from specific knowledge available at certain points or regions of the domain modeled, and they can come from direct local measurements (hard conditioning data) or from other dependent variables (soft data). Such diverse sources of information can be contradictory. For example, one may impose a proportions that are wildly different from those observed in the training image, or conditioning data that are in disagreement with a prescribed spatial continuity. Another case is when the level of knowledge is so rich that it results in very small variability between realizations. To identify such situations, specific methods are used to validate the models produced against available local data, and also to assess the simulation results with respect to the training image (i.e., are the relevant features of the training image retained in the realizations?). The goal of training image-based geostatistics can be seen as to generate random fields (denoted realizations) that incorporate training image–related information.

Part II contains mostly small illustrative examples, whereas Part III contains the actual applications. An accompanying website contains many of the examples in this Part II for researchers to further develop this field. Please visit http://www.trainingimages.org.

CHAPTER 2

The algorithmic building blocks

Before we introduce any algorithms or applications related to MPS, it is important for the reader to be familiar with what we call the *building blocks* of MPS algorithms. These are essential concepts that pertain to not just implementation but also understanding of the inner workings of the algorithms presented in the following chapters. In computer science language, these building blocks would be implemented as functions, routines, objects, and structures that are called in different ways by the various flavors of MPS simulation approaches. At this stage, we provide an empirical description of the various components and technical terms that lie at the basis of the algorithms described in Chapter II.3.

In this chapter, we adopt a generic approach that consists of covering the algorithmic elements that are common to most MPS simulation methods. This way, the discussion will remain valid in regard to future advances that may still use some of these concepts.

It is noteworthy that many of the concepts introduced here, such as the grid organization of the sequential simulation, are not unique to MPS but are also used with semivariogram-based methods. Although they are often not formulated as such, traditional semivariogram-based geostatistical simulation methods can also be seen as assemblages of elementary blocks, or functions. Some examples of building blocks used in classical geostatistics are:

- The definition of neighborhoods (using a minimum number of samples per quadrant): the concept is used throughout simulation and estimation methods;
- The kriging of one single point (in various), which is used in either simulation or estimation methods;
- The histogram transformation (normal-score transforms, and anamorphosis based on Hermite polynomials), which is universally used;
- The method of conditioning by kriging, which is applicable to a range of simulation methods (turning bands and fast Fourier transform based).

The notions introduced are used in Chapter II.3, where we show how the MPS building blocks are assembled to construct simulation algorithms. Although numerous MPS simulation algorithms have been developed, they essentially all use a common ensemble of basic elements that are assembled in various ways to suit specific applications or needs.

Multiple-point Geostatistics: Stochastic Modeling with Training Images, First Edition. Gregoire Mariethoz and Jef Caers.
© 2015 John Wiley & Sons, Ltd. Published 2015 by John Wiley & Sons, Ltd.
Companion website: www.wiley.com/go/caers/multiplepointgeostatistics

2.1 Grid and pointset representations

In geoscience, models are based on localized data consisting of physical quantities measured at specific locations in a coordinate system. What we define as a *pointset format* is measured data in a raw form: the location of each datum is explicitly specified, and the order in which data points are organized is irrelevant. In such a format, each data point needs to be characterized by $D+1$ numbers, where D is the dimensionality of the domain. The first D numbers characterize the location of the measurement in space and time, and the last number specifies the value v measured at that location:

One dimension: $x\ v$

Two dimensions: $x\ y\ v$

Three dimensions: $x\ y\ z\ v$

Four dimensions (e.g., the first three dimensions (3D) of height, length, and depth, plus time): $x\ y\ z\ t\ v$

In cases where more than one variable is measured at a given location, a value is attached for each variable:

Four dimensions with three variables: $x\ y\ z\ t\ v_1\ v_2\ v_3$

Although the pointset representation is appropriate for measurements, modeling often takes place on grids. Grids inform one or several spatially distributed variables on the entire modeling domain considered. Again, this domain can be 1D (e.g., a time series), 2D (e.g., a satellite image), 3D (e.g., a geological formation), or of higher dimensions if both spatial and temporal dimensions are considered simultaneously (Figure II.2.1).

A grid is necessarily spatially complete, meaning that it contains information on every spatiotemporal unit of the area modeled. Such units, called "nodes", are equally spaced and of fixed dimensions. This type of structured grid is called "Cartesian" and can be entirely characterized by:

1 its origin,

2 the size of the nodes,

3 the number of nodes in each dimension.

For example, let us consider a spatial property on a 2D domain of 10 km × 5 km, divided in nodes whose size is 0.5 km × 0.5 km. The origin is considered as the coordinates of the center of the node placed on the lower-left corner of the grid. Note that some conventions can differ, for example the coordinates of the nodes can represent an angle of the node instead of its center, or the origin of the grid can be located in the top-left corner. This should be given particular attention when passing data between software that may use different grid conventions.

In our example, considering a given reference system, the grid origin is located at coordinate 350 km along the X axis and 60 km along the Y axis. These values, summarized in Table II.2.1, entirely define the grid topology.

The entire grid, displayed in Figure II.2.2, contains 200 nodes, each representing the spatial property at the corresponding location. In this space, each nodal

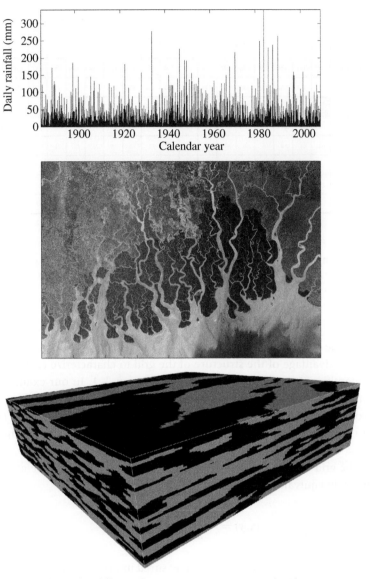

Figure II.2.1 Regular grid in different dimensions. Top: 1D temporal grid representing 120 years of daily rainfall measures in Sydney, Australia. Middle: 2D grid of a satellite image of the Sundarbans region, Bangladesh. Bottom: 3D grid representing the hydrofacies in an alluvial aquifer in the Maules Creek valley, Australia.

Table II.2.1 Topology of the grid considered in the example

Number of nodes	Size of nodes	Origin
$n_x = 20$	$d_x = 0.5$	$o_x = 350$
$n_y = 10$	$d_y = 0.5$	$o_y = 60$

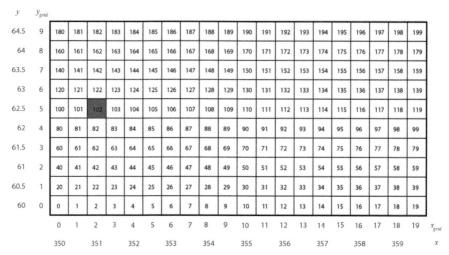

Figure II.2.2 Illustration of the grid representation using IDs. The number inside each cell is the node ID. Blue axis represents the real coordinate system. Red axis represents the coordinate system in units of nodes.

location can be characterized by its coordinates (x, y). However, one can conveniently take advantage of the structure of the grid to characterize each node by a single identifier ID, numbered from 0 to $N = n_x \times n_y - 1$. In our example, the node with $ID = 0$ is at the lower-left corner of the grid, whereas the lower-right corner corresponds to the node with $ID = 19$. The highest ID is 199 and corresponds to the upper-right corner of the grid, and the top-left corner has the $ID = 199 - 19 = 180$.

Once the grid topology is defined, the coordinates (x, y) of a node and its ID are completely equivalent. In 2D, the ID of a node is defined as

$$ID(x, y) = \frac{(x - o_x)}{d_x} + \frac{(y - o_y)}{d_y} n_x \tag{II.2.1}$$

Note that n_y is not needed. If a point measured in 2D has to be migrated on a grid, the coordinates of this point will first have to be rounded such that it falls on the center of a grid node. This rounded coordinate is then used in Equation (II.2.1). Similarly for 3D grids, we have

$$ID(x, y, z) = \frac{(x - o_x)}{d_x} + \frac{(y - o_y)}{d_y} n_x + \frac{(z - o_z)}{d_z} n_x n_y \tag{II.2.2}$$

where, again, n_z is not used. The same principle can be extended to nD grids.

This representation using IDs is more compact and convenient than the pointset representation, although strictly equivalent. The coordinates can be retrieved for any *ID* using modulo operations, according to the following formula for 2D grids:

$$x(ID) = (ID \bmod n_x)d_x + o_x$$

$$y(ID) = \left\lfloor \frac{ID}{n_x} \right\rfloor d_y + o_y \qquad (II.2.3)$$

where $a \bmod b$ is the remainder of the division a/b, or modulus, and $\lfloor \rfloor$ signifies a flooring operation. A similar conversion can be used for 3D grids and extended to *n*D grids:

$$x(ID) = \left[(ID \bmod n_x n_y) \bmod n_x \right] d_x + o_x$$

$$y(ID) = \left\lfloor \frac{(ID \bmod n_x n_y)}{n_x} \right\rfloor d_y + o_y \qquad (II.2.4)$$

$$z(ID) = \left\lfloor \frac{ID}{n_x n_y} \right\rfloor d_z + o_z$$

In Figure II.2.2, the coordinates of each node are indicated with the blue axis labels. However, for most algorithms, the grid is converted to a coordinate system in units of nodes rather than in meters or kilometers. These coordinates in units of nodes are shown with the red axis in Figure II.2.2. In our example, storing the full grid in the pointset format would necessitate storing 200 x coordinates, 200 y coordinates, and 200 property values v, hence a total of 600 numbers.

In the grid format, it takes the six numbers of Table II.2.1 to describe the grid topology, plus 200 properties values, stored in the order of increasing *IDs*. The coordinates do not need to be stored as they are embedded in the grid topology, and neither do the *IDs* that are simply the order in which the data are given. Hence, the same information can be stored with a total of only 206 numbers. Note that in the grid representation, the order in which the values are stored does matter, as opposed to in the pointset format where this order is not important.

An important note here is that in several cases, the domain modeled is not rectangular. However, the topological framework defined above and the definition of regular grids imply that one needs to work with rectangular grids (or boxes in 3D). This issue is usually dealt with by using masks. A mask is a portion of the grid that exists numerically (i.e., the mask is made of nodes that have *IDs* or corresponding coordinates) but that does not have any physical existence and should therefore be ignored when considering spatial relationships. It can be implemented by assigning to all nodes in the mask a specific value that is flagged as representing masked nodes (e.g., the string "N/A" or a numerical value such as −99999). Figure II.2.3 shows the mask that is used for the case study related to climate modeling in Chapter III.3. In this application, we only consider variables related to surface processes; therefore, all areas that are in the ocean are masked. For example, the soil moisture loses its physical meaning in a water body. In theory, the value of soil moisture in a water body should be 1 (corresponding to 100% water content). However, using this value for all nodes covering the sea

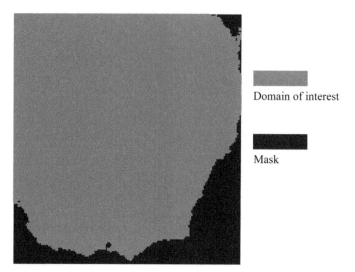

Domain of interest

Mask

Figure II.2.3 Mask applied over southeastern Australia to limit the modeled area to the landmass and exclude the water bodies.

may cause spurious and nonphysical effects near the edges of the model. Masking those nodes is, therefore, preferable and also saves computational effort because the corresponding areas do not need to be simulated.

Most MPS algorithms are based on the definition of neighborhoods (see Section II.2.3). For data stored in a pointset format, finding the neighbors of a node can be difficult. A general solution is to compute the distance between the point of interest and all other points in the dataset, and define the closest neighbors as the nodes with the smallest distance. This approach can be very costly for large data sets, especially because all distances have to be computed anew if another node is considered. Optimized search methods, such as the quad-tree algorithm, can greatly accelerate the search for neighbors in the case of data organized as a pointset; however, it remains an expensive task (Finkel and Bentley, 1974).

In comparison, the search for neighbors is greatly simplified when considering data stored in the grid format. In the 2D grid example, consider the node with $ID = 102$, which is colored in red in Figure II.2.2, corresponding to $x = 351$ and $y = 62.5$. Using the grid organization only, one observes that nodes on the left and on the right have IDs of 101 and 103, respectively, and the nodes on the top and bottom have IDs of 122 and 82, respectively. Jumping further by steps of 20 IDs, one can directly identify the nodes that are one row further as having IDs 142 and 62. This way of using offsets to jump through the grid allows algorithms to conveniently and efficiently accomplish tasks such as searching for neighbors or determining a location to interpolate.

In geological modeling, unstructured grids are used that follow stratigraphic contacts and interfaces. Because the concept of unstructured grids is not specific

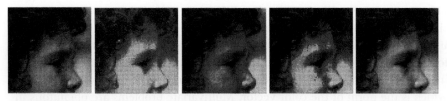

Figure II.2.4 From left to right: original image, red component, green component, blue component, and the intensity of all three components (corresponding to the norm of a 3D vector), which results in grayscale intensity values.

to MPS, it is not dealt with in detail in this chapter; it is, however, discussed in Chapter III.1 on reservoir forecasting.

2.2   Multivariate grids

An example of a grid that is encountered daily by most people is a digital photograph, where the unit is the pixel that defines a small spatial area. Typical digital images represent visual information expressed in the color spectrum. All visible colors can be represented as a mixture of three elementary tones: red, green, and blue. Hence, when describing a color image such as in Figure II.2.4, one must be aware that at each pixel, there is not one but three numbers that inform the amount of each basic tone for this particular pixel. The combined intensity of all three basic tones corresponds to the color image (the leftmost plot of Figure II.2.4). Some areas will tend to be dominated by a certain color, such as the red tone dominating the area of the character's cheek.

This example illustrates that each spatial unit (pixel) may be informed by more than one variable. The color image is a human representation of an object whose numerical description needs three variables to be fully captured. The ensemble of all possible color values for a pixel can be represented as a 3D space, and each color corresponds to a position in this 3D space. Figure II.2.5 represents the color space for a RGB domain (Figure II.2.5a) as well as for a grayscale space

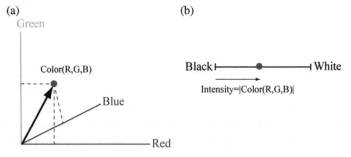

Figure II.2.5 Representation of colorspaces: (a) RGB colorspace that requires three components; and (b) grayscale space, 1D.

(1D; Figure II.2.5b). The multivariate RGB representation can be reduced to a single variable by considering only the norm of the color (R, G, or B) in Figure II.2.5a instead of its individual components. The color information is then lost and replaced by intensity values. In the example of Figure II.2.4, a grayscale image is obtained by taking, for each pixel, the norm of the three basic tones, which is the norm of the color vector. Multivariate problems are ubiquitous in Earth science, such as in subsurface flow problems where each node needs to inform hydraulic conductivity, storativity, and porosity. However, in such a case, averaging these three variables would not yield a meaningful result, and therefore any interpolation or simulation exercise needs to be carried out jointly on all three variables.

2.3   Neighborhoods

At a very fundamental level, characterizing spatially distributed variables consists of defining how the values at certain grid nodes are related to the values at other grid nodes. These relationships are often formulated as being stronger between nodes that are close to each other and weaker between nodes that are more distant. This is reflected by covariance functions that are generally decreasing with increased lags (or, inversely, semivariograms showing $\gamma(h)$ values increasing). In linear geostatistics, this behavior is called "spatial correlation" (with the term "correlation" implying a linear relationship).

When complex spatial structures are considered, the spatial relationships cannot be characterized in such a simple manner. For example, it can be the case that locations far away are less influential than nearby locations, except for a specific category of high values where distant locations become more important. Another example is that the relationship between two values A and B can be affected by the value at a third nearby location, C. Therefore, in those cases, the spatial continuity cannot be fully described by metrics measuring the dependence between pairs of values, but one needs to consider sets of more than two points. The term "multiple-point geostatistics" (MPS) precisely takes its origin in this problem of characterizing spatial continuity over more than two points.

In the case of such spatial dependences, the term "spatial correlation" is no longer appropriate; hence, one tends to speak of spatial continuity or spatial dependence. "Textural coherence" is the term used in computer graphics to designate the fact that textural elements cannot be placed randomly, but need to respect certain relationships with their neighbors. In all cases, it is acknowledged that the relationship between one node and its surrounding neighbors is the basis for the definition of a spatial phenomenon. It is therefore not surprising that an important element in all MPS methods is the definition of a neighborhood.

In semivariogram-based geostatistics, neighborhoods are used to define the extension until which other points should be considered in a kriging system. In the context of methods such as sequential Gaussian simulation (SGS) or sequential indicator simulation (SIS), the definition of a limited neighborhood can be seen as an approximation in faithfulness to a chosen spatial model, which is a price to pay for computational efficiency. When computational cost is not an issue, a "global neighborhood" is used, meaning that all available neighbors are used for the estimation or simulation of a location, based on comparisons of pairs of values.

The role of neighborhoods is different with MPS than in semivariogram-based geostatistics. In the semivariogram-based framework, the "neighborhood" essentially relates to the spatial extension of the influence of one location over other locations. A poorly defined neighborhood may result in artifacts, but reproduction of statistics is not strongly dependent on the chosen neighborhood. In MPS, the neighborhood also controls the radius of influence of values, but it does much more than that. It controls the order of the statistics considered. Whereas two-point dependences are always used in semivariogram-based approaches, MPS uses n-point dependences, where n is solely defined by the neighborhood.

The example above, where two locations A and B depend on a third location C, would need to be described with neighborhoods of at least three nodes. If the neighborhood contains five nodes, up to five-point relationships will be used. With a neighborhood of 20 nodes, up to 20-point relationships are taken into account, resulting in different, higher-order features being considered. Therefore, the definition of the nature of the neighborhood forms an integral part of the spatial model definition; hence, its use is more than just a matter of gaining computational efficiency. A corollary is that there is no equivalence to a unique or global neighborhood in MPS methods, because then the only pattern available to a simulation algorithm would consist of the entire training image (TI), or a subset of the TI if it is larger than the simulation. Any resulting set of realizations would then consist of a tiling of the TI, with very low variability (spatial uncertainty) and poor conditioning capability. At the other end of the spectrum, a neighborhood consisting of zero neighbors would be blind to any spatial dependence and would then at best only reproduce the marginal distribution of the TI. The choice of neighborhood influences what characteristics of a TI are to be retained in the realizations.

Several MPS methods define neighborhoods as *templates*. A template is a fixed set of locations around a central node that are considered for the simulation of the value at the central node. The use of fixed locations is an algorithmic and computational necessity to store conditional probabilities for a large number of combinations of category values (see Section II.2.4). Figure II.2.6 illustrates templates consisting of four, eight, and 12 nodes with a central value (in red). Note that, as illustrated in the rightmost figure, it is not a necessity that the central node is located in the geometrical center of the template; therefore, the term "central node" is only employed by convention.

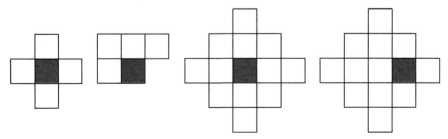

Figure II.2.6 Different types of templates.

In principle, the use of MPS is not restricted to fixed templates, and it is found that more flexibility in the neighborhood is needed to model spatial features at different scales. In practice, even methods that use fixed templates do it by nesting several templates representing different scales, a process that is known as "multiple grids" (see Section II.2.7).

Because the template size directly affects the type of structures that can be modeled, choosing a template is an integral part of a modeling workflow and should be considered an important modeling decision. One approach to template inference is to analyze the type of structures present in the TI and to use a comparable template size. However, the effect of the template choice will also depend on decisions made on other parameters; therefore, one commonly used approach is to look at a-posterior simulations and evaluate if the desired spatial dependence is present. The inference of the template, along with other parameters, is therefore iterative (see more on the parameterization of MPS methods in Chapter II.3). Another practical factor to consider is that the template size usually has a strong influence on the computation cost of MPS simulation.

The alternative to templates and multiple grids is to use flexible neighborhoods, which have long been used in semivariogram-based methods such as SGS and SIS. One possibility is to define the neighborhood of a node as the n closest informed nodes to the central node. This ensures that the order of the statistics considered remains constant (i.e., it does not tend to a unique neighborhood) and also relates the spatial extension of the neighborhood to the density of informed neighbors, which allows considering patterns at several scales. This neighborhood definition is discussed in more detail in the section on the direct sampling (DS) simulation method (see Section II.3.2.3).

A "data event" is the combination of the locations of the neighbor nodes considered, and the values at these nodes. The number of nodes in a data event is denoted n. The value at the central node of a neighborhood has to be determined such that it is somehow spatially coherent with data values at the locations in the neighborhood (this spatial coherence will be formally defined later). In most simulation methods, the aim is to compute the probability distribution function (pdf) for a node located in $\mathbf{x}$. This node is the central node of a data event (here, $\mathbf{x}$

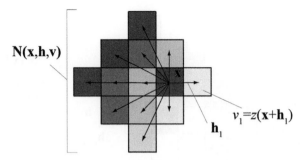

Figure II.2.7 A data event corresponding to the rightmost template of Figure II.2.6.

is a set of coordinates $\mathbf{x} = [x, y]$ in two dimensions, but it is often implemented as a grid *ID*). In Part I of this book and in some other publications, the data event is denoted as $\mathbf{dev}(\mathbf{x})$. Here, we use an alternate notation using N (Neighborhood); let $\mathbf{N}(\mathbf{x}, \mathbf{h}, \mathbf{v})$ therefore be a data event. It is a function of the lag vectors $\mathbf{h}_i$ that describes the position of each neighbor $\mathbf{x}_i$ with respect to the central node $\mathbf{x}$. In 2D, these lag vectors are expressed as $\mathbf{h}_i = [x - x_i, y - y_i]$. It is also a function of the values of the neighbor nodes $v_i = z(\mathbf{x} + \mathbf{h}_i)$, z being the variable of interest. The different elements composing a data event are illustrated in Figure II.2.7. For simplicity, we denote data events $\mathbf{N}$; however, it should be kept in mind that $\mathbf{N}$ is a complex object. The pdf for node $\mathbf{x}$ can then be written as a conditional probability of the type

$$f(\mathbf{x}) = \mathrm{Prob}\,[Z(\mathbf{x})|\mathbf{N}] \qquad (\text{II.2.5})$$

It is possible to extend the concept of a data event to multiple variables. In this case, a template is defined for each variable, hence there are M partial data events, $\mathbf{N}_1, \dots, \mathbf{N}_M$, corresponding to the M variables present, $Z_1, Z_2, \dots, Z_M$. The data event to consider for the conditional pdf in Equation (II.2.5) is the union of all M neighborhoods:

$$\mathbf{N} = \bigcup_{k=1}^{M} \mathbf{N}_k \qquad (\text{II.2.6})$$

Although $\mathbf{N}$ is a composite of several subneighborhoods, it still has a single central node, and all lag vectors $\mathbf{h}$ will be defined with respect to this central node. Figure II.2.8 illustrates such a multivariate data event. Note that in this figure, the data event has the same configuration and number of nodes for each variable, but this need not be the case. Therefore, we denote the number of nodes for each variable differently: n_k designates the number of nodes in the data event for variable k.

The data event shown in Figure II.2.8 is 2D and multivariate; therefore, three components are required in the lag vectors that characterize the position of a neighbor node relative to the central node: $\mathbf{h}_i = [x - x_i, y - y_i, k - k_{central}]$. For the

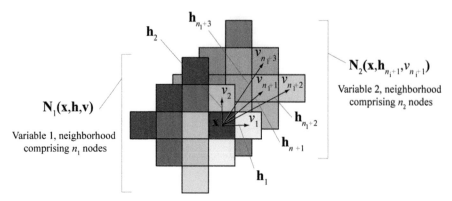

Figure II.2.8 Multivariate template and corresponding multivariate data event.

neighbors that are on the same variable $k_{central}$ as the central node, the third component of the lag vectors is zero. In our example, the term k-$k_{central}$ is equal to zero for nodes 1 to n_1 because they are on the same variable as the central node (these are all the nodes represented in gray). For the nodes n_1+1 to n_1+n_2, represented in brown, the lag vectors have a third component equal to 1, representing the offset between $k_{central}$ and the variable where the neighbor node is situated.

Note that the neighbor node $\mathbf{x}_{n_1+1}$ is collocated with the central node, meaning that it is at the same location in space, but on a different variable (in a sense, it can be seen as living in a parallel dimension). In semivariogram-based geostatistics, a number of multivariate methods aim at performing co-kriging or collocated cosimulation. In MPS, to capture complex relationships, we consider the full multiple-point and multivariate relationship between $\mathbf{x}$ and $\mathbf{x}_{n_k+j}$, with $j = 1 \ldots n_k$ and $k = 1 \ldots M$. Hence all neighbors, on all variables, provide their share of information on what occurs at $\mathbf{x}$. The disadvantage is that such complex spatial models generally require the specification of TIs that are sufficiently large and complete.

As a summary, Table II.2.2 provides an overview of the definitions and nomenclature used in this chapter.

2.4 Storage and restitution of data events

MPS deals with possibly large, multiple and multivariate TIs, grids of considerable size, as well as possibly complex neighborhoods in 3D or 4D. Hence, although the principles formulated in MPS methods and algorithms are generally straightforward and intuitive, their practical implementation presents caveats.

In dealing with the large amount of information and complexity, a learning and restitution approach is taken. In an initial learning step, the large amount of data events present in the TI are learnt by an algorithm – this is the phase

Table II.2.2 Summary of terminology

Neighborhood	Spatial context of a grid node **x**, generally characterizing its local surroundings. It is used as a generic term that does not have a formal definition.
Data event	Set of values **v** and lag vectors **h**, which are defined with respect to a central node **x**.
Template	Set of predetermined and ordered lag vectors **h**, which are defined with respect to a central node, and are intended to be used for organizing a database (see Section II.2.4).
Pattern	Generic term used to informally designate a spatial arrangement of nodes or pixels on a subset of a grid.
Patch	Square or rectangular pattern made of several nodes, which is used as a spatial unit in certain types of simulation methods. Seen in opposition to pixel-based methods, for which the spatial unit is a single node.

when information is stored. In a second step, this learnt knowledge is used to restitute simulations that present desired characteristics of the TI. Note that if certain characteristics of the TI are not desired in the simulation, one approach to filter out these characteristics is to avoid learning (storing) them in the initial phase.

In the following sections, we review the possible approaches for both the storage and restitution steps.

2.4.1   Raw storage of training image

The simplest way for an algorithm to store the properties of the TI is to keep the entire TI in memory, and to keep it under the form of an image. This is what we call a raw version of the TI, in contrast with other storage methods where the TI is not kept in its original (unprocessed) form, but transferred into a database from which data events can be stored and retrieved. This form of information storage was used in the earliest MPS approaches developed (Guardiano and Srivastava, 1993) and is still relied upon by several simulation approaches (see Chapter II.3 on simulation algorithms).

2.4.1.1   Calculating conditional frequencies using convolutions

One drawback of raw storage is that it does not allow selecting which properties of the TI are stored and which others are not. Therefore, if certain characteristics need to be filtered out, this has to be done at the restitution step. In addition, having a single large object in memory is not optimal for restituting the conditional probabilities of data events. With this form of storage, obtaining a probability of a nodal value conditionally to a data event **N** can be computationally demanding for large TIs. The procedure consists of analyzing the TI and counting all occurrences of node configurations equal to **N**. Such an exhaustive analysis of the TI is effectively a form of convolution. Then, the conditional probability of finding category c at the central node is obtained by the ratio of the number of occurrences with this category over the total number of occurrences with any category:

$$P[Z(\mathbf{x}) = c|\mathbf{N}] = \frac{\# \left(Z_{TI} = c|\mathbf{N}\right)}{\sum\limits_{b=1}^{C} \# \left(Z_{TI} = b|\mathbf{N}\right)}, \tag{II.2.7}$$

where C is the total number of categories. This procedure has to be repeated each time a conditional probability is needed, which means at each simulated node. Hence, the number of data events to be analyzed is of the order of $\# nodes_{TI} \times \# nodes_{SIM}$. Note that Equation (II.2.7) is general for the computation of conditional frequencies and will be used again further in this chapter.

Consider in more detail the notion of "convolution" in this context. Given a data event $\mathbf{N_x}$, we analyze the TI to determine if the pattern centered on each TI node $\mathbf{y}$ is identical to $\mathbf{N}$. The result is a logical variable of size identical to the eroded TI, informing whether each TI neighborhood $\mathbf{N}\{TI(\mathbf{y})\}$ hosts the same spatial pattern as $\mathbf{N}$:

$$TI(\mathbf{y}) * \mathbf{N} = \begin{cases} 0 & \text{if } \mathbf{N}[TI(\mathbf{y})] \neq \mathbf{N} \\ 1 & \text{if } \mathbf{N}[TI(\mathbf{y})] = \mathbf{N} \end{cases} \tag{II.2.8}$$

Figure II.2.9 illustrates this type of convolution for a simple data event $\mathbf{N}$ consisting of four nodes, all of them black, including the central node. Five occurrences of this data event are identified in the TI, resulting in a convolution grid $TI(\mathbf{y}) * \mathbf{N}$, the star (*) being the conventional symbol for convolution. The resulting grid mostly consists of zero values, except for five locations where the data event is matched. Note that the external (eroded) area of the TI, depicted in orange, can never contain a matching data event. In this example, the non-eroded part of the TI contains 36 pixels, and therefore the frequency of occurrence of the data event $\mathbf{N}$ is $\#(Z_{TI}|\mathbf{N}) = \frac{5}{36}$. By performing such a convolution, one can determine each of the counters in Equation (II.2.7) and then determine the conditional frequency of Z at a given grid node. This idea was employed in the very first MPS algorithm (see Section II.3.2.1), and modern patch-based algorithms still rely on this convolution framework.

2.4.1.2 L2-based convolution
The principle of convolution can be applied to continuous variables by extending the pattern matching as an error term instead of a binary choice of match–no match. The convolution has then the form

$$TI(\mathbf{y}) * \mathbf{N} = \sum_{\mathbf{N}} |\mathbf{N}[TI(\mathbf{y})] - \mathbf{N}|^p \tag{II.2.9}$$

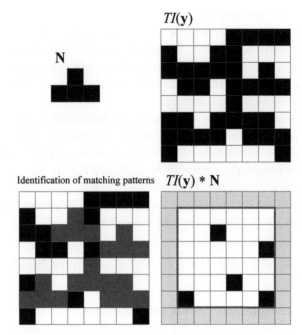

Figure II.2.9 Illustration of a convolution for a binary variable.

where p is an error norm, usually equal to 2, resulting in a Euclidean norm. Other types of distances can be used with the same convolution principle (see Section II.2.5 for an overview).

Figure II.2.10 illustrates such a convolution with a continuous variable. The pattern shown in the center of Figure II.2.10 depicts a gradient from dark to light in a diagonal direction. This pattern is compared to every location in the image of the child's face using Equation (II.2.9). For each pixel y, the convolution value is represented. Because the size of N is 5×5 pixels, convolution values cannot

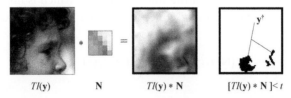

Figure II.2.10 Illustration of a convolution for a continuous variable. From left to right: original image, pattern sought for, convolution result, and threshold-based selection. The two right images are the result of a convolution of the left image using the pattern N. N is enlarged 30 times for better visualization. On the convolution figure, dark values represent lower convolution values (hence, high similarity with N).

be computed for the two-pixel area at the edge of the image. The darker values correspond to better similarity with **N** (lower convolution values), for example on the edge of the character's nose, where the transition between dark and light colors corresponds to what is observed in **N**. Conversely, the hair area has dark values, occurring in a range that is different from the values in N and resulting in high convolution values in this region.

Similar to the categorical case, a conditional probability for a continuous value can be based on the convolution results. One way to proceed is to define a threshold t in the convolution values. All locations $\mathbf{y}^{\dagger}$ whose value falls under the threshold can be considered similar to **N**. In Figure II.2.10, this is represented by the black areas in the rightmost figure. The conditional cumulative distribution function (ccdf) of $Z(\mathbf{x})$ in the simulation can then be approximated by the experimental distribution of all TI values at locations $\mathbf{y}^{\dagger}$:

$$F[Z(\mathbf{x}) \leq z|\mathbf{N}] \cong F[Z_{TI}(\mathbf{y}^{\dagger}) \leq z], \text{ with } \mathbf{y}^{\dagger} : \{[TI(\mathbf{y}^{\dagger}) * \mathbf{N}] < t\} \quad (\text{II.2.10})$$

2.4.2 Cross-correlation based convolution

Although being intuitively straightforward, Equation (II.2.9) can be computationally demanding due to the calculation of the exponent p. In the case of $p = 2$, it has been proposed to calculate instead the correlation coefficient as a faster alternative (Theodoridis and Koutroumbas, 2008; Tahmasebi et al., 2012).

Taking again the definition of a Euclidean norm in Equation (II.2.9), we get a sum of squared differences that can be developed as follows:

$$\sum_{\mathbf{N}} |\mathbf{N}[TI(\mathbf{y})] - \mathbf{N}|^2 = \sum_{\mathbf{N}} \mathbf{N}[TI(\mathbf{y})]^2 + \sum_{\mathbf{N}} (\mathbf{N})^2 - 2 \sum_{\mathbf{N}} \mathbf{N}[TI(\mathbf{y})] \cdot \mathbf{N}. \quad (\text{II.2.11})$$

The first term can be approximated as constant if we assume that the TI is stationary and that the spatial variability occurs at a scale smaller than the size of the data event **N**. The second term is also constant for a given pattern that is searched for. This leaves us with the third term only, which is nothing but the expression of the cross-correlation sequence between $TI(\mathbf{y}) * \mathbf{N}$ and **N**:

$$CC\{\mathbf{N}[TI(\mathbf{y})], \mathbf{N}\} = \sum_{\mathbf{N}} \mathbf{N}[TI(\mathbf{y})] \cdot \mathbf{N}. \quad (\text{II.2.12})$$

Equation (II.2.12) represents a lighter computational load than Equation (II.2.11) because the square term has been dropped. However, because of the strong reliance on the stationary assumption, this measure can be very sensitive to local variations in the TI that are at a scale larger than **N**. One solution is to choose a very large **N**, but this can be impractical. Another solution is to use a normalized version of Equation (II.2.12), but then the computational advantage is lost:

$$CC_N\{\mathbf{N}[TI(\mathbf{y})], \mathbf{N}\} = \frac{CC\{\mathbf{N}[TI(\mathbf{y})], \mathbf{N}\}}{\sqrt{\sum_N \mathbf{N}[TI(\mathbf{y})]^2 \sum_N \mathbf{N}^2}}. \qquad (\text{II.2.13})$$

2.4.3   Partial convolution

Alternative approaches have been proposed to avoid performing the full convolution in Figure II.2.10, based on the fact that the exhaustive map of $TI(\mathbf{y}) * \mathbf{N}$ is often not required. One such alternative is the DS method (Mariethoz et al., 2010) which is based on the same principle as a convolution, but interrupts the convolution as soon as a data event honoring the condition $\mathbf{N}\{TI(\mathbf{y})\} \cong \mathbf{N}$ is found. This approach, which can be seen as a partial convolution, is also based on raw TI storage. It is described in detail in Section II.3.2.3.

Another attractive way of proceeding is by using a hierarchical multiscale search. The TI then needs to be available at several resolutions, the coarser resolutions being obtained by subsampling or averaging the fine ones. The data event searched for is also transformed in the same coarse resolutions. Then, performing a convolution of the coarsest resolution of the TI with the coarsest data event is very fast. The best matching location is then selected, and the corresponding area at the finer scale is identified. This subarea is then used to refine the search at the next resolution, but only considering the subset of the TI that was identified at the coarsest scale. Repeating this operation for all refinement levels very quickly yields a matching data event (Theodoridis and Koutroumbas, 2008). One drawback with this approach is that there is no guarantee that the final match is the best one in the entire TI, because often, the area determined at the coarsest levels may not contain the best matching event for the fine levels. Another drawback is that the approach yields a single matching data event, and may therefore result in reduced stochasticity in the data events retrieved. In contrast, methods that perform single-scale convolutions yield several candidate locations distributed over different areas of the TI, which can then be sampled (Figure II.2.10).

A similar way of proceeding is by using a multiscale gradient search where for each scale a gradient-driven walker roams through the TI seeking the best matching data events (Abdollahifard and Faez, 2013). The walker gets trapped in local minima, but this can be alleviated by using multiple walkers. Last, we mention the principle of logarithmic search, which works at a single resolution but sequentially narrows down the search to the most promising part of the TI.

2.4.4   Tree storage

A more elaborate way of storing data events is in a tree-like logical structure (Strebelle, 2002). Tree structures are a general way of classifying database elements in a hierarchical manner. The general organization of a tree goes from general (the trunk of the tree) to particular (the leaves). Consider first the example of living species that have traditionally been classified in such a hierarchical way. Whereas modern-day classifications are highly complex, an eighteenth-century classification is a good illustration of the principle. In these systems, any life form could be classified as plant, animal, or protist (Figure II.2.11), which corresponds to the highest taxonomic rank. Lower ranks allow refining the classification further by dividing at each level the number of species that can be included in a category. Animals are divided into vertebrates and invertebrates; vertebrates are further divided into fish, amphibians, reptiles, and mammals; and so on. The highest ranks are kingdoms, which are further divided into phyla, classes, orders, and so on until the species level.

Now imagine that a new species is discovered. The existing organization of the tree can be used to store it jointly with similar species that were previously known. The tree structure allows one to easily verify that the species is not already known, because one only needs to compare with the small group of species that are in the last leaf level.

Another useful property of the tree structure is that each branch contains some information about all the levels that come below it. All species that come under the vertebrates' branch have common characteristics because they possess bones. Although a large variety of animals are found in this category, the existence of a skeleton is a common characteristic that unifies this vast ensemble. Similarly, it is possible to represent a prototype of the mammal branch that would reflect a sort of average of all subsequent tree leaves (here we can imagine a warm-blooded, viviparous, four-legged furry creature). Such prototypes can be defined for each level of the tree. The key to classification is to find common characteristics that can be assessed as true or false for all elements of the database (e.g., have a skeleton? yes/no).

Determining the species of a life form can be done very rapidly because the identification of these essential characteristics (or the comparison with a prototype) allows one to dramatically reduce the search space. Without the tree classification, the approach would consist of systematically comparing the unknown creature with all known life forms until a match is found – clearly, such an exhaustive search is a very inefficient way of proceeding. In MPS, tree-based patterns classification will rely on similar principles as taxonomy. When classifying living things, it can be challenging to identify common characteristics that can be assessed as true or false for each creature. When dealing with patterns, the question of identifying such classification criteria is greatly simplified, thanks to the notions of template and data events that we have established in Section II.2.3.

Consider the simple template represented in Figure II.2.12 consisting of four nodes: a central node and three neighbor nodes. As a starting point, we will look at a binary variable; therefore, the value at each node can only be 0 or 1. By convention, white pixels have a value of 0, and black pixels have a value of 1. The

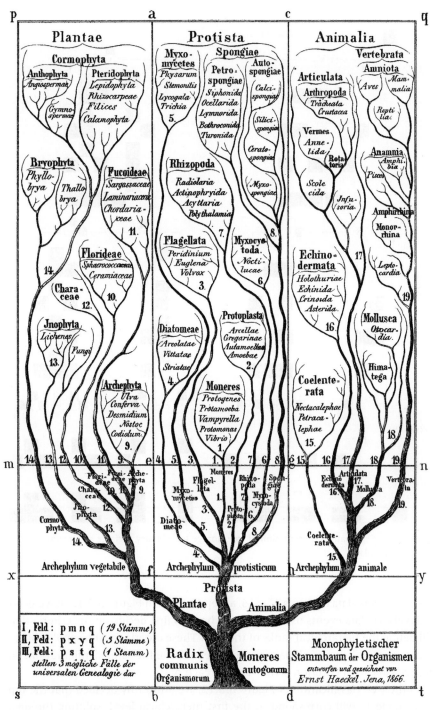

Figure II.2.11 Tree-based classification of the different life forms. From Haeckel, E. (1866). *Generelle Morphologie der Organismen: allgemeine Grundzüge der organischen Formen-Wissenschaft, mechanisch begründet durch die von C. Darwin reformirte Decendenz-Theorie*, Berlin, downloaded from http://en.wikipedia.org/wiki/Ernst_Haeckel

Figure II.2.12 Schematic representation of tree storage for a binary variable and a simple four-node template.

total number of data events is $2^4 = 16$, as represented in Figure II.2.12. If we want to classify 16 data events using a tree, we have to first decide on the criteria that will define the hierarchical ranks of the tree. These criteria must be applicable to all data events. Because all patterns considered have the same configuration, an ideal classification criterion is the value at each of the template nodes.

In the tree classification of Figure II.2.12, we arbitrarily decided that the top node (node 1) will correspond to the first hierarchical level, splitting the data event population in two. Then, nodes 2 and 3 (the left and right nodes in the template) correspond to the second and third levels. With these three levels, the tree has eight leaves, each comprising of two data events: one with a value of 0 at the central node, and another with a value of 1. Any data event corresponding to the defined template can then be classified in this tree.

If more than two categories are considered, the concept is the same, except that instead of having two branches at each level, there are C branches, where C is the number of categories. In total, the tree has n levels (n is the number of nodes in the template), and the number of branches is multiplied by C at each level.

Imagine now that data events from a TI are stored in a tree. Because each data event can be located in the tree, it is easy to keep track of the occurrence of each category c at the central node of each possible data event $\mathbf{N}$ with a counter, $\#[Z(\mathbf{x}) = c|\mathbf{N}]$. In the example of Figure II.2.12, we consider the leftmost bottom data event: all three neighbor nodes are white (category 0). The central node can be either white or black. When one such configuration of neighbors is encountered with a white value in the central node, the counter $\#[Z(\mathbf{x}) = 0|\mathbf{N}]$ is incremented. When the central value is black, the counter $\#[Z(\mathbf{x}) = 1|\mathbf{N}]$ is incremented. Based on the counters, the probability of having a white value at the central node can be approximated as the frequency of occurrences:

$$P[Z(\mathbf{x}) = 0|\mathbf{N}] \frac{\#(Z_{TI} = 0|\mathbf{N})}{\#(Z_{TI} = 0|\mathbf{N}) + \#(Z_{TI} = 1|\mathbf{N})}. \qquad (II.2.14)$$

The general principle is that once all data events present in a TI have been incorporated into the tree, the TI is no longer needed. As in the case of raw storage of the TI, the conditional frequencies are based on Equation (II.2.7). However, the retrieval of the counters is done differently: here, the counters are not obtained by a convolution operation, but by a lookup of the number of occurrences that have been pre-computed in the tree.

With this template (and therefore the same search tree), we now analyze the TI of Figure II.2.13. The first step is to consider the eroded TI. This zone is delineated in red on Figure II.2.13 and comprises 12 nodes, each of them being the central node of a data event. The corresponding 12 data events are displayed in the lower part of Figure II.2.13. This implies that out of the 16 possible data events with this template, four are not present in the TI. The 12 data events present are then stored in the tree, and this is done simply by updating the counters that are at each branch and each leaf of the tree. Figure II.2.14 illustrates the tree with the counters updated so that they store the 12 data events. The initial root level of the tree has a counter of 12, indicating that a total of 12 data events are considered, and each level further subdivides these data events, with the sum of counters for each level being always 12. Note that there are several branches with zero occurrences.

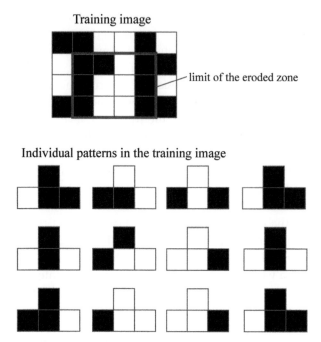

Figure II.2.13 Example TI and all of the individual data events that compose it, considering a simple four-node template.

One general requirement for using a tree to store patterns is that data events are defined on fixed templates. The template has to be the same for all data events considered. In the categorical case, the classification is based on the numbering of the template nodes, and changing the location or numeration of the nodes would destroy the entire classification. In the continuous case, all data events considered need to have the same configuration as the prototypes, otherwise it would not be possible to carry out a comparison between data events and prototypes. This requirement becomes problematic when a mix of large and small data events is needed to represent multiscale structures. Therefore, the solution is to construct multiple trees with one tree per scale considered (see Section II.2.7 on multiple grids), with each tree storing patterns belonging to multiple-scale representations.

Often in multiple-point simulation, incomplete data events occur (i.e., some node values in the template are known, and others have not yet been simulated). In these cases, a recursive search has to be carried out throughout the tree to sum the counters corresponding only to the known nodes. Computationally, such a search is efficient because at each branch, there is a counter that informs how many data event occurrences lie in all the leaves subordinate to the branch. Summing the corresponding counters then allows for the calculation of conditional

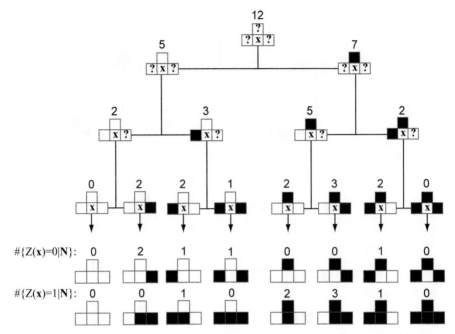

Figure II.2.14 Complete search tree with all counters corresponding to the storage of the TI shown in Figure II.2.13.

frequencies in Equation (II.2.7). The approach is computationally efficient, but the drawback is that the tree storage requires large amounts of memory space.

The theoretical number of branches (and therefore counters) in a tree with N levels and C branches at each level has an exponential form:

$$\#\text{counters} = \sum_{n=1}^{N} C^n, \tag{II.2.15}$$

which means that the tree storage can be impractical in cases with large templates and numerous categories (>4). For example, consider $N = 60$ nodes and $C = 5$; according to Equation (II.2.15), the corresponding tree would have over 10^{42} counters. One solution is to only store in the tree the data events that are actually present in the TI, or in other words not to represent the branches where the occurrence counters are zero. The tree size of Equation (II.2.15) is therefore a theoretical maximum, assuming that the TI contains all the possible data events with the template considered. This usually drastically reduces the tree size. However, for large and pattern-rich images, the storage of search trees can still pose a real problem. Figure II.2.15 shows the same tree as previous, but with all branches and leaves that have zero occurrences removed.

Further reduction in the size of the search tree can be accomplished by collapsing the singleton branches of the tree (Zhang et al., 2012). A singleton branch

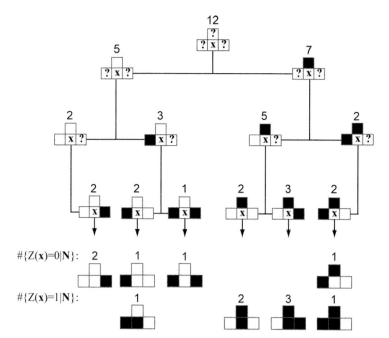

Figure II.2.15 Search tree omitting zero-occurrence counters, corresponding to the storage of the TI shown in Figure II.2.13.

is a series of levels where all but one of the counters is zero. If this occurs for a succession of several levels, then all these branches can be collapsed into a single counter. Figure II.2.16 shows the same tree, but now applying the collapsing of singleton branches. In this case, only the branch corresponding to the rightmost pattern (shown in red) is collapsed; however, for complex TIs and large trees, most of the tree often consists of singleton branches. In such cases, the collapsing can reduce the tree size by a factor 10 to 100. The advantages of such an approach are twofold because, on one hand, they allow reducing the size of the tree, and, on the other hand, the search procedure is accelerated because the tree is simplified.

To conclude on the topic of tree-based storage, we visit the case of continuous variables. The nature of tree structures is very appropriate for storing categorical variable patterns because, at each branch of the tree, a discrete choice has to be made. However, the tree can also accommodate continuous variables through the use of prototypes. If we look again at the case of the classification of living organisms, some classification choices are discrete (skeleton–no skeleton), whereas others are more subtle and do not rely on a unique criterion. For example, to define a mammal, one has to look at multiple criteria because there are exceptions, such as platypuses laying eggs. In a similar way, it is possible to define a group, or cluster of similar continuous patterns, by means of a prototype

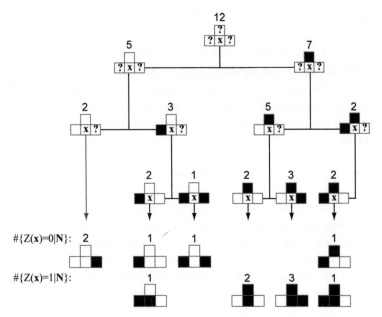

Figure II.2.16 Compact search tree with collapsing of singleton branches, corresponding to the storage of the TI shown in Figure II.2.13.

consisting of the average of all patterns in that group. At each level of the tree, the decision to classify a given pattern in a certain category is based on the comparison between this pattern and the prototypes corresponding to the different choices possible. Although this type of continuous-variable tree has not yet been investigated in the MPS literature, it is commonly used in computer graphics applications (Wei and Levoy, 2000). In this field, the principle of Tree-Structured Vector Quantization (TSVQ) was first introduced in the field of texture synthesis. The basic principle of the TSVQ is a binary tree. Given a set of training data events, the first step is to compute the centroid of the set (the average pattern) and use it as a prototype for the root level of the tree. Then, the set of patterns is split in two groups using a Lloyd algorithm (Gersho and Gray, 1992), which consists of alternations between the centroid computation and nearest centroid partition. This is then carried out for each level of the tree, and each time a prototype represents all patterns in the subtree below a branch. The tree-building process terminates when the number of prototypes exceeds a certain size or when the prototypes are very similar to the TI data events. To improve the efficiency of searching in the tree, the construction process also ensures that the tree is balanced (i.e., that there is a similar number of data events on each side of a crossing). With a well-balanced tree, searching for a data event in the tree can be accomplished using $\log(N_{TI})$ operations, where N_{TI} is the number of different data events present in the TI.

2.4.5 List storage

An alternative to tree storage is to store the TI patterns in the form of a list (Straubhaar et al., 2011). The advantage of a list compared to a tree is that it takes less memory space. For a given TI and template, the list of patterns is exactly equivalent to the last level of the tree (the leaves). The gain in memory lies in the fact that the counters of all upper levels of the tree are not present. For a categorical variable, the theoretical maximal size of the list is the total number of data events:

$$\#\text{counters} = C^n, \tag{II.2.16}$$

which is necessarily smaller than in Equation (II.2.15). The total size is generally smaller than this value because the elements with zero occurrences can be removed. The gain in memory, however, comes at the price of a slower search in the patterns database, because when searching for a pattern the entire list has to be scanned: N_{TI} operations are required for searching a single data event. In the case of the tree, the counters present at each branch can be summed without needing to reach the leaf level. This loss of CPU efficiency can, however, be offset with parallelization of the search in the list. It consists of dividing the list in P subsets, where P is the number of available processors. To obtain the number of occurrences for a given data event, each processor searches its own partial list and adds up the counters corresponding to the data event searched for. In a second step, all P partial list counters are summed to obtain the total number of occurrences to be used in Equation (II.2.7).

If we consider the same example as in Figure II.2.14, there are 16 possible data events and as many counters. Figure II.2.17 shows the structure of the list in this case, with all counters represented (including the counters with no occurrences).

A concept similar to the storage of patterns in a list is the multiple-point histogram (MPH). The classical histogram is simple and does not measure any spatial property because it is only based on counting the frequencies of occurrence of values in given ranges (or bins).

The multiple point histogram works similarly as the traditional histogram, but it counts the frequency of entire patterns instead of the frequency of values. It is therefore a spatial extension of the histogram (Deutsch and Gringarten, 2000). Within a given fixed template, it records the number of occurrences of all possible pattern configurations in a given image. These occurrences are then stored in a table that is a counting distribution of the patterns for that template configuration. To create a meaningful counting distribution, one has to use a small template (e.g., 4×4 and, for 3D, 3×3×2); otherwise, each bin would contain either one or zero patterns. Additionally, the amount of possible configurations grows rapidly as the size of the template and the number of categories increase. The amount of empty bins increases rapidly as well (Boisvert et al., 2007). For these reasons, the MPH is only useful for small cases with few categories.

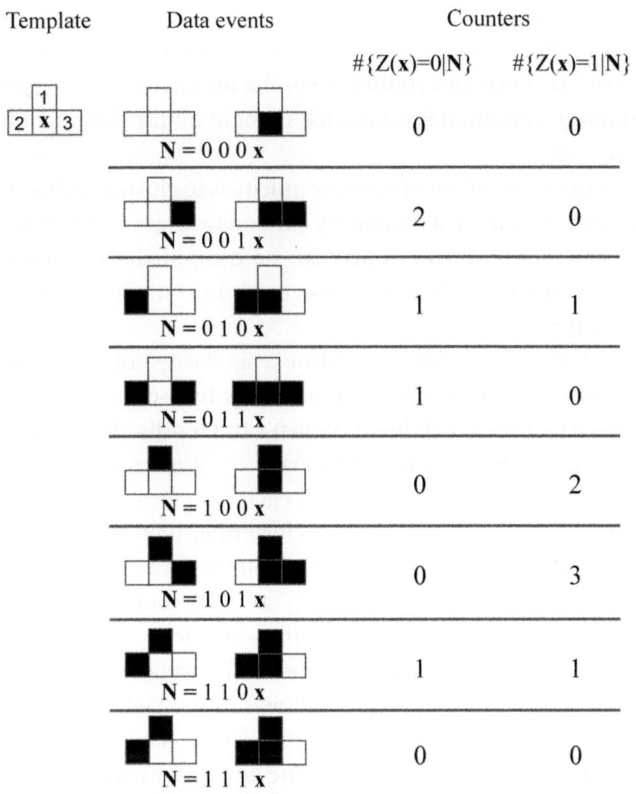

Template	Data events	Counters	
		$\#\{Z(\mathbf{x})=0\|\mathbf{N}\}$	$\#\{Z(\mathbf{x})=1\|\mathbf{N}\}$
$\begin{array}{c}\boxed{1}\\ \boxed{2}\,\boxed{\mathbf{x}}\,\boxed{3}\end{array}$	$\mathbf{N}=0\,0\,0\,\mathbf{x}$	0	0
	$\mathbf{N}=0\,0\,1\,\mathbf{x}$	2	0
	$\mathbf{N}=0\,1\,0\,\mathbf{x}$	1	1
	$\mathbf{N}=0\,1\,1\,\mathbf{x}$	1	0
	$\mathbf{N}=1\,0\,0\,\mathbf{x}$	0	2
	$\mathbf{N}=1\,0\,1\,\mathbf{x}$	0	3
	$\mathbf{N}=1\,1\,0\,\mathbf{x}$	1	1
	$\mathbf{N}=1\,1\,1\,\mathbf{x}$	0	0

Figure II.2.17 Data events storage in a list, corresponding to the storage of the TI shown in Figure II.2.13.

In essence, list storage and multiple-point histogram are the same thing: the MPH of the image represented in Figure II.2.13 consists of the patterns and values indicated in Figure II.2.17. The main difference lies in their usage: whereas a list is used to store a large number of patterns in order to restitute them using a simulation algorithm, a MPH is used to characterize a spatial variable and extract some of its basic characteristics.

2.4.6 Clustering of patterns

In the case of simulation by patches, specific storage strategies have to be designed. These methods populate a grid proceeding with groups of nodes simultaneously (see Section II.2.6.3). Each group of nodes is termed a patch or a block, consisting of n nodes. For categorical variables, the simplest way of storing patches is in a list identical to that in Figure II.2.17 (Arpat and Caers, 2007). The same strategy is also applicable for continuous variables, such as in the SIMPAT algorithm (Section II.3.3.1). A drawback, however, for the continuous case is that the nature of a continuous variable defined in $\mathbb{R}$ implies that there are no two identical data events in a given TI. As a result, all counters in the list would be equal to 1 and there would be N_{TI} elements: as many elements as there are data events in the TI. Notwithstanding the problems related to storing such a list, using it to compute conditional frequencies would be problematic knowing all frequencies are $1/N_{TI}$.

The solution to this problem of counter unicity is to cluster similar data events in categories. Then, because each category contains a larger number of individual data events, frequencies of occurrence can be computed. Once categories are defined, an easy way to classify one data event is to compare it to prototypes at each subtree segment.

The key to pattern clustering is to define a similarity criterion between data events. One approach to defining such criteria is to use filter scores. In a first step, N_f elementary filters are defined, which can identify the basic shapes that are expected to occur. Then, each TI data event is compared to these filters, and the result of this comparison is a score. For each data event, a vector of N_f filter scores is obtained and used as a basis for clustering, typically using a k-means algorithm. Figure II.2.18 illustrates the use of filters. As a result, the data events are classified into k categories, with $k \ll N_{TI}$, and the prototype of a category is the average of all data events contained in this category.

At the time of simulation, comparing the neighborhood of the location to be simulated, namely **N**, with all prototypes allows finding which category is most compatible with the neighborhood. One data event is then chosen randomly in this category and used for the simulation. The advantages are:

• Because there are many occurrences in each category, the counters for each category are higher than 1, and therefore stochasticity can be introduced in the simulation;

• The list to scan is much smaller than the number of data events present in the TI;

• The use of filters allows one to customize the classification to specifically represent patterns relevant for a given application. This flexibility can also be seen as a disadvantage because, in many applications, one may not know in advance what are the most appropriate filters to use.

Filters can be seen as a way of reducing the dimensionality of the patterns from n (the number of nodes in a pattern) to N_f (the number of filters used). An alternative way of reducing the dimensionality of patterns is to use wavelet coefficients instead of filters (Chatterjee and Dimitrakopoulos, 2012; Chatterjee et al., 2012). Each wavelet coefficient characterizes the frequency component of a patch. The k-means classification of data events is then identical, except that it is based on the wavelet coefficients instead of the filter scores. Similar to filters, the wavelets approach necessitates choosing a wavelet basis function as well as the scale of the wavelet decomposition.

A general way of framing the classification of patterns is by using a distance-based approach that relies on multidimensional scaling (MDS). Given a set of

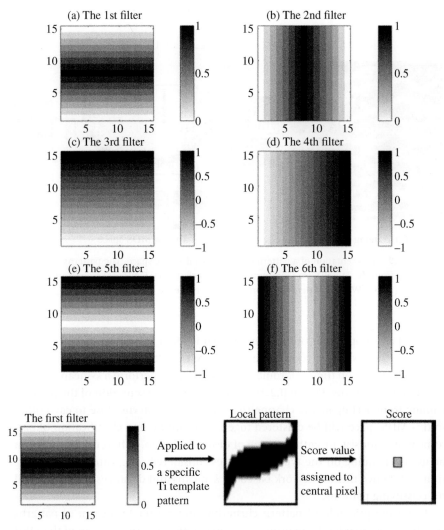

Figure II.2.18 Representation of the filters used in the FILTERSIM algorithm and illustration of the computation of a filter score for a given data pattern. Modified from Zhang, T., et al. (2006). With kind permission from Springer Science and Business Media.

training data events, one can compute a distance matrix **D** that describes the dissimilarity between each data event and all other data events. The approach is general because this distance can be based on a norm, filters, wavelets, or any other relevant metric (see Section II.2.5). MDS then translates **D** into a configuration of points defined in an n-dimensional Euclidean space $\Re$ (Cox and Cox, 1994; Borg and Groenen, 1997; Caers, 2011). The points in this spatial representation are arranged in such a way that their Euclidean distances in $\Re$ correspond as much as possible to the dissimilarities of the objects. The MDS algorithm can

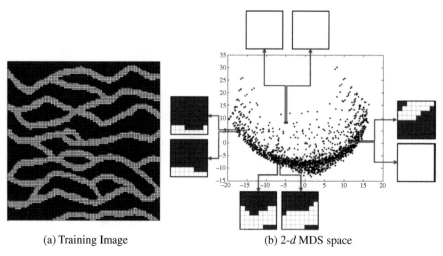

(a) Training Image   (b) 2-*d* MDS space

Figure II.2.19 Representation of a pattern database in a distance space. The patterns originating from the TI (left) are represented in 2D space (right). Figure (a) from Honarkhah and Caers (2010). With kind permission from Springer Science and Business Media.

be posed as an optimization problem, which is often solved using least-squares methods. The result is a coordinates matrix $\mathbf{X}$ that defines the position of each point in $\mathfrak{R}$. Based on these coordinates, it is possible to perform a k-means classification of the patterns. Figure II.2.19 shows a MDS representation of the patterns coming from a TI depicting channelized structures. Patterns close to each other in this MDS space will be clustered together, resulting in a classification of patterns that is consistent with the type of distance chosen. Different distances can be used depending on the type of variables considered (e.g., categorical or continuous), hence this framework based on k-means and distances is very general for classifying patterns.

Using the principle of pattern clustering, the cluster-histogram of patterns (CHP) is a tool that has been proposed as an extension of the concept of the MPH (Honarkhah and Caers, 2010), in the DISPAT algorithm. Instead of counting the frequencies of each pattern, which often collapse to zero, the CPH constructs distributions based on the number of patterns within each cluster (Tan et al., 2013). To achieve this, the patterns are clustered into groups that are defined by a meaningful distance. For each cluster, the number of patterns is recorded and the prototype is calculated that defines the cluster (in a similar way as for the continuous-variable trees described in this chapter). This prototype can be seen as representative of each cluster. Typically, the prototype could be the mean of the patterns within that cluster or a medoid pattern. Figure II.2.20 shows an example of a set of realizations generated with three different simulation algorithms and the input TI (these algorithms are described in detail in Chapter II.3). Forty-eight clusters were generated based on the TI.

Training image

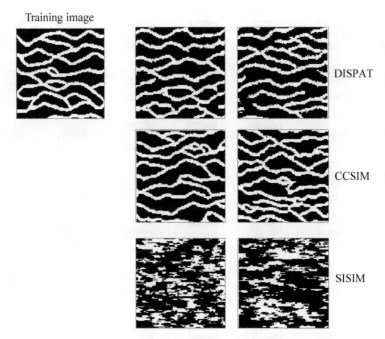

DISPAT

CCSIM

SISIM

Figure II.2.20 A TI and two realizations (from a total of 50) generated with three algorithms: DISPAT, CCSIM (see Chapter II.3), and sequential indicator simulations (SISIMs; Deutsch and Journel, 1992).

Figure II.2.21 shows the number of patterns that occur in each of these clusters for the four cases considered (TI and three simulation algorithms). Figure II.2.21 represents the CHP, which is then a summary of the pattern frequencies in the images. Comparing the different CHPs allows determining if the simulations correctly reproduce the pattern frequencies present in the TI.

2.4.7   Parametric representation of patterns

A completely different way of storing the patterns of a TI is by using a parametric model. Semivariogram-based geostatistics is based on the parametric expression of the spatial relationship between two locations. This principle has been extended to interactions between more than two points, in a mathematical framework using spatial cumulants (Dimitrakopoulos et al., 2010).

In probability theory, cumulants define a probability distribution in an equivalent way as moments, and the formulation of cumulants is based on a combination of moments. Their principle is to describe in a probabilistic sense the interactions between multiple locations, and hence they can be seen as a generalization of the covariance to higher orders. Because higher moments are considered, using cumulants allows one to better capture complex geological structures, non-Gaussian features, and connectivity patterns present in the TI.

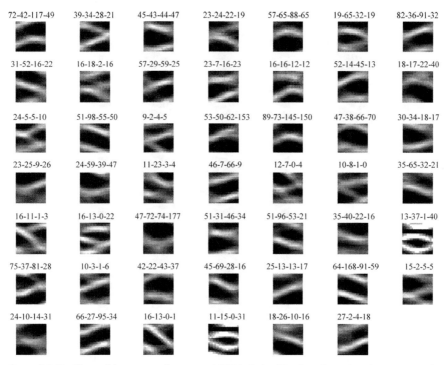

Figure II.2.21 Cluster-histograms of patterns (CHPs) derived by first clustering the patterns of the TI, then classifying the patterns of the realizations based on these clusters. Each image is a prototype representing a cluster of patterns, and the numbers above each prototype are the frequencies associated to this cluster for each of the four cases considered (TI, DISPAT, CCSIM, and SISIM).

For a random variable Z, the nth-order cumulant is defined as a function of the relative positions of the n points considered. The second-order cumulant, which describes the interaction of two points separated by a vector $\mathbf{h}$, is defined as

$$c_2(\mathbf{h}) = E[Z(\mathbf{x})Z(\mathbf{x}+\mathbf{h})] - E[Z(\mathbf{x})]^2. \qquad (\text{II}.2.17)$$

Higher-order cumulants require the definition of a template (i.e., a set of fixed lag vectors). Once these lag vectors are defined, one can compute high-order cumulants. For example, the third-order cumulant is given by:

$$\begin{aligned}
c_3(\mathbf{h}_1, \mathbf{h}_2) = {}& E[Z(\mathbf{x})Z(\mathbf{x}+\mathbf{h}_1)Z(\mathbf{x}+\mathbf{h}_2)] \\
& -E[Z(\mathbf{x})]E[Z(\mathbf{x}+\mathbf{h}_1)Z(\mathbf{x}+\mathbf{h}_2)] \\
& -E[Z(\mathbf{x})]E[Z(\mathbf{x}+\mathbf{h}_1)Z(\mathbf{x}+\mathbf{h}_3)] \\
& -E[Z(\mathbf{x})]E[Z(\mathbf{x}+\mathbf{h}_2)Z(\mathbf{x}+\mathbf{h}_3)] + 2E[Z(\mathbf{x})]^3
\end{aligned} \qquad (\text{II}.2.18)$$

General formulations for higher moments can be found in Dimitrakopoulos et al. (2010); however, orders higher than 5 are rarely used in practice. In practice, the nth order cumulant can be computed from a TI or a densely informed data set,

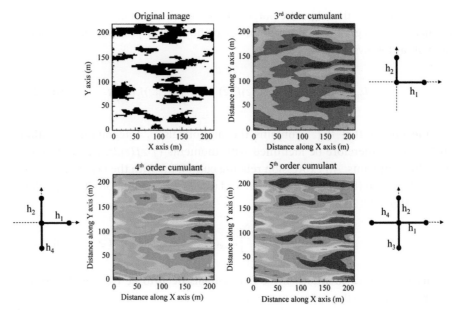

Figure II.2.22 Spatial cumulants of orders 3, 4, and 5 computed on a simple binary image. Next to each cumulant map, the corresponding lag vectors are displayed. Reprinted from Mustapha, H. and R. Dimitrakopoulos (2010). With permission from Elsevier.

given a set of locations $\mathbf{x}_1, \ldots, \mathbf{x}_n$, whose relative locations are defined by the lag vectors $\mathbf{h}_1, \ldots, \mathbf{h}_{n-1}$. Figure II.2.22 illustrates the cumulants of orders 3, 4, and 5 computed on a simple 2D image.

Cumulants are therefore obtained by scanning a TI for a range of lag vector configurations. The cumulant value for each configuration of nodes can then be treated in a similar way as an experimental frequency (Equation (II.2.7)), and stored in a tree or in a list.

Cumulants can be used to derive expressions for probability density functions. Consider first the univariate case and density $f(z)$ with moments m_i and cumulants $c_i, i = 1, \ldots \infty$. One way of approximating this density, knowing all of its cumulants, is by means of (orthogonal) Legendre polynomials:

$$f(z) = \sum_{m=0}^{\infty} L_m \frac{P_m(z)}{\|P_m(z)\|} \quad -1 \leq z \leq 1, \tag{II.2.19}$$

where $P_m(z)$ is the mth-order Legendre polynomial (Abramowitz and Stegun, 1965):

$$P_m(z) = \frac{1}{2^m m!} \left(\frac{d}{dz}\right) [(z^2 - 1)^m] = \sum_{i=0}^{m} a_{i,m} z^i. \tag{II.2.20}$$

The coefficients of the Legendre polynomials are calculated in the usual fashion by means of integration and based on the orthogonality property of such polynomials as follows:

$$L_m = \int_{-1}^{1} \sqrt{\frac{2m+1}{2}} P_m(z) f_z(z)\, dz = \int_{-1}^{1} \bar{P}_m(z) f_z(z)\, dz. \qquad (\text{II}.2.21)$$

One notices that $P_m(z)$ contains the monomials z^i. Integration (Equation (II.2.21)) can therefore be identified with moments ($\int z f(z) dz$, etc) and therefore also cumulants. An expression relating cumulants with the Legendre coefficients (Equation (II.2.21)) has been derived (see Mustapha and Dimitrakopoulos, 2010a, Appendix B) as:

$$L_m = \sqrt{\frac{2m+1}{2}} \sum_{i=0}^{m} \sum_{j=0}^{i} \binom{i}{j} a_{i,m} c_{i-j} m_j \quad i = 0, \dots m; \, m \geq 0 \qquad (\text{II}.2.22)$$

In any practical calculations, these polynomials need to be truncated at some order k, leading to the approximation:

$$f(z) \simeq f_k(z) = \sum_{m=0}^{k} L_m \frac{P_m(z)}{\|P_m(z)\|} \quad -1 \leq z \leq 1. \qquad (\text{II}.2.23)$$

In higher dimensions (spatial modeling), one is interested in approximating either multivariate distributions $f(z_0, z_1 \dots z_n)$ or conditional distributions $f(z_0 | z_1 \dots z_n)$. The latter can be used as a building block in sequential simulation (see Section II.2.6). Consider, for example, a bivariate distribution expanded as:

$$f(z_0, z_1) = \sum_{m=0}^{\infty} \sum_{n=0}^{\infty} L_{m,n} \bar{P}_m(z_0) \bar{P}_n(z_1) \quad -1 \leq z_0, z_1 \leq 1, \qquad (\text{II}.2.24)$$

with the Legendre coefficients expressed as:

$$L_{m,n} = \int_{-1}^{1} \int_{-1}^{1} \bar{P}_m(z_0) \bar{P}_n(z_1) f(z_0, z_1)\, dz, \qquad (\text{II}.2.25)$$

for which expressions similar to Equation (II.2.21) can be calculated as a function of cumulants. In order to derive a conditional distribution, one relies on the following expression:

$$f(z_0 | z_1) = \frac{f(z_0, z_1)}{\int_{-1}^{1} f(z_0, z_1)\, dz_0} \qquad (\text{II}.2.26)$$

Mustapha and Dimitrakopoulos (2010a) provide the developments of any higher-order density, whether multivariate or conditional.

2.5 Computing distances

Most MPS algorithms require computing distances between patterns at one stage or another, such as when building a patterns database, where the classification of a pattern is based on a similarity (or distance) with a prototype, or when simulated patterns need to be compared with those in the database in order to retrieve frequencies.

Mathematically, a distance $d(.)$ between two objects $\mathbf{A}$ and $\mathbf{B}$ is a scalar quantity, defined by three fundamental properties:

1 It must be positive-definite:

$$d(\mathbf{A}, \mathbf{B}) \geq 0 \quad and \quad d(\mathbf{A}, \mathbf{B}) = 0 \Leftrightarrow \mathbf{A} = \mathbf{B} \qquad \text{(II.2.27)}$$

2 It must be symmetric:

$$d(\mathbf{A}, \mathbf{B}) = d(\mathbf{B}, \mathbf{A}) \qquad \text{(II.2.28)}$$

3 And it has to respect the triangular inequality, meaning that the trajectory between $\mathbf{A}$ and $\mathbf{B}$ cannot be shortened if it is forced to pass through a third point $\mathbf{C}$:

$$d(\mathbf{A}, \mathbf{B}) \leq d(\mathbf{A}, \mathbf{C}) + d(\mathbf{B}, \mathbf{C}). \qquad \text{(II.2.29)}$$

In the next sections, we review distances that honor these properties and are commonly used for comparing patterns.

2.5.1 Norms

In this section, we discuss the possible norms used to compare two data events $\mathbf{N}(\mathbf{x}_1, \mathbf{h}, \mathbf{v})$ and $\mathbf{N}(\mathbf{x}_2, \mathbf{h}, \mathbf{v})$. In order to simplify the notations, we denote each of the compared data events with a superscript 1 or 2, and each individual node in these data events with a subscript:

$\mathbf{N}^1 = \mathbf{N}(\mathbf{x}_1, \mathbf{h}, \mathbf{v})$, with the value at each of the n nodes in $\mathbf{N}^1$ denoted $\mathbf{N}_1^1$, $\mathbf{N}_2^1, ..., \mathbf{N}_n^1$

and

$\mathbf{N}^2 = \mathbf{N}(\mathbf{x}_2, \mathbf{h}, \mathbf{v})$, with the value at each of the n nodes in $\mathbf{N}^2$ denoted $\mathbf{N}_1^2$, $\mathbf{N}_2^2, ..., \mathbf{N}_n^2$.

In doing so, we imply that the lag vectors $\mathbf{h}$ are identical between both compared data events and the number of nodes n in each data event. For the case where data events have different numbers of elements, specific dissimilarity measures have been developed in the pattern recognition literature such as the Edit distance or dynamic programming-based techniques; however, we do not cover them here because they are typically not used in the context of MPS.

Because data events are complex objects, it is convenient to break down the operations required to compare them. Three levels can be considered: (1) comparison of the value of a node with the value of another node; (2) comparison of two univariate data events, which are groups of nodes; and (3) comparison of multivariate data events, which are groups of univariate data events. A distance function that compares univariate data events is denoted as $d(.)$, and a distance function that compares groups of univariate data events is denoted as $D(.)$.

The simplest distance between data events is a norm consisting of the mean errors on each node, possibly put to an exponent p. It is expressed as:

$$d(\mathbf{N}^1, \mathbf{N}^2) = \frac{1}{n} \sum_{i=1}^{n} \left| \mathbf{N}_i^1 - \mathbf{N}_i^2 \right|^p. \qquad \text{(II.2.30)}$$

The exponent p determines how large errors will be treated: a large p value results in a large distance when there are only a few nodes that have a high mismatch. Conversely, small p values tend to give equal importance to all error magnitudes. The most commonly used norms are the case with $p = 1$ corresponding to the Manhattan distance, and the case of $p = 2$, which is the Euclidean distance. If patterns contain extreme values, some norms may become unstable. This instability is particularly marked when a large p is used. One solution is to perform a uniform score transform of the variable prior to simulation. The values of the TIs are first ranked from 1 to N_{TI}, then rescaled in the interval [0 1]. This new variable is then simulated using any of the norms discussed above. The final simulations are then back-transformed such that they have the same marginal distribution as the TI.

The norms above are appropriate when dealing with continuous variables; however, they cannot be used with categorical variables. Because categorical variables have no order relationship, there is no possible nuance in the degree of similarity between categories compared. For example, compare the following two continuous vectors:

1 1 1 2 1 3 4
2 1 2 4 4 3 4

Here, we can state that the distance between 1 and 4 is larger than the distance between 1 and 2. Therefore, using a Manhattan distance, we obtain the following errors:

  0 1 2 3 0 0

Consider the categorical equivalent of the example, with alphabet letters (which are categories) replacing numbers:

1 A A B A C D
2 A B D D C D

The absence of an (numerical) order relationship means that the distance between A and D is no different from the distance between A and B. We can only state that values are of the same category or not. Hence, the appropriate distance consists of the proportion of matching nodes:

$$d(\mathbf{N}^1, \mathbf{N}^2) = \frac{1}{n} \sum_{i=1}^{n} a_i, \quad a_i = \begin{cases} 0 & \text{if } \mathbf{N}_i^1 = \mathbf{N}_i^2 \\ 1 & \text{if } \mathbf{N}_i^1 \neq \mathbf{N}_i^2 \end{cases}. \tag{II.2.31}$$

With this categorical distance, the errors between both strings of letters would be as follows:

  0 1 1 1 0 0

We have seen in Section II.2.3 that data events can be multivariate (Figure II.2.8). Such multivariate data events may consist of several continuous variables, several categorical variables, or a combination of both. One general way of obtaining a distance between such complex data events is by separately computing the distance between each variable of each data event, and then to integrate the distances:

$$D(\mathbf{N}^1, \mathbf{N}^2) = \frac{1}{M} \frac{1}{n} \sum_{k=1}^{M} d^k(\mathbf{N}^{1_k}, \mathbf{N}^{2_k}), \tag{II.2.32}$$

where $\mathbf{N}^{1_k}$ and $\mathbf{N}^{2_k}$ denote the kth variable of the first and second compared data events, respectively; and $d^k(.)$ is the distance that is specific to each variable considered, for example Equation (II.2.30) for continuous variables and Equation (II.2.31) for categorical variables. This averaged distance can be tweaked by adding weights for each variable:

$$D\left(\mathbf{N}^1, \mathbf{N}^2\right) = \frac{1}{\sum\limits_{k=1}^{M} w^k} \frac{1}{n} \sum_{k=1}^{M} w^k d^k \left(\mathbf{N}^{1_k}, \mathbf{N}^{2_k}\right). \tag{II.2.33}$$

2.5.2 Hausdorff distance

The Hausdorff distance is popular in the image analysis literature for measuring the dissimilarity between two sets of data **A** and **B** (Huttenlocher et al., 1993). Individual elements of **A** are denoted a, and individual elements of **B** are denoted b. Its particularity lies in the fact that it does not compare, one by one, every element a with each element b, but instead considers both sets in their entirety. It is considered that **A** and **B** are similar if each point in one set is close to all points in the other set.

To describe the rationale of the Hausdorff distance between sets, we consider two situations that are illustrated in Figure II.2.23:

1 Case 1: the distance $d(a, \mathbf{B})$ between one point a and a curve **B**
2 Case 2: the distance $d(\mathbf{A}, \mathbf{B})$ between a curve **A** and another curve **B**.

The first case is straightforward: one computes the distance between a and each point on **B**, denoted b_i, and $d(a, \mathbf{B})$ is the minimum of these distances:

$$d(a, \mathbf{B}) = \min_i d(a, b_i) \tag{II.2.34}$$

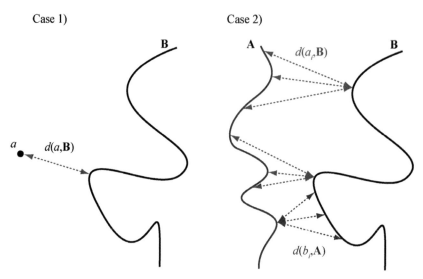

Figure II.2.23 Illustration of the Hausdorff distance between two sets A and B.

For the second case, we can consider many points in **A**, and similarly as in case 1 compute the distance $d(a_i, \mathbf{B})$ for each point in **A**. At this point, we have a large number of distances but not yet a unique measure of how dissimilar **A** and **B** are. This unique measure is obtained by integrating those different measures into a single metric. This can be accomplished by taking the mean, the minimum, the maximum, or any quantile of the point-to-set distances. Dubuisson and Jain (1994) have shown that when computing distances between spatial patterns, an integration by taking the average point-to-set distance provides the best option:

$$d\left(\mathbf{A}, \mathbf{B}\right) = \frac{1}{J} \sum_{j=1}^{J} \min_{i} d(a_j, b_i). \qquad \text{(II.2.35)}$$

One issue is that the resulting distance is not symmetric, that is, in most cases, $d\left(\mathbf{A}, \mathbf{B}\right) \neq d\left(\mathbf{B}, \mathbf{A}\right)$. This is clearly illustrated in Figure II.2.23: in case 2, the red arrows and the blue arrows converge differently. Hence, at this stage, it cannot be called a distance because it violates the very definition of a distance, which requires symmetry. This pseudo-distance can, however, be made symmetric by integrating $d(\mathbf{A}, \mathbf{B})$ and $d(\mathbf{B}, \mathbf{A})$, and similarly as before there are many ways of doing it, such as taking the maximum of $d(\mathbf{A}, \mathbf{B})$ and $d(\mathbf{B}, \mathbf{A})$, the minimum, the mean, the median, and so on. In this case again, Dubuisson and Jain (1994) studied which integration operation is the most appropriate when computing distances between spatial patterns, and found that it is the maximum of $d(\mathbf{A}, \mathbf{B})$

and $d(\mathbf{B}, \mathbf{A})$. Hence, the type of Hausdorff distance to be used with spatial patterns is computed by nesting min and max operations:

$$d(\mathbf{A}, \mathbf{B}) = d(\mathbf{B}, \mathbf{A}) = \max\left[\frac{1}{J}\sum_{j=1}^{J}\min_i d(a_j, b_i), \frac{1}{I}\sum_{i=1}^{I}\min_j d(b_i, a_j)\right]. \qquad (\text{II.2.36})$$

In order to compute this Hausdorff distance on two data events $\mathbf{X}$ and $\mathbf{Y}$, the templates need to be first converted to individual points (see Section II.2.1), and then each data event is considered as a set of points (i.e., x_i is an element of $\mathbf{X}$). The comparison of data events then proceeds with Equation (II.2.36).

2.5.3 Invariant distances

So far, we have considered that in order for two data events to be perfectly similar, identical values need to perfectly match spatially, as defined by the lag vectors $\mathbf{h}$. In some cases, however, it may be desirable to provide some tolerance in either matching value or matching the position of nodes. Data events are then compared *up to* a given characteristic; an exact one-to-one match is not required.

A first possibility is to compare data events up to a translation in their value. To this end, the values of both $\mathbf{N}^1$ and $\mathbf{N}^2$ are centered on zero before being compared. This results in a mean-invariant distance:

$$d^{MI}(\mathbf{N}^1, \mathbf{N}^2) = d(\mathbf{N}^1 - \overline{\mathbf{N}^1}, \mathbf{N}^2 - \overline{\mathbf{N}^2}), \qquad (\text{II.2.37})$$

where $\overline{\mathbf{N}^1}$ and $\overline{\mathbf{N}^2}$ represent the mean of all values in $\mathbf{N}^1$ and $\mathbf{N}^2$. Going one step further, we can compare the normalized values using a normalization-invariant distance:

$$d^{NI}(\mathbf{N}^1, \mathbf{N}^2) = d\left(\frac{\mathbf{N}^1 - \overline{\mathbf{N}^1}}{\sigma_{\mathbf{N}^1}}, \frac{\mathbf{N}^2 - \overline{\mathbf{N}^2}}{\sigma_{\mathbf{N}^2}}\right). \qquad (\text{II.2.38})$$

Next, we consider the comparison of data events up to a spatial transformation. Define a geometrical parametric transformation T_p applied on $\mathbf{N}$, where p is a parameter affecting the transformation. Several types of geometrical transformations can be considered. Consider two types of transformations: (1) rotations centered on $\mathbf{x}$, denoted $R(.)$; and (2) affinity (or homothetic) transforms, denoted $A(.)$. Note that we consider only geometrical transforms affecting the lag vectors $\mathbf{h}$, whereas the values $Z(\mathbf{x}_i)$ remain unchanged. For rotations, the parameter p corresponds to the rotation angle. For affinity transforms, p corresponds to the affinity ratio. Figure II.2.24 shows a few examples of transformations of a simple data event.

A geometry-invariant distance is then defined as:

$$d^{GI}(\mathbf{N}^1, \mathbf{N}^2) = \min\left[d\left(T_p(\mathbf{N}^1), \mathbf{N}^2\right)\right] \quad : \forall p \text{ in } [a, b] \qquad (\text{II.2.39})$$

The formulation of this distance is similar to case 1 of the example we used for the Hausdorff distance, except that here the parameter p comes into play. Such

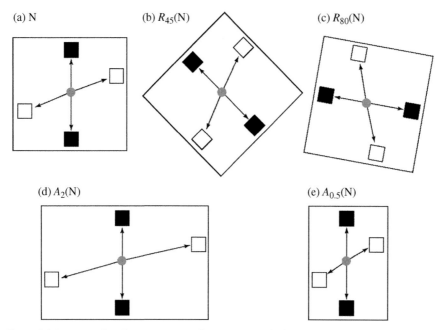

Figure II.2.24 Examples of geometric transformations applied to N. (a) Original data event N. (b) 45° rotation. (c) 80° rotation. (d) Factor 2 affinity along the X-axis. (e) Factor 0.5 affinity along the X-axis. Modified from Huysmans, M. and A. Dassargues (2011). With kind permission from Springer Science and Business Media.

a metric compares $\mathbf{N}^2$ with the ensemble of possible data events resulting from geometric transforms of $\mathbf{N}^1$. Comparing both data events up to a transformation entails that if the distance is small, there exists a transformation $T(.)$ that renders $T(\mathbf{N}^2)$ similar to $\mathbf{N}^1$. When considering geometry-invariant distances, Figure II.2.24a, Figure II.2.24b, and Figure II.2.24c are identical up to a rotation, and Figure II.2.24a, Figure II.2.24d, and Figure II.2.24e are identical up to an affinity transform.

One advantage of using geometry-invariant distances in multiple-point simulations is that it becomes possible to consider not only the data events present in the TI, but also the ensemble of their parametric transformations with p in the range $[a,b]$. Hence, it amounts to searching in a larger pool of possible spatial patterns than is actually present in the TI.

2.5.4 Change of variable

It is clear that a large number of candidate distances can be selected for comparing patterns and/or data events. The choice of distance will depend on the nature of the spatial variables. In some cases, however, direct application of the distances listed may not lead to satisfactory results. Next, we therefore propose to transform a "difficult" variable into another variable that is more amenable to storage in a pattern database or to the computation of meaningful distances between patterns.

2.5.4.1 Proximity transform

When considering categorical variables, a simple norm as described in Equation (II.2.30) may result in a large distance between patterns that are visually and structurally very similar. An example is given in Figure II.2.25a, where two categorical patterns N_1 and N_2 are compared, consisting of a diagonal linear structure. Although both patterns present the same diagonal feature, the feature does not exactly overlap, resulting in most nodes being dissimilar.

One solution consists of transforming the categorical variable into a continuous variable that informs the distance to the edge of the feature of interest (Arpat, 2005). Such a "distance map" is shown in Figure II.2.25b. It results in a transformed variable $T(N_1)$ where the grayscale levels indicate the distance to the diagonal black object. The computation of a distance between both distance maps, using for example a Euclidean norm, is then more meaningful than the distance between categorical patterns. If the objects in both patterns are at the same location, the distance will be small. If the objects have a different shape or

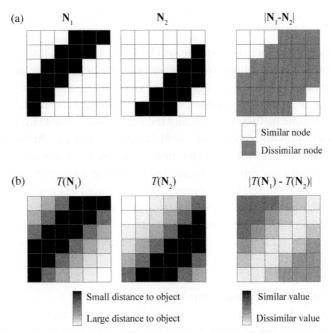

Figure II.2.25 Illustration of the use of proximity transform as a more relevant comparison between patterns. (a) Patterns compared by counting the number of mismatching nodes. (b) The same patterns after proximity transform can be compared using an Euclidean norm, which is more appropriate to distinguish the distance between the objects compared.

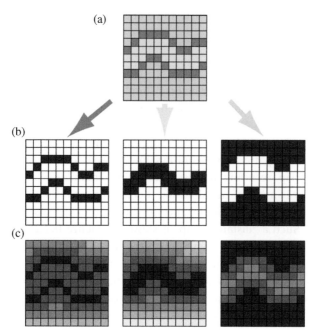

Figure II.2.26 Separation in indicators to compute proximity transforms on patterns having more than two categories. (a) Multicategory pattern, (b) transformation into indicators, and (c) proximity transform on indicators (modified from Arpat, 2005).

are at a different location, the distance will be large. However, if the objects are similar but only slightly shifted, the distance will now be relatively small, which may in some cases be a more interesting way of comparing patterns.

In the case of multiple categories, the computation of the proximity transform has to be done by first transforming the categorical variable into a series of indicators corresponding to each of the M categories. The proximity transform is then applied on each indicator separately, resulting in M distance maps. The comparison of patterns can then be carried out using a multivariate distance, such as in Equation (II.2.32). This procedure is illustrated in Figure II.2.26, with a three-category pattern in (a), the indicator variable for each of the three categories in (b), and the corresponding distance transform in (c). This triplet of distance maps can then be compared with the equivalent triplet created based on another pattern.

2.5.4.2 Edge property transform

In applications related to fluid flow, the most important feature to model is the connectivity of the permeable and impermeable bodies. One challenge arises when these bodies are very thin. If the grid nodes are larger than these thin features, they cannot be represented correctly, resulting in excessive disconnectivity. One solution is to use very high-resolution grids, but this can

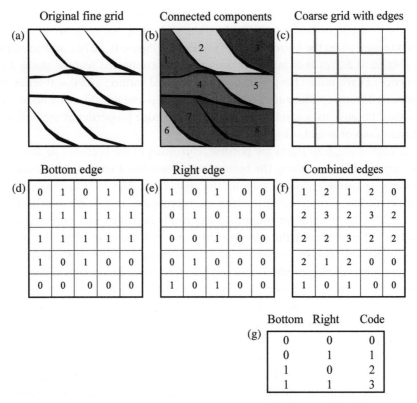

Figure II.2.27 (a) Fine-scale shale layers, (b) connected components, (c) derived coarse grid with edges, (d) indicator of presence of bottom edge, (e) indicator of presence of right edge, (f) combined edges code, (g) coding of edges. Modified from Huysmans and Dassargues (2011).

be computationally challenging. The alternative, first proposed by Stright et al. (2006) and then further investigated by Huysmans and Dassargues (2011), is the edge property transform.

It consists of using a coarse grid where information on the thin connected bodies is encoded using higher dimensionality that is expressed as additional categories. In the type of categorical image considered (Figure II.2.27a), the areas between the thin connected features can be conceptualized as disconnected blocs. Using mathematical morphology (Serra and Soille, 1994), it is easy to isolate the connected components of the nonconnected phase (Figure II.2.27b). A coarse grid is then defined where the thin connected features are represented at the boundaries between nodes. If the connected component indices of the two nodes differ, then the edge between these two nodes is flagged as 1 (dark on Figure II.2.27c). Otherwise, the edge between the two nodes is flagged as 0, indicating that no thin connected feature is present.

Figure II.2.27c contains the essential connectivity properties of the original fine-scale image; however, at this stage, this information is not represented on a grid, but rather on the edges of the nodes. This edge information is then encoded on the coarse grid nodes by two matrices: one indicating the presence of a connected feature along the bottom edge of each node (Figure II.2.27d), and another one (Figure II.2.27e) indicating the presence of a connected feature along the right edge of each node (1 = presence of a connected feature, and 0 = absence of a connected feature). The edge properties for the top and left sides of a node are not stored because they can be obtained from the edge properties at the bottom and right sides of neighboring nodes.

For each node, four states are possible, consisting of the presence or absence of a connected feature along the bottom edge, associated with the presence or absence of a connected feature along the right edge. This information can be summarized into four categories, which are represented in a single matrix (Figure II.2.27f). The coding for the four categories is described in Figure II.2.27g. Based on a TI representing the edge property, one then simulates new edge properties, which can then be converted to real thin properties in a flow simulator.

Using this encoding method, one can simulate on a coarse grid a categorical variable that contains the essential properties of a much finer grid, which has potentially one or two orders of magnitude more nodes. The encoded variable has four instead of three categories, but this can be handled easily with most MPS simulation methods. Effectively, this method trades off resolution for dimensionality: the fine-scale information is conveniently expressed as additional categories. An interesting parallel can be made with the method of appearance–space texture synthesis, which uses a similar approach to encode information on continuous patterns as additional variables (Lefebvre and Hoppe, 2006). The increase in dimensionality allows using smaller neighborhoods while enhancing the reproduction of patterns.

2.5.5 Distances between distributions

In some cases, it may be useful to compare patterns or entire images based on their statistical properties. Patterns are now no longer compared locally; rather, two patterns are considered similar if they present similar statistical properties. For example, two patterns may have a similar histogram. One can also compare patterns or images in terms of their MPH or CHP.

The statistical literature offers various methods to test whether two distributions are statistically significantly different. Most of these methods rely on defining a metric for such a comparison. Consider two counting distributions summarized with vectors $\mathbf{p}$ and $\mathbf{q}$, which have an equal number J of elements. The following measures of dissimilarity are commonly applied to such cases:

The Jensen–Shannon divergence:

$$JSD(\mathbf{p}||\mathbf{q}) = \frac{1}{2}K(\mathbf{p}||\mathbf{m}) + \frac{1}{2}K(\mathbf{q}||\mathbf{m}) \qquad \text{(II.2.40)}$$

where $\mathbf{m} = (\mathbf{p} + \mathbf{q})/2$ and $K(\mathbf{p}||\mathbf{m})$ is the Kullback–Liebler divergence, defined as:

$$K(\mathbf{p}||\mathbf{m}) = \sum_{j=1}^{J} \ln \frac{\mathbf{p}_j}{\mathbf{m}_j} \mathbf{p}_j \tag{II.2.41}$$

The earth mover distance (EMD): with the EMD, the pdfs to compare are seen as two piles of material. The EMD is then defined as the minimum cost of turning one pile into the other. The cost is defined as the amount of material moved times the distance by which it is moved.

The Bhattacharyya distance:

$$BC(\mathbf{p}||\mathbf{q}) = -\ln \left(\sum_{j=1}^{J} \sqrt{\mathbf{p}_j \mathbf{q}_j} \right) \tag{II.2.42}$$

The chi-squared distance:

$$\chi^2(\mathbf{p}||\mathbf{m}) = \frac{1}{2} \sum_{j=1}^{J} \frac{(\mathbf{p}_j - \mathbf{m}_j)^2}{(\mathbf{p}_j + \mathbf{m}_j)} \tag{II.2.43}$$

Norms: the Euclidean and Manhattan norms seen in this chapter can also be used to compare distribution functions.

2.6   Sequential simulation

Most MPS simulation methods can be conceptualized as having an empty grid as input (it can also be partially empty depending on the nature of the data available) and a full grid as output. In an empty grid, the grid topology is defined, but the values attributed to each *ID* are not determined and can be represented by a standard code, such as −99999. Note that undefined nodes are different from masked nodes. Undefined nodes do not have a defined value attributed in the inputs, but they have a value in the output. Masked nodes have neither input nor output values; they are simply ignored during the simulation.

In the output grid, the values should be organized such that they have a particular spatial relationship with their neighbors, and this relationship is based on the TI. In this sense, the value at each node depends on the value of its neighbors, and the relationship goes both ways: a node also influences the values at the neighboring locations. If the value of a single node is changed, it is not sufficient to locally adapt its neighborhood. This chain of interdependences can be expressed in a probabilistic form. Consider the variable Z on a simulation grid SG having N_{SG} nodes. The probability of a value occurring at location $\mathbf{x}$ is dependent on all other values of the grid, either known or unknown. One can imagine an N_{SG}-dimensional joint cumulative probability distribution (cdf), such that each sample from it corresponds to a set of values that present the desired spatial relationships. For a single, continuous variable, this cdf is expressed as:

$$F(z,\mathbf{x}) = \text{Prob}[Z(\mathbf{x}_1) \le z_1, \; Z(\mathbf{x}_2) \le z_2, \; \dots, \; Z(\mathbf{x}_{N_{SG}}) \le z_{N_{SG}}]. \quad \text{(II.2.44)}$$

Such a distribution can be difficult if not impossible to formulate, especially if the relations between the different locations are complex and cannot be written in an analytical form. This is typically the case when the spatial dependence is expressed by a TI, where the complete multivariate distribution is analytically intractable. In such cases, numerical approaches are adopted. The most commonly used approach in this regard relies on the principle of sequential decomposition, which consists in decomposing the above cdf as follows (Deutsch and Journel, 1992):

$$F(z,\mathbf{x}) = \text{Prob}\,[Z(\mathbf{x}_1) \le z] \cdot \text{Prob}[Z(\mathbf{x}_2) \le z|z(\mathbf{x}_1)] \cdots \\ \text{Prob}\,[Z(\mathbf{x}_N) \le z|z(\mathbf{x}_1), \dots, z(\mathbf{x}_{N-1})] \quad \text{(II.2.45)}$$

In practice, this decomposition is used to sequentially simulate values along a path, namely, first sampling the value of a first location, say $Z(\mathbf{x}_1)$, then sampling the value of a second location $Z(\mathbf{x}_2)$ conditional to $Z(\mathbf{x}_1)$, based on the local cdf $\text{Prob}\,[Z(\mathbf{x}_2) \le z|z(\mathbf{x}_1)]$. The third location $Z(\mathbf{x}_3)$ is determined conditionally to both $Z(\mathbf{x}_1)$ and $Z(\mathbf{x}_2)$, and so on, until all nodes have been visited, as illustrated in Figure II.2.28. If conditioning data are present, the method is identical. In that case, certain nodes are initially informed, hence the decomposition does not start with $Z(\mathbf{x}_1)$, but with $Z(\mathbf{x}_{p+1})$, where p is the number of conditioning data.

Each local cdf depends on the ensemble of all previously determined locations. If the principle of sequential simulation is strictly followed, then there are $N{-}1$ conditioning terms when simulating the last node, which again can lead to an intractable high-dimensional cdf. The solution generally adopted is to consider only a limited neighborhood (i.e., a subset of locations) of size n for determining local ccdfs, with $n{\ll}N_{SG}$. This is related to the discussion on the impossibility of using a global neighborhood in the context of MPS.

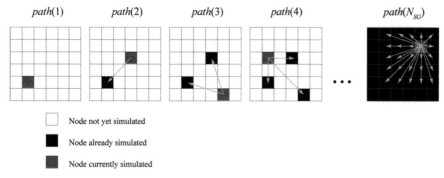

path(1)   *path*(2)   *path*(3)   *path*(4)   *path*(N_{SG})

☐   Node not yet simulated

■   Node already simulated

▨   Node currently simulated

Figure II.2.28 Steps of a sequential simulation with a random path.

An important point with the sequential decomposition is that the final result depends on the order in which the nodes are visited. This order is called a *simulation path*, and several possibilities have been investigated in the literature. Some of the most common simulation paths are described in the remainder of this section.

2.6.1   Random path

The random path is the most commonly used path for visiting nodes in sequential simulation. It consists of visiting all uninformed nodes of the grid in a random order. It is the path type traditionally used with SGS. Using a random path entails that the first nodes simulated will tend to be further away from their neighbors, and therefore large-scale features will be simulated first. Then, as the grid becomes denser, the frequency of cases with close neighbors increases, informing small-scale structures. All areas of the grid are populated simultaneously (in a statistical sense), and the grid gets filled progressively. An analogy would be a Polaroid photograph that slowly appears on the argentic paper (see Figure II.2.29). Obtaining different realizations is accomplished by generating a new path for each realization, using another random seed.

Algorithmically, defining a random path requires generating a vector containing the *ID*s of each uninformed node in the grid, ordered randomly. Starting from an initial ordered vector $ID(i) = 1 \ldots N$, the random path $ID\_random(i)$ is obtained by the following pseudo-code:

For each element i in the path:

Choose r as an integer random number in the interval $[i, N]$.

Define $ID\_random(i) = ID(i)$.

Define $ID\_random(r) = ID(r)$.

The general idea is that the initial ordered path is mixed by performing permutations of each *ID* with another *ID* randomly chosen at a location further in the path. The fact that only permutations are performed ensures that each *ID* appears only once in the path.

A simple variant of the random path can be made, ensuring that each simulated node has some nearby informed neighbors. This is accomplished by a rejection procedure that only accepts simulating a node if one of its direct neighbors

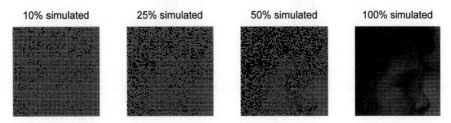

Figure II.2.29 Illustration of the random simulation path.

is informed. If no neighbor is informed, the node is rejected and another random candidate node is tested. The simulation then grows from the conditioning data. A simulation front progresses, and the rejected candidate nodes are eventually simulated as the entire grid is covered. This type of path is called the "random-neighbor path".

2.6.2 Unilateral path

An alternative approach is to visit the grid nodes in a linear manner, starting in one corner of the grid and then progressing along one of the grid dimensions at a time. This type of path, termed a "unilateral path", has the advantage that the sequential decomposition (Equation (II.2.45)) is not approximated. The linear nature of the simulation implies that it can be assimilated to a series of Markovian dependences (Daly, 2004). Each node only needs to be conditioned to the previous neighbors (see Figure II.2.30) because the chain of dependences from one node to the next is unbroken from the beginning to the end of the path.

This approach has, however, drawbacks. It may be difficult to honor conditioning data that lie ahead of the path, and therefore some structures built prior to simulating the data locations may be incompatible with the data values. This issue has been in large part addressed by splitting the template in two parts: a causal part that considers the previously simulated nodes that constitute the "past" of the simulation, and the noncausal part of the template that looks ahead and is usually empty (Daly, 2004; Parra and Ortiz, 2011). The noncausal part of the template has an influence on the simulation only when data points are encountered, and thus it allows for conditioning the simulated node to these data.

Another alternative to condition models when a unilateral path is used, is to update the local conditional probability (Equation (II.2.7)) with a kriging estimate (Section II.4.2.2). Although the kriging estimate does not account for multiple-point statistics, it has the advantage of considering data points

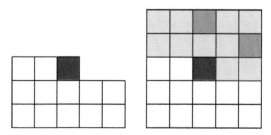

Figure II.2.30 Left: typical L-shaped template used for unilateral simulation starting in the lower-left corner. Right: conditioning with a unilateral path. The causal part of the template is in white, and the noncausal part is in pink. Informed nodes (conditioning data) are shown in green.

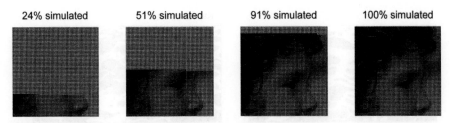

Figure II.2.31 Illustration of the unilateral simulation path with the direction $+X +Y$.

located further than the spatial extension of the template (Kjønsberg and Kolbjørnsen, 2008).

A feature of the unilateral path is that it can be run along different directions. For example, assuming the origin of the grid in the lower-left corner, one can choose a path that first simulates nodes along the X direction from left to right. When the first row is simulated, it proceeds to the next row along the Y direction. When the first XY plane is completed, it increments Z to go on with the next XY plane, and so on until the entire grid is simulated. Such a path can be represented as $+X+Y+Z$ (see Figure II.2.31).

One could also use a path that first simulates columns along the Y direction and then increments X when it arrives at the end of a column. This path would be denoted $+Y+X+Z$. It is also possible to run through the dimensions in the reverse order: starting at the upper-right corner of the grid, one takes the first row from right to left, then simulates XY planes from the top to the bottom. This is represented by using the minus sign: $-X-Y-Z$.

A large number of combinations are possible, and a drawback of the unilateral path is that different choices of path directions generally influence the simulation results. It has been observed, for example, that in some cases the simulated structures tend to be parallel to the first path direction. The occurrence of such artifacts depends on the algorithm used as well as the parameters of this algorithm. Nevertheless, compared to a random path, the use of a unilateral path often leads to better reproduction of TI patterns.

The directional character of the unilateral path can be extended to multivariate grids. In a case comprising several variables, an obvious possibility is to simulate one variable at the time $(+X +Y +Z +V)$, but this is not necessarily optimal. In this case, the structures that have been simulated for the first variable will control all other variables. In this sense, the first simulated variable has more importance than the others. One possibility to avoid this hierarchy in the variables is to use a path that visits the variables in a random order, although it can visit the nodes according to a unilateral path: $+X +Y +Z$ *random V*. Similarly, one could use a fully random path through the nodes and the variables (*random X, random Y, random Z* and *random V*), or a random path through the nodes with a unilateral order in the variables (*random X, random Y, random Z*, and $+V$).

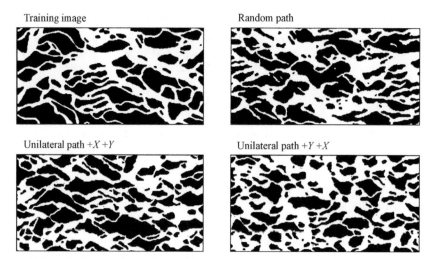

Training image   Random path

Unilateral path $+X+Y$ Unilateral path $+Y+X$

Figure II.2.32 The effect of the path on the simulation. Training image (Ganges delta, Bangladesh), and simulation with random and two unilateral paths in different directions. Simulations obtained with the direct sampling algorithm.

The simulation path therefore is often a modeling choice. Choosing different types of paths may yield different realizations as well as results in sets of realizations with varying spatial uncertainty. Such choices have to be made for the path, but also for most of the parameters governing MPS: for example, the choice of a neighborhood or various other parameters that are specific to each simulation method. To illustrate this point, Figure II.2.32 shows realizations obtained with the same TI and parameters, except for the path used. Realizations corresponding to a random and two unilateral paths are displayed. They show that the use of a random path tends to result in improperly represented thin connected structures, whereas the unilateral path represents the connectivity better but preferentially produces elongated features in the path direction.

2.6.3 Patch-based methods

The patch-based approach (also termed the "pattern-based approach") is a more recent development in MPS and proceeds by simulating entire regions, or patches, at a time. Instead of sampling a value conditional to a data event, it consists of sampling an entire region, using neighborhoods consisting of the zones overlapping the previously simulated patches (Arpat and Caers, 2007; El Ouassini et al., 2008). In this context, the different simulation paths reviewed above still apply: the simulation domain can be covered with patches in a random order, according to a unilateral order or in any possible order. The only difference is that the simulation elementary unit is a patch instead of a node, with the patch itself consisting of a number of nodes that are simultaneously pasted in the simulation grid (Figure II.2.33).

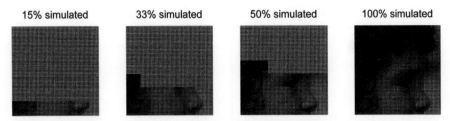

Figure II.2.33 Illustration of a patch-based simulation with a unilateral path.

2.6.4 Patch carving

With patch-based methods, it is common that two simulated patches do not coincide perfectly, resulting in possible vertical and horizontal artifacts reflecting the patch size. This occurs because the new patch is superimposed on the previously simulated ones. One alternative to simple superposition of patches is to find the cut that results in minimal error in the transition between patches.

In 2D, such a minimum boundary cut can be obtained very efficiently using dynamic programming (Dijkstra, 1959), and it results in significantly improved pattern continuity. This minimum-cut approach has already been discussed in part I (Figure I.5.9) and will be described in detail in Section II.3.3.4.

2.7 Multiple grids

MPS simulation requires data events to have a size or extent comparable to those of the spatial structures present. The data event size can intuitively be considered as equivalent to the search radius in semivariogram-based methods (see the discussion in Section I.3.6). In the case of very large structures, the data event size required may become impractical. The solution proposed is to use sparse nested paths and corresponding sparse data events. The approach, called the "multiple-grids approach", is able to better capture the structures at different scales (Strebelle, 2002).

The use of multiple grids can be seen as a specific type of simulation path, where the simulation grid is divided into G multigrids denoted g, with $g = 1 \ldots G$. The first multiple grid is very sparse, and then each multiple grid covers the space in a progressively denser fashion. For each multiple grid, the spacing between simulated nodes is scaled by a factor 2^{g-1}, until the last level encompasses all nodes in the grid. The main principle is that the larger search templates of the coarse multiple grids capture the large-scale structures of the TI, and then the finer grids simulate the smaller structures. The nodes simulated in the previous multiple-grid levels are then frozen and used as conditioning data in the finer grids.

Figure II.2.34 illustrates the use of four multigrid levels, with each level represented in a different color. The simulation starts with a large template, which

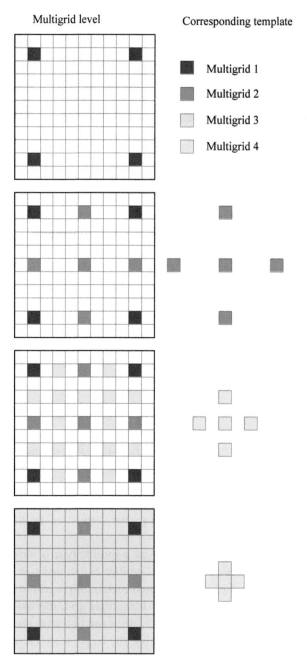

Figure II.2.34 Principle of multiple grids. The left column shows a simulation grid with the nodes belonging to each multigrid level in a different color. The right column shows the five-node templates corresponding to each level (template for the first level is not shown).

Training image

1 multigrid

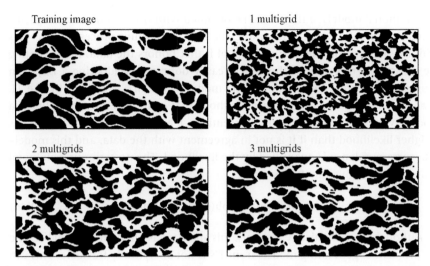

2 multigrids

3 multigrids

Figure II.2.35 Illustration of the effect of using or not using multiple grids, with the same TI as in Figure II.2.32. Simulations obtained with the SNESIM algorithm.

is reduced as the grids become denser (the largest template is not shown because it would be larger than the simulation grid). This template size reduction is necessary for the template nodes to fall on unsimulated grid nodes. Note that the shape of the template does not need to be the same for each multiple grid. The use of different templates requires a separate database to be constructed for each multiple-grid level (recall that a database, whether a tree or list, is defined with respect to a given template).

Without using multiple grids, the reproduction of the large-scale structures can be significantly affected. Figure II.2.35 shows this by comparing the results of using a single grid or up to three multiple grids.

2.8 Conditioning

Geostatistical models are stochastic representations of natural phenomena that need to incorporate both global and local information. With MPS, the global information is given by the TI: the spatial continuity, the histogram or proportion of the categories considered, the type of patterns, and so on. Geostatistics is not the only discipline that considers such information. Other disciplines, such as computer graphics, have developed methods to produce models honoring this kind of information. The challenge specific to geostatistics is that these models also need to be conditioned to local information.

One approach to conditioning is to adopt a Bayesian point of view where the TI is seen as defining a prior ensemble of models. This prior then consists of all unconditional models M that could be obtained with this TI and using a specific simulation method with given parameters. The conditioning process updates this prior by sampling or selecting a subset of it that honors the data. Incorporating the measurement errors of the data is easily done when adopting a Bayesian approach. The prior distribution of the models, $\text{Prob}(M)$, is defined by the TI. In addition, one has to formulate a likelihood that quantifies the probability of a model given the data $\text{Prob}(\text{data}|M)$. Essentially, if a model honors the data, it has a higher likelihood than if it is not in agreement with the data, and this model–data correspondence is quantified by the likelihood. The posterior probability of a model is then given by Bayes' rule:

$$\text{Prob}(M|\text{data}) \sim \text{Prob}(\text{data}|M)\text{Prob}(M) \qquad (\text{II.2.46})$$

Although mathematically rigorous, this approach is often impractical because the problem of generating models from TIs is not analytically tractable. This can be tackled with Monte Carlo methods, but then, sampling from the posterior can be extremely computationally demanding. It requires using a Metropolis sampler, a Gibbs algorithm, or a rejection sampler, which all necessitate building a large number of models that are rejected in the sampling process, the remaining ones being sampled from the posterior. The rejection rates with these methods are generally very high, resulting in heavy computational cost (this is extensively discussed in Chapter II.9 on inverse problems with TIs).

It is therefore more practical to seek models that are approximately conditioned using faster methods, such that they reasonably correspond to the models that would have been obtained from a Bayesian approach. The best example of such approximate conditioning is hard data (i.e., isolated locations where the value to simulate is known). A practical approach often used for hard data is to condition models by construction. In a broad sense, it can be seen as building models around the data, using the hard data as a starting point and then building the rest of the model such that it embeds them. Certain types of data, however, are not easily amenable to such approximations, and an indirect conditioning has to be adopted, such as Bayesian approaches.

The remainder of this section reviews the most common types of data that are available in spatial models, and how conditioning in MPS to such data is approached.

2.8.1 The different types of data

There exists a large diversity in measurement techniques for recording natural processes. As a result, data come in various forms and shapes. The way data are integrated in geostatistical models is specific to the type of data and to the type of problem to be solved.

A first point to consider is whether a given measurement directly informs the process of interest, or if it is only indirectly related to the modeled variable.

Whereas the value of direct data can be locally assigned to the simulation grid, indirect data need to be integrated through a relation that can be either simple or complex. For example, a relatively simple relationship is a target local probability, which can be integrated by locally perturbing the cdf of the simulated variable. A more complex relationship is when there is a possibly nonlinear dependence between the measured data and the simulated variable. The solution adopted lies in using a multivariate TI that explicitly represents this dependence and for algorithms to then reproduce it. The most difficult data are related to the variable of interest through complex physical processes that need to be described with a forward model. A typical example is when using head measurements to infer the hydraulic conductivity distribution in an aquifer. This is usually addressed through inverse problems (Chapter II.9) that can be computationally challenging.

Another factor to take into consideration is the support of a measurement (i.e., area or volume). The measured value can correspond to a point. It may also be a value integrated on a larger support. If this support coincides with the volume of a grid node, the measured data can simply be migrated to the grid. If this support is different from the grid node volume, appropriate upscaling or disaggregation techniques need to be used to determine what this measurement becomes when it is translated at the scale of a node.

Also important is the question of whether a measurement is exact (i.e., if there is negligible uncertainty associated with it) or if it is inexact. Strictly speaking, all measurements are inexact because measurement devices typically have errors associated with them. However, in many cases, this error can be neglected and the data are then imposed as fixed values in the simulation grid. Data that are exact and at the same support than the modeling volume are often termed "hard data". Hard data are usually easy to integrate in MPS simulation because it is relatively straightforward to design algorithms that use such data as a starting point for building spatial models.

As opposed to hard data, the term "soft data" is used to designate data that are inexact and/or indirect. When data are uncertain, it is not possible to build the models starting from the data. One approach that can be used in this case is to first create realizations of the data that reflect the measurement uncertainty. These randomized data realizations can then be considered as hard data for MPS models. Because the MPS models locally reproduce the hard data values, the uncertainty impaired to the data will be reproduced in the ensemble of model realizations.

Finally, the variable measured can be either continuous or categorical. Although all MPS approaches can deal with categorical variables, only some of them also accommodate continuous data, and therefore an appropriate simulation method has to be chosen.

Table II.2.3 provides the summary of a classification that takes into consideration the definitions above.

Table II.2.3 Overview of the different data types

Data property	Dichotomy	Data integration method
Relationship to the property modeled	Direct or indirect	Direct data are assigned to grid nodes. Indirect data can be integrated by local perturbation of statistics (target probability), by reproducing observed multivariate relationships, or through the resolution of an inverse problem.
Support	Point or area (or volume in 3D)	If data support is the volume of a node, its value can be simply migrated to the simulation grid. Otherwise, upscaling or disaggregation to the node support needs to be applied.
Measurement uncertainty	Exact or inexact	Exact data are used as a starting point for generating MPS models. For inexact data, realizations of the data need to be obtained that reflect the data uncertainty.
Data type	Continuous or categorical	Choice of the MPS simulation method. Alternatively, apply a threshold to a continuous variable to convert it to a categorical one.

2.8.2 Different types of data: an example

To better illustrate the different types of data described above, consider the example of the simulation of hydraulic conductivity in an aquifer using MPS. In this problem, the inputs would be a TI of hydraulic conductivity as well as measurements. Typical measurements that are used for aquifer characterization include well tests, geophysics, and possibly the analysis of rock samples. The variable of prime interest is often hydraulic conductivity, which describes the resistance that a porous medium offers to the flow of water.

One way to estimate hydraulic conductivity is to conduct pumping tests that provide, for a given pumping rate, the drawdown in the well as a function of time. Such data are then interpreted by fitting an analytical model to this function, thereby providing, under certain assumptions of homogeneity, an estimate of the hydraulic conductivity integrated over a relatively large volume that depends on the pumping rate and duration. In this case, the hydraulic conductivity obtained from a pumping test is direct because it provides a value for hydraulic conductivity. The data support is a volume, and the measured quantity is continuous. It is a modeling decision to assume the data are exact or inexact, but in all generality they are inexact because they are the result of a fitting procedure that generally implies some degree of uncertainty.

Regarding the notion of exact or inexact data, different degrees of uncertainty can be considered. If the hydraulic conductivity value is assumed to be exact (i.e., hard data) and its support volume is similar to the volume of a grid node, then that value can be simply migrated to the corresponding grid node. If the data are considered inexact, one can assume that this can be represented by a pdf of the hydraulic conductivity at the well location and use the same framework as in the case with exact data, but with a local value that is randomized for each realization using the pdf of the data value. If the assumptions of homogeneity in the analytical drawdown model are considered inacceptable, one can take the route of considering the measured drawdown as indirect data informing hydraulic conductivity. A transient groundwater model is then built that reproduces the pumping test conditions. This model is then used to solve a spatial inverse problem to derive heterogeneities that are consistent with the observed time-drawdown function. See Chapter II.9 for more details on inverse problems in the context of MPS simulation.

Slug tests are another type of measurement that informs hydraulic conductivity. Such tests are a category of well-testing methods where instantaneous injection of a given volume in the well causes a local perturbation of the piezometric level. The recovery from this perturbation can be analyzed, again using an analytical model. This type is very similar to pumping test data, and it can be integrated in similar ways. The main difference is that the resulting hydraulic conductivity is only measured in a small volume around the well with a slug test. Therefore, the value has a smaller support than a pumping test. In cases where both well test and slug test measurements are available, it is not possible to choose the size of the grid node such that it corresponds to all measurement volumes, and upscaling and disaggregation methods are necessary to transfer the measured value across different scales.

Any geophysical surveys would provide indirect data. For example, electric tomography provides information on the electric resistivity of the rock materials, which is a continuous variable that only partially reflects hydraulic conductivity. The inference of resistivity is based on an inverse problem that is resolved on a grid. It allows estimating resistivity at the same scale as the grid used for simulating hydraulic conductivity. However, the inverse procedure also entails that the outcome bears significant uncertainty.

The main interest in such geophysical measurements for aquifer characterization is that they cover a spatially extensive area and may inform local heterogeneities. If a simple relationship between electrical resistivity and hydraulic conductivity exists, one can convert this resistivity into local probabilities of observing certain geological facies at the survey location. Several MPS algorithms allow integrating such local probability information.

If there is no explicit statistical relationship available, then this relationship can be modeled by first obtaining a TI for resistivity that is locally consistent with the TI of hydraulic conductivity. This can be done, for example, by applying a forward model of resistivity to the hydraulic conductivity TI. This dual TI (or bivariate TI) is then used to simulate hydraulic conductivity conditionally to the observed resistivity values. For such purposes, the concept of multivariate neighborhoods needs to be used, and MPS algorithms designed for multivariate modeling are needed. As in the case of pumping tests, the most general way of integrating data is through the resolution of an inverse problem that uses the MPS algorithm as a regularization model.

The last type of data is the identification of geological facies in core samples. When wells are drilled, the rock type along the well profile is often recorded. Certain rock types have typical hydraulic conductivity ranges, providing important detailed local information. In most cases, such core samples are much smaller than the model grid nodes, and there may even be several data points falling in the same grid node. In this case, several options are possible: one can, for example, take the most frequently occurring facies and assign it to the corresponding grid node, or assume the data as inaccurate and draw facies values according to their frequency within the grid node.

Facies data are of a different nature than hydraulic conductivity because they provide categorical information (i.e., sand, clay, or carbonate). It is therefore indirect information that providing a range (or a pdf) for the hydraulic conductivity value of a node. In this case, it is relatively straightforward to obtain a TI consisting of the facies that correspond to the hydraulic conductivity TI. As in the case of geophysics, this bivariate TI can be used to jointly simulate the hydraulic conductivity and the facies.

2.8.3 Steering proportions

In many applications of MPS where categorical variables are used, it is desirable to control the simulated proportions. This is particularly the case when modeling geological bodies, where the proportion of each lithofacies may be known. For example, in a mining deposit it is common for a large number of drill holes to inform with a high level of confidence the marginal proportion of composites. Because this proportion directly determines the total amount of ore present in the deposit, it is critical to correctly honor this information. A similar situation occurs in oil reservoirs where geophysical surveys and rock physics analysis provide some information on the volumetric proportion of the porous lithofacies, which control the proportion of sand (or pay facies) and, therefore, the total amount of oil in place. Depending on the amount of wells drilled or the quality of the geophysical survey, the proportion information may vary in precision.

Here, we can distinguish two different situations where proportions need to be constrained:

1 A soft information is available on the presence of a given facies c at a specific location $\mathbf{x}$, which can be expressed as a probability of occurrence $P[Z(\mathbf{x}) = c]$.
2 The global proportion of facies c, $P[Z = c]$, is precisely known, or its uncertainty can be considered negligible.

Here, we review the methods that can be used to address these specific cases.

2.8.3.1   Servosystem

In a typical sequential simulation using the conditional frequencies in Equation (II.2.7), reproduction of the TI histogram is not ensured. The proportion of categories in a simulation not only depends on the TI, but are also driven by a number of other factors and modeling choices that have been described in this section, including the template used, the type of simulation path, the multiple-grid strategy, and so on. Therefore, it is often observed that unconditional simulations have facies proportions that differ from those of the TI. That is, the simulations reproduce some high-order properties of the TI but fail to honor univariate statistics.

The problem can be even more acute for conditional simulations, where the TI histogram may be different from the histogram of the data. This inconsistency in the model inputs entails that the resulting univariate statistics are poorly controlled and a specific mechanism has to be put in place to enforce what is actually desired. If the target histogram is precisely known (e.g., based on numerous drill holes data), it can be challenging to find a TI that exactly matches those proportions (on this topic, see Section II.8.2 on model validation). In data-poor situations, the proportions may not be strongly constrained. In this case, a mismatch between training proportions and (too sparse) data may not need to be explicitly resolved, because the proportions derived from the data are deemed unreliable.

There is a practical need for constraining global proportions of simulations (the histogram in the categorical case). However, matching a target histogram may come at the expense of matching higher-order statistics, and hence a trade-off has to be found. The strategy most often used to control the simulated univariate statistics is a servosystem approach. In this method, univariate statistics such as proportions are calculated on the fly, such that, for each category c, the proportion p_c^s in the (possibly incomplete) simulation remains close to the target proportion for this same category p_c^t. This is accomplished by applying a perturbation to Equation (II.2.7), which becomes

$$P^*[Z(\mathbf{x}) = c|\mathbf{N}] = \frac{\#(Z_{TI} = c|\mathbf{N})}{\displaystyle\sum_{b=1}^{C} \#(Z_{TI} = b|\mathbf{N})} + \frac{w}{1-w}\left(p_c^t - p_c^s\right), \qquad (II.2.47)$$

where $w \in [0, 1]$ is the servosystem weight factor that quantifies the relative importance of the proportions control against the faithful reproduction of the TI structures (Remy et al., 2009). If $w = 0$, no correction is applied. Conversely, as w tends to 1, the target proportion is exactly reproduced, but the resulting probabilities can be completely unrelated to those obtained using Equation (II.2.7).

One issue is that the more complete the simulation becomes, the more difficult it becomes to correct the proportions. To illustrate this, consider the application of the servosystem to a 1D random walk (i.e., a coin-tossing function),

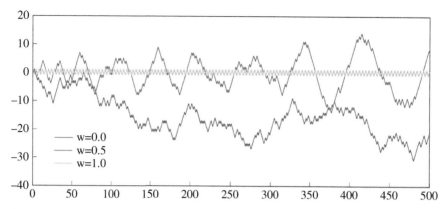

Figure II.2.36 Servosystem with different intensities applied to a 1D Brownian motion.

where at each step there is an equal probability of an increase of 1 or a decrease of 1:

$$b(x) = \begin{cases} b(x-1) + 1 & \text{if } r \le 0 \\ b(x-1) - 1 & \text{if } r > 0 \end{cases}, \quad r : U\{-1, 1\}. \tag{II.2.48}$$

The blue line on Figure II.2.36 shows the results of such a function without the use of a servosystem. One characteristic of random walks is that they tend to be nonstationary on a finite domain: the longer the function, the larger the deviations are from the zero value. The green line shows the same function, but now applying a servosystem with $w = 1$. Because the servosystem attempts to push back the average toward zero, a perfect oscillation between -1 and 1 is observed. The red line shows the intermediate case, with $w = 0.5$. Although the servosystem ensures that the function does not deviate from a zero mean, it results in oscillations of increasing amplitude. These are due to the inertia effect of the mean: the average is computed on the integral of the previously simulated nodes; hence, when this integral is large, an equally large excursion is required to apply a correction. This demonstrates that the servosystem may in certain cases create artifacts. It corrects the univariate proportions, but in doing so it may affect higher-order statistics, and such effects are not well controlled. In this example, the desired proportions are imposed, but this is accomplished at the cost of creating artifacts that manifest themselves by the low-frequency fluctuations.

2.8.3.2 Probability aggregation

In many cases, auxiliary data are provided to constrain the proportions locally. Such data, termed "soft data", are typically obtained by remote sensing or, when considering geological applications, by geophysical surveys. Soft data provide exhaustive but low-resolution information, which can be translated into local probability distributions. At each node of the simulation grid $\mathbf{x}$, some other variable $Y(\mathbf{x})$ is considered to be available, resulting in the conditional probability:

$$f^2(\mathbf{x}) = P[Z(\mathbf{x}) = c|Y(\mathbf{x})]. \qquad (\text{II.2.49})$$

This only provides information related to the soft data. The information coming from the neighborhood, based on the experimental pattern frequencies derived from Equation (II.2.7), can be expressed with another local pdf:

$$f^1(\mathbf{x}) = P[Z(\mathbf{x}) = c|\mathbf{N}]. \qquad (\text{II.2.50})$$

These two local pdfs are not independent. The rationale of probability aggregation methods (Allard et al., 2012) is to somehow combine Equations (II.2.49) and (II.2.50) using an aggregation operator, denoted $\wedge$, such as to obtain a combined pdf that is an approximation of the probability conditional to the joint event $\{\mathbf{N}, Y(\mathbf{x})\}$.

$$f^a(\mathbf{x}) = f^1(\mathbf{x})^{\wedge} f^2(\mathbf{x}) \approx P\left[Z(\mathbf{x}) = c|\mathbf{N}, Y(\mathbf{x})\right], \qquad (\text{II.2.51})$$

In the case where $\mathbf{N}$ and $Y(\mathbf{x})$ are independent, the aggregation collapses to a simple product of probabilities. In the more general case, one will need to account for information redundancy, and several approaches have been proposed. For a detailed discussion on the developments of these approaches, we refer the reader to Allard et al. (2012).

Posed in general terms, probability aggregation can be seen a determining the probability of an event, denoted A, conditional on the occurrence of a set of events, D_i, $i = 1, \ldots, N$. The most usual case is, however, to aggregate only two pdfs. We wish to approximate the probability $P(A|D_1, \ldots, D_N)$ on the basis of simultaneous knowledge of the N conditional probabilities $P(A|D_i)$.

We can then distinguish three methods of probability aggregation that are used in geostatistics:

Bordley's formula (Bordley, 1982)

The method is based on the use of odd ratios, which are defined as:

$$O(A) = \frac{P(A)}{1 - P(A)}, \quad 0 \leq P(A) < 1. \qquad (\text{II.2.52})$$

Then:

$$O_G(A) = O_0(A)^{w_0} \prod_{k=1}^{N} \left[\frac{O_i(A)}{O_0(A)}\right]^{w_k}, \quad w_k \in (0, \infty), \qquad (\text{II.2.53})$$

where O_0 is the odd ratio of the marginal (unconditional) probability $P[Z(\mathbf{x}) = c]$. The weights w_k can be seen as a way of quantifying redundancy or as a confidence factor for each source of information.

Note that Bordley's formula can be extended to the continuous case:

$$f^a(\mathbf{x}) = \frac{1}{\eta} f^0(\mathbf{x}) \prod_{k=1}^{N} \left[\frac{f^k(\mathbf{x})}{f^0(\mathbf{x})} \right]^{w_k}, \quad w_k \in (0, \infty), \tag{II.2.54}$$

where η is a normalization constant and f^0 is the marginal pdf.

The tau model (Journel, 2002; Krishnan, 2008)

The tau model is essentially equivalent to Bordley's formula for the binary case, with $w_k = 1$ for $k = 1, \dots, N$. It is also based on odd ratios and expressed by the following equations:

$$f^a(\mathbf{x}) = \frac{1}{1+x}, \tag{II.2.55}$$

with:

$$\frac{x}{x_0} = \left(\frac{x_1}{x_0} \right)^{\tau_1} \left(\frac{x_2}{x_0} \right)^{\tau_2}, \quad \tau_1, \tau_2 \in (-\infty, \infty), \tag{II.2.56}$$

and the different terms defined as:

$$x_0 = \frac{1 - P\left[Z(\mathbf{x}) = c \right]}{P\left[Z(\mathbf{x}) = c \right]}, \tag{II.2.57}$$

$$x_1 = \frac{1 - P\left[Z(\mathbf{x}) = c | \mathbf{N} \right]}{P\left[Z(\mathbf{x}) = c | \mathbf{N} \right]}, \tag{II.2.58}$$

$$x_2 = \frac{1 - P\left[Z(\mathbf{x}) = c | Y(\mathbf{x}) \right]}{P\left[Z(\mathbf{x}) = c | Y(\mathbf{x}) \right]}. \tag{II.2.59}$$

The values often used in the tau model are $\tau_1 = \tau_2 = 1$, assuming some standardized form of conditional independence.

References

Abdollahifard, M. J. & Faez, K. 2013. Fast direct sampling for multiple-point stochastic simulation. *Arabian Journal of Geosciences*, 1–13.

Abramowitz, M. & Stegun, I. 1965. *Handbook of Mathematical Functions with Formulas, Graphs, and Mathematical Tables*, Dover, New York.

Allard, D., Comunian, A. & Renard, P. 2012. Probability aggregation methods in geoscience. *Mathematical Geosciences*, 44, 545–581.

Arpat, B. 2005. *Sequential Simulation with Patterns*, PhD diss., Stanford University.

Arpat, B. & Caers, J. 2007. Conditional simulations with patterns. *Mathematical Geology*, 39, 177–203.

Boisvert, J. B., Pyrcz, M. J. & Deutsch, C. V. 2007. Multiple-point statistics for training image selection. *Natural Resources Research*, 16, 313–321.

Bordley, R. 1982. A multiplicative formula for aggregating probability assessments. *Management Science*, 28, 1137–1148.

Chatterjee, S. & Dimitrakopoulos, R. 2012. Multi-scale stochastic simulation with a wavelet-based approach. *Computers and Geosciences*, 45, 177–189.

Chatterjee, S., Dimitrakopoulos, R. & Mustapha, H. 2012. Dimensional reduction of pattern-based simulation using wavelet analysis. *Mathematical Geosciences*, 44, 343–374.

Daly, C. 2004. Higher order models using entropy, Markov random fields and sequential simulation. *In:* Leuangthong, O. & Deutsch, C. V. (eds.), *Geostatistics Banff 2004*, Kluwer Academic, Banff, AB, 215–224.

Deutsch, C. & Gringarten, E. 2000. Accounting for multiple-point continuity in geostatistical modeling. *In: 6th International Geostatistics Congress of Southern Africa, 2000 Cape Town*, 156–165.

Deutsch, C. & Journel, A. 1992. *GSLIB: Geostatistical Software Library*, Oxford University Press, New York.

Dijkstra, E. 1959. A note on two problems in connexion with graphs. *Numerische Mathematik*, 1, 269–271.

Dimitrakopoulos, R., Mustapha, H. & Gloaguen, E. 2010. High-order statistics of spatial random fields: Exploring spatial cumulants for modeling complex non-Gaussian and non-linear phenomena. *Mathematical Geosciences*, 42, 65–99.

Dubuisson, M. & Jain, A. 1994. A modified Hausdorff distance for object matching. In: *International Conference on Pattern Recognition*. Jerusalem, Isarel.

EL Ouassini, A., Saucier, A., Marcotte, D. & Favis, B. D. 2008. A patchwork approach to stochastic simulation: A route towards the analysis of morphology in multiphase systems. *Chaos, Solitons and Fractals*, 36, 418–436.

Finkel, R. A. & Bentley, J. L. 1974. Quad trees a data structure for retrieval on composite keys. *Acta Informatica*, 4, 1–9.

Gersho, A. & Gray, R. 1992. *Vector Quantization and Signal Compression*, Kluwer Academic, Boston.

Guardiano, F. & Srivastava, M. 1993. Multivariate geostatistics: Beyond bivariate moments. *In:* Soares, A. (ed.) *Geostatistics-Troia*, Kluwer Academic, Dordrecht.

Honarkhah, M. & Caers, J. 2010. Stochastic simulation of patterns using distance-based pattern modeling. *Mathematical Geosciences*, 42, 487–517.

Huttenlocher, D. P., Klanderman, G. A. & Rucklidge, W. J. 1993. Comparing images using the Hausdorff distance. *IEEE Transactions on Pattern Analysis and Machine Intelligence*, 15, 850–863.

Huysmans, M. & Dassargues, A. 2011. Direct multiple-point geostatistical simulation of edge properties for modeling thin irregularly shaped surfaces. *Mathematical Geosciences*, 43, 521–536.

Journel, A. 2002. Combining knowledge from diverse sources: An alternative to traditional data independence hypotheses. *Mathematical Geology*, 34, 573–596.

Kjønsberg, H. & Kolbjørnsen, O. 2008. Markov mesh simulations with data conditioning through indicator kriging. *In:* Ortiz, J. M. & Emery, X. (Eds.), *Proceedings of the Eighth International Geostatistics Congress 2008, 1–5 Dec. 2008*.

Krishnan, S. 2008. The tau model for data redundancy and information combination in earth sciences: Theory and application. *Mathematical Geosciences*, 40, 705–727.

Lefebvre, S. & Hoppe, H. 2006. Appearance-space texture synthesis. *ACM Transactions on Graphics*, 25, 541–548.

Mariethoz, G., Renard, P. & Straubhaar, J. 2010. The direct sampling method to perform multiple-point geostatistical simulations. *Water Resources Research*, 46.

Mustapha, H. & Dimitrakopoulos, R. 2010a. High-order stochastic simulation of complex spatially distributed natural phenomena. *Mathematical Geosciences*, 42, 457–485.

Mustapha, H. & Dimitrakopoulos, R. 2010b. A new approach for geological pattern recognition using high-order spatial cumulants. *Computers and Geosciences*, 36, 313–334.

Parra, A. & Ortiz, J. M. 2011. Adapting a texture synthesis algorithm for conditional multiple point geostatistical simulation. *Stochastic Environmental Research and Risk Assessment*, 25, 1101–1111.

Remy, N., Boucher, A. & Wu, J. 2009. *Applied Geostatistics with SGeMS: A User's Guide*, Cambridge University Press, Cambridge.

Serra, J. & Soille, P. 1994. *Mathematical Morphology and Its Applications to Image Processing*, Kluwer Academic, Dordrecht.

Straubhaar, J., Renard, P., Mariethoz, G., Froidevaux, R. & Besson, O. 2011. An improved parallel multiple-point algorithm using a list approach. *Mathematical Geosciences*, 43, 305–328.

Strebelle, S. 2002. Conditional simulation of complex geological structures using multiple-point statistics. *Mathematical Geology*, 34, 1–22.

Stright, L., Caers, J. & Li, H. 2006. Coupled geological modeling and history matching of fine-scale curvilinear flow barriers. Paper presented at the 19th SCRF annual meeting, Stanford Center for Reservoir Forecasting, Stanford, CA, May.

Tahmasebi, P., Hezarkhani, A. & Sahimi, M. 2012. Multiple-point geostatistical modeling based on the cross-correlation functions. *Computational Geosciences*, 16, 779–797.

Tan, X., Tahmasebi, P. & Caers, J. 2013. Comparing training-image based algorithms using an analysis of distance. *Mathematical Geosciences*, 46, 149–169.

Theodoridis, S. & Koutroumbas, K. 2008. *Pattern Recognition*, Academic Press, New York.

Wei, L. & Levoy, M. 2000. Fast texture synthesis using tree-structured vector quantization. In: *SIGGRAPH '00: 27th Annual Conference on Computer Graphics and Interactive Techniques, 2000 New Orleans*. ACM Press/Addison-Wesley, Reading, MA.

Zhang, T., Pedersen, S. I., Knudby, C. & Mccormick, D. 2012. Memory-efficient categorical multi-point statistics algorithms based on compact search trees. *Mathematical Geosciences*, 44, 863–879.

Zhang, T., Switzer, P. & Journel, A. 2006. Filter-based classification of training image patterns for spatial simulation. *Mathematical Geology*, 38, 63–80.

CHAPTER 3

Multiple-point geostatistics algorithms

A large number of multiple-point geostatistics (MPS) algorithms currently exist, and new ones are developed regularly. These algorithms use various ways of defining, storing and restituting patterns. The profusion of methods can be daunting for anyone wanting to start using or developing MPS algorithms. In this chapter, we review the approaches one by one, going into the details of each algorithm.

It is our view that many algorithmic elements are common to different simulation methods. For example, the use of a simulation path, the calculation and sampling from local probability distributions, and the use of multiple grids only need to be understood once. These various components or ingredients that are part of each simulation method have been detailed in Chapter II.2. Here, we start from the point of view that these notions are known, and we use them in the description of the simulation techniques, in a similar way as a program would call generic functions to perform specific tasks. With this presentation of the existing methods, we also want to highlight how similar many of them are, and the fact that some methods often only differ by a slight algorithmic component. This is highlighted by the fact that we can establish a generic template for MPS algorithms that captures their essential elements.

We do not discuss in which case the individual algorithms should be applied; instead, Part III will serve as illustrations of what method is suitable in what particular circumstances. This chapter simply focuses on the theoretical considerations and technical details of the simulation algorithms.

Broadly speaking, simulation methods are often classified in pixel-based and pattern-based methods, and both categories have advantages and drawbacks. In the description, we focus on functionality by identifying which algorithm can do what, or which constraints can be imposed, in order to give the reader guidelines and an overview of the capabilities of each method.

The Markov random field (MRF)-based methods are dealt with separately (Chapter II.4) because they are based on a philosophy that is different from MPS.

Multiple-point Geostatistics: Stochastic Modeling with Training Images, First Edition. Gregoire Mariethoz and Jef Caers.
© 2015 John Wiley & Sons, Ltd. Published 2015 by John Wiley & Sons, Ltd.
Companion website: www.wiley.com/go/caers/multiplepointgeostatistics

3.1   Archetypal MPS algorithm

Before going into the detailed description and the specifics of each algorithm, we introduce here a generic template that synthesizes the elements that are common to most algorithms. This will allow a discussion of each algorithmic element separately while remaining general. This generic template is described here:

Inputs

A At least one variable Z to simulate
B Simulation grid SG, with nodes (or patches) denoted $\mathbf{x}$
C Training image TI
D [*optional*] Set of conditioning data points
E [*optional*] Local probability map L
F Set of algorithm-specific parameters

Algorithm steps

 1 Migrate conditioning points to the corresponding grid nodes.
 2 Analyze TI to derive patterns database DB.
 3 Define a simulation path.
 4 **Do** until stopping criterion is met:
 5   Define the node $\mathbf{x}$ to be simulated.
 6   Find $\mathbf{N}_x$, the neighborhood of $\mathbf{x}$.
 7   Determine the conditional distribution of $Z(\mathbf{x})$ conditional to $\mathbf{N}_x$ and to $L(\mathbf{x})$.
 8   Sample a value from this distribution.
 9   Place this value in the SG.
10 **End.**
11 Postprocessing

Outputs

A SG, with variable Z known at all nodes

This overall archetypal organization remains relatively consistent and is valid, with specific adaptations, for all algorithms in this book, and it is likely to remain valid for future algorithms as well.

The different algorithms offer different versions of these basic steps, but this general scheme can be seen as a template. Some examples of variations that can occur between the different existing algorithms are:

- For patch-based algorithms, $\mathbf{x}$ represents patches instead of nodes.
- The use of multigrids can be assimilated to a specific type of path, and a different neighborhood for each grid.
- Different ways of defining neighborhoods are possible (see Section II.2.3).
- The DB can be defined and organized in a number of ways (see Section II.2.4).
- The way the conditional distributions are computed (point 7) can vary.
- For methods based on sequential simulation, the stopping criterion of step 4 is met when all nodes of the simulation grid have been simulated. With iterative

simulation methods, some convergence criterion is formulated as a stopping criterion for the algorithm.

- Z can be multivariate, and in this case a set of variables is simulated: Z_1, $Z_2, \ldots Z_m$.
- The postprocessing is mostly performed independent of the simulation algorithm used, and therefore will be described separately in Section II.3.3.

Note that parameterization of MPS algorithms is an important part of the modeling process, with the algorithmic parameters often having a direct influence on the final ensemble of realizations, and therefore on the final quantification of spatial uncertainty. The general approach recommended is to perform a quick sensitivity analysis prior to real-world modeling, by doing unconditional realizations on a small subdomain. Validation techniques described in Section II.8.3 can then be used to select the appropriate range of parameters for the chosen simulation algorithm. As a starting point, we provide guidelines and references, when available, on the parameterizations of some of the most common MPS algorithms.

3.2 Pixel-based algorithms

3.2.1 ENESIM (Guardiano and Srivastava, 1993)

Inputs

A One categorical variable Z to simulate
B Simulation grid SG, with nodes denoted $\mathbf{x}$
C Categorical training image TI
D Set of conditioning data points
E Definition of a template

Algorithm steps

1 Migrate conditioning points to the corresponding grid nodes.
2 Define a simulation path.
3 **Do** until all SG nodes have been visited:
4   Define the node $\mathbf{x}$ to be simulated.
5   Find $\mathbf{N}_x$, the data event at $\mathbf{x}$.
6   Scan the entire TI for all occurrences of $\mathbf{N}_x$.
7   Determine the conditional distribution of $Z(\mathbf{x})$ as the histogram of the central node values for all occurrences.
8   Sample a value from this distribution.
9   Place this value in the SG.
10 **End.**

Outputs

A SG, with variable Z known at all nodes

Remarks

- ENESIM is the earliest MPS implementation (Guardiano and Srivastava, 1993). It is not used anymore in most practical cases because of its high CPU cost.
- An algorithm that is similar to ENESIM, but specifically aimed at continuous variables, was proposed by Efros and Leung (1999).

3.2.2   SNESIM (Strebelle, 2002)

Inputs

A One categorical variable Z to simulate
B Simulation grid SG, with nodes denoted **x**
C Categorical training image TI
D Set of conditioning data points
E Local probability map L
F Parameters: number of multigrids G, search template **T**, minimum number of replicates r, tau weights, and global proportion for each category p_c^t

Algorithm steps

1 Migrate conditioning points to the corresponding grid nodes.
2 **For** each multiple grid g:
3   Define an expanded template **T**$_g$.
4   Construction of search tree from analysis of the TI at level g. Events with a number of occurrences in the TI lower than r are ignored in the tree.
5   Define a simulation path for grid g.
6   **Do** until all nodes in grid g have been visited:
7   Define the node **x** to be simulated.
8   Find **N**$_x$, the neighborhood of **x**.
9   Determine the conditional distribution of $Z(\mathbf{x})$ conditional to **N**$_x$ based on the search tree.
10   Aggregate this distribution with the local probability $L(\mathbf{x})$, using the tau model.
11   Sample a value from the resulting distribution.
12   Place this value in the SG.
13   **End.**
14 **End.**

Outputs

A Simulation grid SG, with variable Z known at all nodes

Parameterization

- A complete guide to the parameterization of SNESIM is given in Liu (2006).

The main parameters of the algorithm are the template definition, essentially relying on the number of neighbors n, the minimum number of replicates r, and the number of multiple grids G. A value of $n = 30$–50 nodes is generally regarded as appropriate, although values up to $n = 80$ or 100 nodes can be required to model complex 3D structures. Most software implementations allow modifying the shape of the search neighborhood;

however, an isotropic search neighborhood (circular in 2D or spherical in 3D) often yields the best results.

It is recommended to use at least three levels of multiple grids, and more if large-scale structures are present. The algorithm is not very sensitive to the minimum number of replicates, and a value at least higher than 1 is recommended to avoid the effect of verbatim copy (see Section II.8.3.2).

Remarks

- The SNESIM algorithm is very widely used in various fields of geoscience because there is an efficient implementation freely available (SGeMS). The use of tree storage can, however, make it RAM intensive. It is restricted to categorical, univariate problems.
- The IMPALA algorithm is essentially similar to SNESIM, except that in step 4 a list is used instead of a tree to store data events (Straubhaar et al., 2011). As a result, the RAM consumption is decreased, with higher CPU cost. This is compensated by the search in the list, which is parallelized. Further significant acceleration is obtained by having a structure that combines trees and lists (Straubhaar et al., 2013).
- The HOSIM algorithm has similarities with SNESIM, except that it uses spatial cumulants to store patterns instead of frequencies (for details, see Mustapha and Dimitrakopoulos, 2011). The spatial cumulants are also stored in a tree structure.
- GROWTHSIM is another algorithm that is also essentially similar to SNESIM, except that it uses a random-neighbor path (Eskandaridalvand and Srinivasan, 2010).

3.2.3   Direct sampling (Mariethoz et al., 2010)

Inputs

A m variables $Z_1, Z_2, \ldots Z_m$ to simulate
B Simulation grid SG, with nodes denoted **x**
C Training image TI with m categorical or continuous variables, or a mix of types
D Set of conditioning data points
E Maximum number of neighbors n, distance threshold t, maximum fraction of TI scanned f

Algorithm steps

  1 Migrate conditioning points to the corresponding grid nodes.
  2 Define a simulation path.
  3 **Do** until all SG nodes have been visited:
  4   Define the node **x** to be simulated.
  5   Find $\mathbf{N}_x$, the uni- or multivariate neighborhood, made of the n closest neighbors of **x**.
  6   Randomly scan the TI, and for each TI location **y**, compute the distance $d(\mathbf{N}_x, \mathbf{N}_y)$.

---

7   At the first occurrence of $d(\mathbf{N}_x, \mathbf{N}_y) \leq t$, assign $Z(\mathbf{y})$ in the *SG*.

8   If no occurrence is found after having scanned a fraction f of *TI*, use the data event with the smallest distance found, and assign its central value in the *SG*.

9 End.

Outputs

A *SG*, with variables $Z_1, Z_2, \ldots Z_m$ known at all nodes

---

Parameterization

- A complete guide to the parameterization of the direct-sampling (DS) algorithm is given in Meerschman et al. (2013).

  The main parameters of the algorithm are the maximum number of closest neighbors n, the distance threshold t and the fraction of scanned TI f. The recommended value for the number of neighbors is between $n = 10$ for simple cases and at most $n = 60$ for complex 3D structures.

  A threshold value of $t = 0$ is generally not recommended because it slows down the algorithm dramatically and leads to verbatim copy of the TI. A value of $t = 0.1$ can be considered in most cases as an upper bound for the threshold. For categorical variables, the threshold value defines the number of mismatching nodes in the data event (i.e. with $n = 20$ and $t = 0.1$, a data event with two mismatching nodes is still acceptable).

  The fraction of scanned TI f should be smaller than the maximum possible value of $f = 1$ to avoid verbatim copy. It is found that the results are not very sensitive to this parameter, although lower f significantly reduces computing time. If the TI is large, it is therefore recommended to start with $f = 0.1$, and only if the results are not satisfying increase the value up to $f = 0.5$.

Remarks

- No multiple grids are used. The data events, being defined as the n closest nodes, continuously reduce in size as the informed nodes become denser. This results in multiple-scale patterns being taken into account without explicitly specifying multiple grids.
- It is possible to use DS as a patch-based simulation method, by pasting a group of nodes instead of a single node at step 7 (Rezaee et al., 2013).
- The basis of DS is very similar to that of ENESIM. The main difference is that ENESIM explicitly models a probability distribution, whereas DS samples the TI without computing probabilities. ENESIM (1) performs an exhaustive scan of the TI (convolution), (2) builds a conditional cumulative distribution function (cdf), and (3) samples this cdf. In contrast, DS directly samples the TI conditionally to a data event. Computationally, this results in an interrupted convolution (i.e., the scan is stopped at the first matching data event).
- DS strongly relies on the definition of the distance function $d(.)$. When multiple variables are used, d should be chosen appropriately for categorical or continuous variables, or a combination of both (see Section II.2.5).

- It is possible to tweak the distance function to control global proportions or local proportions, or to impose a local average. This is accomplished by adding to the distance function an error term on the local pdf E_p that quantifies the difference between the target probability $L(\mathbf{x})$ and the current proportion $P^*(\mathbf{x})$ in the partially simulated SG. It has the form:

$$E_p \sim \left| P^*(\mathbf{x}) - L(\mathbf{x}) \right|. \qquad (\text{II}.3.1)$$

The distance of a given TI data event is then increased as a function of E_p, resulting in the acceptance of data events that stir the simulation histogram such that it becomes similar to the target local or global histogram. The approach is valid for the histogram in both categorical and continuous cases.

3.2.4 Simulated Annealing (Peredo and Ortiz, 2011)

Inputs

A m variables $Z_1, Z_2, \ldots Z_m$ to simulate

B Simulation grid SG, with nodes denoted $\mathbf{x}$

C Training image TI with m categorical or continuous variables, or a mix of types

D Set of conditioning data points

E Local probability map L

F Perturbation function $Pert$. It can, for example, be used to alter a category value, or to add a random uniform value to a continuous variable.

G A global objective function O that quantifies the mismatch in statistics between patterns frequencies in a realization with the TI, and possibly the mismatch with a probability map, or any other simulation constraint

H Parameters: search template $\mathbf{T}$, cooling parameters, stopping criterion

Algorithm steps

1 Analyze the TI to derive patterns frequencies.

2 Migrate conditioning points to the corresponding grid nodes.

3 Initialize the remaining nodes with values taken from the marginal distribution.

4 Evaluate O.

5 **Do** until the stopping criterion is met:

6   Randomly choose a node $\mathbf{x}$ that is not a conditioning datum.

7   Propose an altered value for $\mathbf{x}$, using the perturbation mechanism $Pert$.

8   Evaluate O.

9   Calculate a probability of acceptance for the altered value with a Boltzmann distribution depending on the current temperature.

10   Accept or reject the perturbation according to this probability.

11 **End.**

Outputs

A SG, with variable Z known at all nodes

Remarks

- The probability of acceptance of step 9 is formulated using a Boltzmann distribution:

$$P(accept) = \begin{cases} 1 & \text{if } O_{new} \leq O_{old} \\ e^{(O_{old}-O_{new})/T} & otherwise \end{cases}. \qquad (II.3.2)$$

where T is the temperature parameter, which is gradually reduced with the iterations.

- The objective function O is very generic and can include the entire range of characteristics to be modeled. Therefore, the annealing approach is very flexible, the drawback being a high computational cost (Deutsch and Wen, 2000).

3.3 Patch-based algorithms

One observation common in MPS is that when comparison between TI and generated realizations is made, some patterns are exactly reproduced or "copied" from the TI to the simulation grid. Such local verbatim copy can be quantified with coherence maps (see Section II.3.2). This is a direct consequence of textural coherence and the finite size of the TI.

The idea of patch-based methods is to take a shortcut: by simulating entire patches, it tends to quilt simulation values that are also next to each other in the TI. One advantage of these patch-based methods is therefore that their implementations are generally faster than pixel-based algorithms.

3.3.1 SIMPAT (Arpat and Caers, 2007)

Inputs

A One variable Z to simulate, either categorical or continuous
B Simulation grid SG, with nodes denoted **x**, each being the centroid of a patch
C Training image TI
D Set of conditioning data points
E Parameters: number of multigrids G, patch template **T**

Algorithm steps

1 Migrate conditioning points to the corresponding grid nodes.
2 **For** each multiple grid g:
3   Define an expanded template $\mathbf{T}_g$.
4   Construct a list of data events from analysis of the TI at level g.
5   Define a random simulation path for grid g.
6   **Do** until all nodes in grid g have been visited:
7   Define the location **x** to be simulated. Locations already simulated in previous multiple-grid levels are resimulated.

8	Find $\mathbf{N}_x$, the data event formed by the informed nodes in the patch centered on **x**.
9	Find the subset of patches in the data events list that match the conditioning data present in the area of the current patch.
10	Among these candidates, find the patch that is most similar to $\mathbf{N}_x$ according to a distance function $d(.)$.
11	Place this patch in the SG.
12	**End.**
13 End.	

Outputs

A *SG*, with variable *Z* known at all nodes

Parameterization

• The most important parameter in patch-based methods is the size of the patch. Honarkhah (2011) has proposed a methodology for automatically determining the optimal patch size. It consists in computing the mean entropy of the patterns in the TI for a series of square templates of different sizes. The mean entropy for a square neighborhood of *w* by *w* nodes is defined as

$$ME(w) = \frac{1}{\#\mathbf{N}_w} \sum_{k=1}^{\#\mathbf{N}_w} entropy \left\{ \mathrm{P} \left[Z(\mathbf{x}) | \mathbf{N}_w^k \right] \right\}, \qquad (\mathrm{II}.3.3)$$

where $\#\mathbf{N}_w$ is the number of patterns of size *w* by *w* in the TI, and the entropy represents a measure of how informative the neighborhood is with respect to the data event central node value. For small templates entropy is low because patterns require short encodings (according to Shannon information theory), while large templates require large encodings, hence resulting in large mean entropy. It is proposed that the optimal patch size is somewhere in between. To find this point, so-called elbow plots are used, which represent the mean entropy versus the patch size. The point of inflexion of the elbow plot is then chosen as the optimal patch size.

3.3.2   FILTERSIM (Zhang et al., 2006)

Inputs

A One variable *Z* to simulate, either categorical or continuous
B Simulation grid *SG*, with nodes denoted **x**, each being the centroid of a patch
C Training image *TI*

D Set of conditioning data points
E Parameters: number of multigrids G, patch template **T**
F Set of filters having the same size as **T**

Algorithm steps

1 Migrate conditioning points to the corresponding grid nodes.
2 **For** each multiple grid g:
3   Define an expanded template $\mathbf{T}_g$.
4   Cluster the TI patterns at level g based on their filter scores, and store a list for each resulting cluster.
5   Compute a prototype for each cluster.
6   Define a random simulation path for grid g.
7   **Do** until all nodes in grid g have been visited:
8   Define the location **x** to be simulated. Locations already simulated in previous multiple-grid levels are resimulated.
9   Find $\mathbf{N}_x$, the data event formed by the informed nodes in the patch centered on **x**.
10   Find the cluster prototype that is the most similar to $\mathbf{N}_x$ according to a distance function $d(.)$.
11   Randomly draw one patch from the corresponding cluster.
12   Place this patch in the SG, and freeze the nodes in a central patch area. Frozen nodes are unfrozen at the end of each multiple grid.
13   **End.**
14 **End.**

Outputs

A SG, with variable Z known at all nodes

Remarks

- The distance function $d(.)$ should be chosen appropriately for categorical or continuous variables. For accurate conditioning to hard data, a higher weight is given in the distance function to the hard data than to the data simulated previously in the path.
- It is possible to integrate local probability data by using a tau model. The cluster prototype of step 10 is then aggregated with the local soft data, resulting in an altered cluster prototype. Then, the database cluster with the closest prototype to this altered prototype is chosen, and a patch is drawn from it (Wu et al., 2008).
- The DISPAT algorithm is essentially similar to FILTERSIM, except that in step 4 the filter scores used for patterns classification are replaced by a kernel k-means applied on an MDS transform of the TI patterns (Honarkhah and Caers, 2010). The WAVESIM algorithm (Chatterjee et al., 2012) is another variation of FILTERSIM, using wavelet coefficients instead of filter scores as the basis for pattern classification.

3.3.3 Patchwork Simulation (El Ouassini et al., 2008)

Inputs

A One binary variable Z to simulate
B Simulation grid SG, with nodes denoted **x**, each being the centroid of a patch
C Training image TI
D Set of conditioning data points
E Parameters: rectangular patch template **T**, amount of overlap o

Algorithm steps

1 Migrate conditioning points to the corresponding grid nodes.
2 Extract all patches from the TI.
3 Construct a graph of transition probabilities between the patches along the main directions of the grid, based on patch transitions observed in the TI.
4 Define a unilateral simulation path where successive patches overlap.
5 **Do** until all nodes have been visited:
6   Define the location **x** of the patch to be simulated.
7   The patch can have 0, 1, 2, or 3 (in 3D) adjacent patches already simulated that overlap the current patch. N_x is defined as these overlap areas.
8   Select the subset of patches that match conditioning data that are present in the area of the current patch.
9   Use the graph of transition probabilities to assign probabilities to the remaining patches.
10   Draw one patch according to those probabilities.
11   Place this patch in the SG.
12 **End.**

Outputs

A SG, with variable Z known at all nodes

Remarks
- Based on transition probabilities, this patchwork method is limited to binary variables. It is, however, the precursor for other patch-based algorithms that are based on a unilateral path. These are described below.
- The method does not require multiple grids because (1) it uses large patches that are able to reproduce large features, and (2) the Markovian properties of the unilateral path ensure that large-scale structures are reproduced.

3.3.4 CCSIM (Tahmasebi et al., 2012)

Inputs

A m variables $Z_1, Z_2, \ldots Z_m$ to simulate
B Simulation grid SG, with nodes denoted **x**, each being the centroid of a patch
C Training image TI with m categorical or continuous variables

D Set of conditioning data points
E Parameters: rectangular patch template **T**, distance threshold t, amount of overlap o

Algorithm steps

1 Migrate conditioning points to the corresponding grid nodes.
2 Define a unilateral simulation path where successive patches overlap.
3 **Do** until all nodes have been visited:
4   Define the location **x** of the patch to be simulated.
5   The patch can have 0, 1, 2, or 3 (in 3D) adjacent patches already simulated that overlap the current patch. $\mathbf{N}_x$ is defined as these overlap areas.
6   Perform a convolution of the TI with $\mathbf{N}_x$ (1, 2, or 3 convolutions required depending on the number of adjacent patches), using a cross-correlation distance function.
7   Find the subset of patches that match conditioning data present in the area of the current patch.
8   If this subset is empty, recursively split the patch in four subpatches that are then simulated (starting from step 4).
9   In this subset, select the patches that correspond to convolution values lower than t.
10   Among the remaining candidates, randomly select one patch.
11   Place this patch in the SG.
12 **End.**

Outputs

A SG, with variable Z known at all nodes

Parameterization

- Mahmud et al. (2014) found that large patches tend to cause verbatim copy, and therefore they do not recommend patch sizes extending over 50 nodes (which is probably a very high value already).
- Empirical tests gave indicative values for the overlap size of 1/4 to 1/3 of the patch size, which is similar to the recommendation of Efros and Freeman (2001).
- A distance threshold of 0 should not be used to avoid verbatim copy. As a rule of thumb, the threshold should be chosen such that there is generally a pool of at least five possible candidate patches in the TI that can be used in the simulation.

Remarks

- Although CCSIM relies on a cross-correlation distance, other distances are possible, which should be chosen appropriately for categorical or continuous variables.
- One variant of the method is to replace the threshold t by a minimum number of candidates (or replicates) ε. In step 9, candidates are then ranked according to their distance, and the ε best ranked candidates are selected. This ensures that the number of candidates is always sufficient to allow for stochasticity in the realizations.

- The method of image quilting (IQ), originally proposed by Efros and Freeman (2001), is identical to CCSIM except that an additional step is added after step 11, consisting of correcting any remaining mismatch in the overlap area by identifying a minimum boundary cut using dynamic programming (see Section II.2.6.4). The procedure for a 2D vertical cut is described below. Its adaptation for a horizontal cut is straightforward.

Inputs
A Two overlapping patches B_1 and B_2 with overlapping areas B_1^{ov} and B_2^{ov}

Algorithm steps
1 Define an error surface $\mathbf{e} = (B_1^{ov} - B_2^{ov})^2$, having a rectangular size p by o.
2 Compute the cumulative minimum error along the cutting direction:
3 **For** each row i in $\mathbf{e}$ (with i = 2 ... p), do:
4   **For** each column j in $\mathbf{e}$, with $j = 1 ... o$, do:
5   Calculate the cumulative minimum error $\mathbf{E}$ using the three closest pixels on the previous row (or the two closest pixels, if on an edge):
6          $\mathbf{E}_{i,j} = \mathbf{e}_{i,j} + \min(\mathbf{E}_{i-1,j-1}, \mathbf{E}_{i-1,j}, \mathbf{E}_{i-1,j+1})$; if $j = 2 ... (o - 1)$
7          $\mathbf{E}_{i,j} = \mathbf{e}_{i,j} + \min(\mathbf{E}_{i-1,j}, \mathbf{E}_{i-1,j+1})$; if $j = 1$
8          $\mathbf{E}_{i,j} = \mathbf{e}_{i,j} + \min(\mathbf{E}_{i-1,j-1}, \mathbf{E}_{i-1,j})$; if $j = o$
9   **End.**
10 **End.**
11 Identify the coordinate k corresponding to the entry with the smallest value on the last row of $\mathbf{E}$. This location corresponds to the arrival point of a path of minimum cost through the error surface.
12 Trace back the minimum values for each row i, going backward (with $i = p - 1 ... 1$), and each time identify the cutting path as $\min(\mathbf{E}_{i,k-1}, \mathbf{E}_{i,k}, \mathbf{E}_{i,k+1})$.

Outputs
A Minimum error boundary cut

3.4 Qualitative comparison of MPS algorithms

A quantitative comparison of each algorithm is very difficult because each method applies to different types of data, has different parameters, and might perform better in some cases than in others. Based on our experience, however, we are able to provide a qualitative comparison of the main characteristics of the MPS algorithms described in this chapter. Table II.3.1 describes these characteristics, mostly focusing on flexibility in terms of data integration and CPU performance. We purposely did not include comparison in terms of pattern reproduction because the results would be very case dependent and subjective.

Table II.3.1 Qualitative comparison of all MPS methods reviewed in this chapter[5]

	Categorical variables	Continuous variables	Multivariate simulation	Hard data	Soft probabilities	Nonstationarity (see Chapter II.6)	CPU performance
ENESIM	✓	✗	✗	✓	✗	✓[3]	Slow
SNESIM	✓	✗	✗	✓	✓	✓[1,2]	Fast
IMPALA	✓	✗	✗	✓	✓	✓[1,2]	Fast
HOSIM	✓	✓	✗	✓	✗	✓[3]	Fast
GROWTHSIM	✓	✗	✗	✓	✓	✓[2]	Unknown
DS	✓	✓	✓	✓	✓	✓[1,2]	Medium
SA	✓	✓	✓	✓	✓	✓[4]	Slow
SIMPAT	✓	✓	✗	✓	✓	✓[3]	Slow
DISPAT	✓	✓	✗	✓	✓	✓[2]	Fast
FILTERSIM	✓	✓	✗	✓	✓	✓[2]	Medium
WAVESIM	✓	✓	✗	✓	✓	✓[2]	Medium
PS	✓	✗	✗	✓	✗	✓[3]	Unknown
CCSIM	✓	✓	✓	✓	✗	✓[3]	Very fast
IQ	✓	✓	✓	✓	✗	✓[3]	Very fast

[1] Nonstationarity included using control maps.

[2] Nonstationarity controlled using soft probabilities.

[3] Nonstationarity can be controlled using zonation.

[4] Nonstationarity included in the objective function.

[5] Markov random fields (MRFs) and Markov mesh models (MMMs) are not included in this diagram. They are described separately in Chapter II.4.

3.5 Postprocessing

3.5.1 Resolution of conflicts with path alterations

Most MPS implementations use a limited neighborhood; therefore, the sequential decomposition is truncated to a set of closest neighbors. This means that the value attributed to a node, conditionally to a certain neighborhood at a given stage of the simulation, can be incompatible with the neighborhood of this same node at a later stage of the simulation, when new values have been populated.

Such incompatible neighborhoods result in difficulties when simulating certain nodes, causing artifacts in the simulations. The patterns that make up a neighborhood may not be present at all in the TIs, making it difficult or impossible to compute a local cdf. This problem is common to all multiple-point simulation methods, and it is more acute when a random path is used in the simulation grid (the Markovian properties of unilateral paths do alleviate this problem). One commonly used strategy is to degenerate the neighborhood considered until the incompatibility disappears (Strebelle, 2002). It is, however, acknowledged that such strategies do not solve all incompatibilities between neighborhoods, but rather enable the algorithm to ignore the problem. Resulting artifacts can be expressed, for example, as a lack of spatial continuity of the simulated structures, such as discontinuous channels where they should be connected.

A number of approaches have been proposed to overcome the issue of incompatible neighborhoods, all of them using the order in which the nodes are visited, or revisited. One strategy is to locate the nodes that were simulated using a degraded neighborhood, and resimulate these nodes after the simulation has been completed (Strebelle and Remy, 2005). The random path is therefore altered during the simulation in an attempt to resolve the neighborhood incompatibilities. One consequence of this approach is that the incompatible values that are temporarily accepted still propagate inconsistencies to nodes that are simulated later. Therefore, if a node is successfully resimulated, it is not guaranteed that all its neighbors are consistent with each other. Alternatively, the method of Stien et al. (2007) deletes the problematic nodes, leaving them to be simulated after all other node values have been determined. Although this approach avoids the propagation of errors caused by incompatible values, the nodes need to be eventually simulated with their current neighborhood, however unsuitable that neighborhood may be.

In the context of simulations using a unilateral path, Suzuki and Strebelle (2007) developed the real-time postprocessing method (RTPP), which consists of walking back the unilateral path when problematic neighborhoods are encountered, and resimulating the most recent nodes until the nodes can be satisfactorily simulated. This method has the advantage of correcting all inconsistencies because it resimulates the neighborhoods of the problematic nodes, not just the problematic nodes themselves. As inconsistencies are resimulated immediately, it avoids propagation to their neighbors. The syn-processing method (Mariethoz et al., 2010) applies the same principle with random path simulation. As with RTPP, it resimulates values as soon as inconsistencies are met. When a neighborhood is encountered that does not exist in the TI, the simulation is halted while one of the nodes in the problematic neighborhood is resimulated. As a result, the problematic neighborhood is changed and may now be free of inconsistencies. The procedure is recursive – meaning that it is possible to resimulate the neighbor of the neighbor, and so on – and it is therefore slow and may not converge.

All of the approaches mentioned here do reduce the presence of artifacts in the simulations. However, none is perfect, and some inconsistencies are expected to occur. The presence of incompatibilities between neighborhoods is a problem inherent to the use of a finite TI.

3.5.2   A posteriori remediation methods

An efficient but possibly time-consuming alternative to postprocessing the realizations is to iterate the simulations on the same grid. This means that once all grid nodes have been visited, a new simulation path is defined and all nodes are resimulated, but this time with entirely informed neighborhoods. This postprocessing operation can be carried out for several postprocessing passes, which iteratively remove residual discrepancies between simulated and TI patterns.

In many cases, these residual discrepancies manifest themselves as noise. Hence, instead of iterative resimulations, one can directly use existing noise removal methods. In the case of categorical variables, mathematical morphology offers erosion–dilation methods (Serra and Soille, 1994). The maximum a posteriori selection method (Deutsch, 1998) also offers an interesting alternative in this regard.

References

Arpat, B. & Caers, J. 2007. Conditional simulations with patterns. *Mathematical Geology*, 39, 177–203.

Chatterjee, S., Dimitrakopoulos, R. & Mustapha, H. 2012. Dimensional reduction of pattern-based simulation using wavelet analysis. *Mathematical Geosciences*, 44, 343–374.

Deutsch, C. V. 1998. Cleaning categorical variable (lithofacies) realizations with maximum a-posteriori selection. *Computers and Geosciences*, 24, 551–562.

Deutsch, C. V. & Wen, X. H. 2000. Integrating large-scale soft data by simulated annealing and probability constraints. *Mathematical Geology*, 32, 49–67.

Efros, A. A. & Freeman, W. T. 2001. Image quilting for texture synthesis and transfer *Proceedings of the ACM SIGGRAPH Conference on Computer Graphics*, 341–346.

Efros, A. & Leung, T. 1999. Texture synthesis by non-parametric sampling. *In: Seventh IEEE International Conference on Computer Vision, 1999, Kerkyra, Greece*, 1033–1038.

El Ouassini, A., Saucier, A., Marcotte, D. & Favis, B. D. 2008. A patchwork approach to stochastic simulation: A route towards the analysis of morphology in multiphase systems. *Chaos, Solitons and Fractals*, 36, 418–436.

Eskandaridalvand, K. & Srinivasan, S. 2010. Reservoir modelling of complex geological systems: A multiple-point perspective. *Journal of Canadian Petroleum Technology*, 49, 59–68.

Guardiano, F. & Srivastava, M. 1993. Multivariate geostatistics: Beyond bivariate moments. *In*: Soares, A. (Ed.), *Geostatistics-Troia*, Kluwer Academic, Dordrecht.

Honarkhah, M. 2011. *Stochastic Simulation of Patterns Using Distance-Based Pattern Modeling*, PhD dissertation, Stanford University, Stanford, CA.

Honarkhah, M. & Caers, J. 2010. Stochastic simulation of patterns using distance-based pattern modeling. *Mathematical Geosciences*, 42, 487–517.

Liu, Y. 2006. Using the SNESIM program for multiple-point statistical simulation. *Computers & Geosciences*, 23, 1544–1563.

Mahmud, K., Mariethoz, G., Caers, J., Tahmasebi, P. & Baker, A. 2014. Simulation of Earth textures by conditional image quilting. *Water Resources Research*, 50(4), 3088–3107.

Mariethoz, G., Renard, P. & Straubhaar, J. 2010. The direct sampling method to perform multiple-point geostatistical simulations. *Water Resources Research*, 46.

Meerschman, E., Pirot, G., Mariethoz, G., Straubhaar, J., Van Meirvenne, M. & Renard, P. 2013. A practical guide to performing multiple-point statistical simulations with the direct sampling algorithm. *Computers and Geosciences*, 52, 307–324.

Mustapha, H. & Dimitrakopoulos, R. 2011. HOSIM: A high-order stochastic simulation algorithm for generating three-dimensional complex geological patterns. *Computers and Geosciences*, 37, 1242–1253.

Peredo, O. & Ortiz, J. M. 2011. Parallel implementation of simulated annealing to reproduce multiple-point statistics. *Computers and Geosciences*, 37, 1110–1121.

Rezaee, H., Mariethoz, G., Koneshloo, M. & Asghari, O. 2013. Multiple-point geostatistical simulation using the bunch-pasting direct sampling method. *Computers and Geosciences*, 54, 293–308.

Serra, J. & Soille, P. 1994. *Mathematical Morphology and Its Applications to Image Processing*, Kluwer Academic, Dordrecht.

Stien, M., Abrahmsen, P., Hauge, R. & Kolbjørnsen, O. 2007. Modification of the SNESIM algorithm. *In: Petroleum Geostatistics 2007*, 10–14 September 2007, Cascais, Portugal, EAGE.

Straubhaar, J., Renard, P., Mariethoz, G., Froidevaux, R. & Besson, O. 2011. An improved parallel multiple-point algorithm using a list approach. *Mathematical Geosciences*, 43, 305–328.

Straubhaar, J., Walgenwitz, A. & Renard, P. 2013. Parallel multiple-point statistics algorithm based on list and tree structures. *Mathematical Geosciences*, 45, 131–147.

Strebelle, S. 2002. Conditional simulation of complex geological structures using multiple-point statistics. *Mathematical Geology*, 34, 1–22.

Strebelle, S. & Remy, N. 2005. Post-processing of multiple-point geostatistical models to improve reproduction of training patterns. *In:* Deutsch, O. L. A. C. V. (Ed.), *Geostatistics Banff 2004*, Springer, Berlin.

Suzuki, S. & Strebelle, S. 2007. Real-time post-processing method to enhance multiple-point statistics simulation. *In: Petroleum Geostatistics 2007*, 10–14 September 2007, Cascais, Portugal, EAGE.

Tahmasebi, P., Hezarkhani, A. & Sahimi, M. 2012. Multiple-point geostatistical modeling based on the cross-correlation functions. *Computational Geosciences*, 16, 779–797.

Wu, J., Zhang, T. & Journel, A. 2008. Fast FILTERSIM simulation with score-based distance. *Mathematical Geosciences*, 40, 773–788.

Zhang, T., Switzer, P. & Journel, A. 2006. Filter-based classification of training image patterns for spatial simulation. *Mathematical Geology*, 38, 63–80.

CHAPTER 4

Markov random fields

4.1 Markov random field model

4.1.1 Definition

Markov random fields (MRFs) are theoretical models formulated within the domain of probability theory to deal with a large variety of problems, including image restoration, image completion, segmentation, texture synthesis and super-resolution modeling (Besag, 1974; Kindermann and Snell, 1980; Geman and Geman, 1984; Clifford, 1990; Cressie, 1993; Winkler, 2003; Li, 2009). Such models are popular in both mathematical statistics and computer science, as well as in application areas such as remote sensing (e.g., Bouman and Shapiro, 1994). The intended application of MRFs in these areas is different in many respects from the type of applications covered in this book. In particular, the seminal MRF application of image-processing problems starts from a complete image; it is two-dimensional (2D) and requires some operation on that specific image. A geostatistical problem is 3D or 4D, often has sparse data, and in addition considers analog or expert information, in this book formulated through training images. The training image does not need processing in this traditional sense; it is instead used as a pattern database whose patterns need to be "borrowed" and "anchored" (in qualitative terms) to the specific measured data. In what follows in this chapter, we will assume that the variable z of interest is binary, although the extension to multicategory is at least methodologically not different.

A MRF is formulated within a traditional mathematical statistical framework. The model is fully probabilistic, namely, a specification of the full spatial law over some regular grid (termed "lattice") is explicitly made. The model contains parameters that will need to be estimated from data, in this case the training image. Then, after the model is fully specified, simulation is performed by sampling from the estimated spatial law. Such sampling often requires iterative Markov chain Monte Carlo methods.

Similar to many of the geostatistical approaches, MRF formulations impart local characteristics on the spatial law. The Markov property enters by specifying that the full conditional distribution can be specified as a conditioning on a local neighborhood $\mathbf{N}(\mathbf{x}_i)$:

$$p(z(\mathbf{x}_i)|\text{all other } z(\mathbf{x}_j)) = p(z(\mathbf{x}_i)|\mathbf{N}(\mathbf{x}_i)) \qquad (\text{II.4.1})$$

Multiple-point Geostatistics: Stochastic Modeling with Training Images, First Edition. Gregoire Mariethoz and Jef Caers.
© 2015 John Wiley & Sons, Ltd. Published 2015 by John Wiley & Sons, Ltd.
Companion website: www.wiley.com/go/caers/multiplepointgeostatistics

The specification must be symmetric, meaning that if $\mathbf{x}_i$ is a neighbor of $\mathbf{x}_j$, then $\mathbf{x}_j$ must also be a neighbor of $\mathbf{x}_i$.

A condition that is often imposed is that the spatial law is positive for all possible outcomes $\mathbf{z}$, that is,

$$p(\mathbf{z}) > 0 \quad \forall \mathbf{z} \tag{II.4.2}$$

In practical terms, this means that any outcome or realization is possible, including those that could be deemed a priori as completely unrealistic. Under this assumption, a theorem has been developed (the Hammersley–Clifford theorem; see, e.g., Besag, 1974; Clifford, 1990) that states that such spatial law can be written as

$$p(\mathbf{z}) = K \exp\left(-\sum_{c \in C} V_c(\mathbf{z}_c) \right) \tag{II.4.3}$$

As an immediate consequence, the conditional distribution can be written as follows:

$$p(z(\mathbf{x}_i)|\mathbf{N}(\mathbf{x}_i)) \sim \exp\left(-\sum_{c:\mathbf{x}_i \in C} V_c(\mathbf{z}_c) \right) \tag{II.4.4}$$

A detailed description of the various components clarifies the notation:

- $\mathbf{z}_c$: denoted a *clique*. A clique is a concept in graph theory. A set of nodes is connected using vertices. The geostatistical grid can be seen as a particular form of a graph. An example is shown in Figure II.4.1(a). A clique is defined as a complete subgraph on the graph shown. The term "complete" means that every node in that clique is also neighbor to all other nodes in the same clique. For example, $\{1,4,5\}$ is a clique, but $\{1,2,3\}$ is not a clique because nodes 1 and 3 are not connected. Figure II.4.1 shows an example of how this can be applied to a 2D grid. Figure II.4.1(b) shows a first-order neighborhood (a 3×3 template) around a certain grid location (black node). The order is taken as spiraling away from the center node $\mathbf{x}_0$. The underlying graph of this neighborhood is shown as well: for the same reason that $z(\mathbf{x}_2)$ is a neighbor of $z(\mathbf{x}_0)$, $z(\mathbf{x}_1)$ is a neighbor of $z(\mathbf{x}_7)$.

- Figure II.4.1(c) shows the 10 possible geometrical configurations of cliques. An example clique is $\mathbf{z}_c = \{z(\mathbf{x}_1), z(\mathbf{x}_7), z(\mathbf{x}_8)\}$. In total, the set C of possible cliques c consists of 50 cliques. Hence, the sum $c : \mathbf{x}_i \in C$ in Equation (II.4.4) goes over 50 such cliques.

- V_c: potential function. These are arbitrary multivariate functions as long as they are finite. Such functions can be predefined or contain parameters, and they may be restricted to pairwise interactions. A few example functions are:

Parameterized: for $\mathbf{z}_c = \{z(\mathbf{x}_1), z(\mathbf{x}_7), z(\mathbf{x}_8)\}$, $V_c(\mathbf{z}_c) = \theta_c z(\mathbf{x}_1) z(\mathbf{x}_7) z(\mathbf{x}_8)$

$$\text{or} \quad \theta_c^1 z(\mathbf{x}_1) + \theta_c^2 z(\mathbf{x}_7) + \theta_c^3 z(\mathbf{x}_8) \tag{II.4.5}$$

Pairwise interaction: for $\mathbf{z}_c = \{z(\mathbf{x}_1), z(\mathbf{x}_7), z(\mathbf{x}_8)\}$,

$$V_c(\mathbf{z}_c) = z(\mathbf{x}_1)z(\mathbf{x}_7) + z(\mathbf{x}_1)z(\mathbf{x}_8) + z(\mathbf{x}_7)z(\mathbf{x}_8) \tag{II.4.6}$$

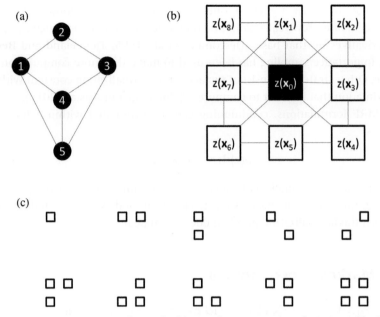

Figure II.4.1 (a) Representation of a graph; (b) a graph representation of a 3×3 neighborhood in the grid; and (c) the various clique configurations of a 3×3 neighborhood.

- The normalization constant K:

$$K = \frac{1}{\sum_{z \in \Omega} \exp\left(-\sum_{c \in C} V_c(\mathbf{z}_c)\right)}$$

(II.4.7)

which represents the sum over all possible outcomes $\mathbf{z}$. Recall that any $\mathbf{z}$ is possible ($p(\mathbf{z}) > 0$). This means, for the simple binary case, that this sum is over all possible combinations over the entire size of the grid (lattice). Clearly, in any real setting, K can never be calculated.

The MRF model is very general, but as a consequence it is very vague in its model specification: there are no specifications regarding what the functions V_C should be; whether all possible clique configurations need to be considered, or only a few; and, if a few cliques are considered, which ones. Estimation of the parameters proceeds through maximum likelihood estimation (Geyer and Thompson, 1995):

$$\hat{\theta}_{MLE} = \max_{\theta} K(\theta) \exp\left(-\sum_{c \in C} V_c(\mathbf{z}_c, \theta_c)\right)$$

(II.4.8)

with θ the set of all parameters consisting of the union of all clique-related parameters θ_c. This maximization problem, however, cannot be solved because the normalization constant cannot be calculated for any arbitrary θ. As an approximation, for each guess θ, one uses a Markov chain Monte Carlo sampling to generate a few realizations and then calculate likelihood function values and derivatives on that basis (Descombes et al., 1995; Tjelmeland and Besag, 1998). Importance sampling has been used to make this more computationally feasible. Other strategies consist of replacing the normalization constant with an approximation that is easier to compute (Tjelmeland and Austad, 2012).

In MRF formulations, conditioning can be formulated within a Bayesian model as

$$p(\mathbf{z}|\mathbf{d}) \sim L(\mathbf{d}|\mathbf{z})p(\mathbf{z}) \qquad \text{(II.4.9)}$$

Again, due to the normalization, such models are impractical for the problems treated in this book. For this reason, subclasses of models have been developed for which the normalization problem does not appear.

4.2   Markov mesh models

4.2.1   Model specification and parameter estimation

MRF models have not seen published 3D application in the fields of study of this book for the reasons of the computational and methodological challenges outlined above. Instead, the focus has been on a subclass of such models, termed "Markov mesh models" (MMMs; Abend et al., 1965; Daly, 2004; Stien and Kolbjørnsen, 2011). MMMs can also be seen as partially ordered MRF models (POMMs; Cressie and Davidson, 1998), where the model construction does not use a general neighborhood model, as in Equation (II.4.1), but orders the grid locations, then formulates the model using this order. This results in an easier parameter estimation problem and also in a more efficient sampling.

Unlike MRFs, MMMs have an asymmetric model specification. However, this asymmetry also makes the parameter estimation and sampling from such models much more convenient. MMMs specify the model through a strict ordering of neighborhoods defined using a raster path over the grid nodes. The reader should not confuse this with simulation with a raster path (Chapter II.2); at this point, we are only interested in model specification. The sequential model specification is illustrated in Figure II.4.2. The following assumption similar to the MRF assumption is made:

$$p(z(\mathbf{x}_i)|\text{all previous } z(\mathbf{x}_i)) = p(z(\mathbf{x}_i)|\mathbf{N}(z(\mathbf{x}_i))) \qquad \text{(II.4.10)}$$

One could therefore consider the MMM to be an extension of a 1D time series to 2D or 3D. Just as in time, an order (i.e., a "history") exists informing what lies ahead; in this context, the model therefore reduces to

$$p(\mathbf{z}) = \prod_{all\,i} p(z(\mathbf{x}_i)|\mathbf{N}(\mathbf{x}_i)) \qquad \text{(II.4.11)}$$

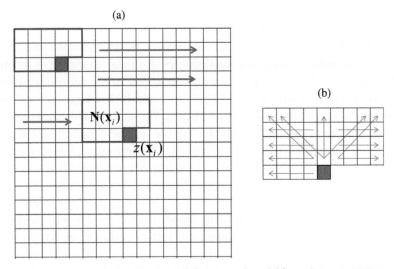

Figure II.4.2 (a) Asymmetric neighborhood definition and model formulation in MMM; and (b) an example of choices on a neighborhood where higher-order interaction is considered: the arrows indicate the direction and the order in which interaction increases (after Stien & Kolbjørnsen, 2011).

In terms of model specification, the same framework of functions and cliques as in MRF is used. One way of formulating the potential functions and their parameterization is by means of generalized linear models (Stien and Kolbjørnsen, 2011). The sum of potential functions in Equation (II.4.4) is replaced by a general linear model (GLM), where the "explanatory" variables are the model variables in the neighborhood and the response or target variable is the variable just ahead of that neighborhood. Consider the neighborhood as explanatory variables, such as in Figure II.4.2:

$$\mathbf{N}(\mathbf{x}_i) = [z(\mathbf{x}_i + \mathbf{h}_1), z(\mathbf{x}_i + \mathbf{h}_2), \dots, z(\mathbf{x}_i + \mathbf{h}_N)] \tag{II.4.12}$$

and the target as $z(\mathbf{x}_i)$. As with any GLM, a nonlinear transformation on the explanatory variables is made:

$$\mathbf{f}_i = \mathbf{f}(\mathbf{N}(\mathbf{x}_i)) = [f_1(\mathbf{N}(\mathbf{x}_i)), \dots, f_M(\mathbf{N}(\mathbf{x}_i))] \tag{II.4.13}$$

using a number M of arbitrary (finite) functions f. The conditional distribution, Equation (II.4.10), can then be written as

$$p(z(\mathbf{x}_i)|\mathbf{N}(\mathbf{x}_i)) = \frac{\exp(\theta^T \mathbf{f}_i z(\mathbf{x}_i))}{\exp(\theta^T \mathbf{f}_i) + 1} \tag{II.4.14}$$

where the parameters θ of the GLM become parameters of the MMM. Note that this model specification is similar to a neural network model specification of the local conditional distribution (Caers and Journel, 1998; Caers and Ma, 2002), except for the ordering of the neighborhood along a raster path. As with the neural network specification, the normalization is part of the formulation, hence the problem of normalization is avoided.

Maximum likelihood estimation of the parameters in the MMM formulation is solved using iterative weighted least squares. A similar challenge as in MRF occurs in the model specification, namely, the choice of the functions f (functions V in MRF) and what the "size" M of the model is. Too large a model may lead to overfitting, and too small to underfitting. The latter would mean that certain salient features of the training image are not represented in the model and hence will not be reproduced through sampling. In this context, it is proposed to use an indicator function for f, namely, a function that indicates the absence and presence of some feature in the entire neighborhood $\mathbf{N}(\mathbf{x}_i)$ in comparison with the indicator value at $\mathbf{z}(\mathbf{x}_i)$. For example, one can include all pairwise interactions with the reference cell. Additionally, the modeler can choose to include certain higher-order interactions deemed important, such as particular directions along which high connectivity of certain categories is present. Figure II.4.2(b) shows such an example. Note that such choices need to be explicitly made by the modeler, based on intuition and possibly visual appreciation of the training image. If the number M becomes too large, principal component analysis can be used to reduce the size of M by retaining only a few components.

After the model has been specified and the parameters have been estimated, any sampling from the unconditional spatial law proceeds directly along a raster path. One particular challenge, also observed in MRF, is the reproduction of the training image marginal proportion in MMM. GLMs rely on assumptions of independence that do not hold in the spatial context. This means that certain properties of the estimated model exhibit biases. One such bias is on the marginal distribution (the proportions in the categorical case). Maximum likelikhood estimates produce models that, when sampled, have a proportion different from the training image proportions. It is also possible that the particular choice of indicator functions causes this bias. For this reason, proportion "steering" (very similar to the servo system described in Section II.2.8.3.1) is proposed by adjusting, through trial and error, the maximum likelihood estimates until the marginal distribution is reproduced.

Figure II.4.3 shows an application of an MMM model to a 2D binary training image (the "Strebelle" training image). A two-point interaction neighborhood of size 5×5 pixels as well as higher-point interactions of a maximum size of eight pixels in all directions is used in the model specification. The resulting realizations show good channel connectivity.

4.2.2   Conditioning in Markov mesh models

The above formulation represents an unconditional (prior) model formulation and is completely stationary in nature: the local neighborhood formulation defines the model, and that same neighborhood model is used through the domain. At least in the formulations above, any nonstationary character needs to

Training image

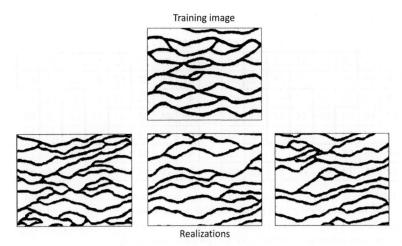

Realizations

Figure II.4.3 Application of an MMM model to a simple 2D binary training image. Results courtesy of Odd Kolbjørnsen.

come from data. Conditioning to hard data is challenging in MMM because some of the data lie ahead of the path; hence, when reaching those data, the partially created simulation may be inconsistent with them. A similar problem occurs in raster-based MPS methods, such as cross-correlation simulation or image quilting (see Section II.3.3.4). Consider for now only hard data, denoting any such hard data ahead of the current $\mathbf{x}_i$ as $\mathbf{hd}(\mathbf{x}_i)$. The neighborhood-conditional posterior distribution is written as

$$p(z(\mathbf{x}_i)|\mathbf{N}(\mathbf{x}_i),\mathbf{hd}(\mathbf{x}_i)) = \frac{p(z(\mathbf{x}_i)|\mathbf{N}(\mathbf{x}_i),\mathbf{hd}(\mathbf{x}_i))}{p(z(\mathbf{x}_i)|\mathbf{N}(\mathbf{x}_i))}p(z(\mathbf{x}_i)|\mathbf{N}(\mathbf{x}_i)) \quad \text{(II.4.15)}$$

It is proposed in Kjønsberg and Kolbjørnsen (2008) to estimate the ratio in this equation using traditional geostatistics, for example indicator kriging, where the variogram is borrowed from the training image. A possible alternative would be to use search trees and calculate the ratio of these two probabilities directly from the multiple-point statistics of the training image.

4.3 Multigrid formulations

4.3.1 Multigrid MMMs

Because of the raster path in MMMs, the neighborhood or template chosen can be quite small yet still be efficient in reproducing large-scale features. This was also observed in raster path MPS algorithms: the size of the neighborhood can be much smaller than the largest-scale feature that one attempts to model and simulate. However, in 3D, with large-scale connected features such as channels, the neighborhood still needs to be considerably large, making the model specification as well as the estimation of MMM model parameters tedious. To address this

1	2	3	4	5	6	7	8	9
10	11	12	13	14	15	16	17	18
19	20	21	22	23	24	25	26	27
28	29	30	31	32	33	34	35	36
37	38	39	40	41	42	43	44	45
46	47	48	49	50	51	52	53	54
55	56	57	58	59	60	61	62	63
64	65	66	67	68	69	70	71	72
73	74	75	76	77	78	79	80	81

1	46	10	47	2	48	11	49	3
26	50	27	51	28	52	29	53	30
16	54	17	55	18	56	19	57	20
31	58	32	59	33	60	34	61	35
4	62	12	63	5	64	13	65	6
36	66	37	67	38	68	39	69	40
21	70	22	71	23	72	24	73	25
41	74	42	75	43	76	44	77	45
7	78	14	79	8	80	15	81	9

Figure II.4.4 Single-grid (left) versus multigrid (right) ordering of the cells for the MMM formulation. In the coarsest grid of the multigrid ordering, nodes 1 to 9 are colored green. Note that the same neighborhood configuration can be maintained, rendering parameter estimations more practical (modified from Kolbjørnsen et al., 2014).

issue, the multigrid approach used in MPS methods is adopted within an MMM framework (Kolbjørnsen et al., 2014). Note, however, that any MRF formulation requires explicit inclusion of the path in the model. In a multigrid-based formulation, the path has now two structures: the level of the grid and the raster. As shown in Section II.2.7, in a multigrid simulation, the coarse grid cells are visited first, followed by grid cells on a consecutively refined grid. This induces an ordering or hierarchy in the visited cells. Figure II.4.4 shows an example of such order. The MMM formulation needs now some adjusting to reflect this order. At the same time, the neighborhood for each cell should have the same geometry, rendering the estimation of parameters through maximum likelihood efficient. Otherwise, a number of different parameter sets for each neighborhood needs to be estimated. A simple solution is to put a box around each cell, as shown in Figure II.4.4, and then exclude those cells belonging to the "future" (i.e., those cells of lesser order). Due to the regularity of the multigrids, the shape of the remaining neighborhood remains the same.

Except for the geometry of the neighborhood, there is no methodological difference between the multiple-grid and single-grid MMM approaches: user-defined neighborhood functions of the within-neighborhood nodes are defined, a parameter is associated with each such function, and a generalized linear model is formulated and estimated. For the binary case, the number of parameters is now equal to the number of multigrid levels times the number of categories times the number of functions retained. Compared to the single-grid approach, the neighborhood is no longer convex (compare Figure II.4.2b with Figure II.4.4); this nonconvexity needs to be accounted for in the calculations of the neighborhood functions (Kolbjørnsen et al., 2014).

Figure II.4.5 shows an example, now in 3D, using the neighborhood definition of Figure II.4.4. Again, we find good channel connectivity in the resulting realizations.

Training image

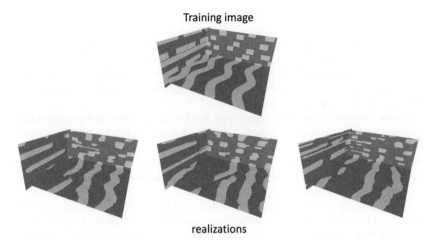

realizations

Figure II.4.5 Training image and simulated MMM models. Results courtesy of Odd Kolbjørnsen (see also Kolbjørnsen et al., 2014).

4.3.2 Multigrid MRF

In a more general setting, the multigrid approach can also be included in an MRF (Toftaker and Tjelmeland, 2013). Recall that in the more general MRF model formulation, no raster path exists. If one is splitting the grid into multiple levels, then each level- or subgrid-specific MRF model needs to be formulated conditional to the coarser levels. However, the problem of normalization in MRF makes this infeasible, at least in a rigorous sense. It is therefore assumed that the multigrid injects an order into the model and the MRF is approximated with a POMM, the order being defined by a multigrid representation only (a multigrid MMM includes the raster path). Unlike MMM, there is no totally complete ordering in a POMM, only a partial ordering. Using the forward-backward algorithm, Toftaker and Tjelmeland (2013) approximate POMMs that allow for large neighborhoods and maximum likelihood formulations on the clique parameters through importance sampling. Because of this approximate formulation, sampling can proceed using a noniterative sequential simulation.

References

Abend, K., Harley, T. & Kanal, L. N. 1965. Classification of binary random patterns. *IEEE Transactions on Information Theory*, 11, 538–544.

Besag, J. 1974. Spatial interaction and the statistical analysis of lattice systems. *Journal of the Royal Statistical Society. Series B (Methodological)*, 192–236.

Bouman, C. A. & Shapiro, M. 1994. A multiscale random field model for Bayesian image segmentation. *IEEE Transactions on Image Processing*, 3, 162–177.

Caers, J. & Journel, A. G. 1998. Stochastic reservoir simulation using neural networks trained on outcrop data. *SPE Annual Technical Conference and Exhibition*, no. 49026, 321–336.

Caers, J. & Ma, X. 2002. Modeling conditional distributions of facies from seismic using neural nets. *Mathematical Geology*, 34, 143–167.

Clifford, P. 1990. Markov random fields in statistics. *In:* Grimmett, G. R. & Welsh, D. J. A. (eds.), *Disorder in Physical Systems*, Oxford University Press, Oxford, 19–32.

Cressie, N. A. 1993. *Statistics for Spatial Data*, revised edition, John Wiley & Sons, Inc., New York.

Cressie, N. & Davidson, J. L. 1998. Image analysis with partially ordered Markov models. *Computational Statistics & Data Analysis*, 29, 1–26.

Daly, C. 2004. Higher order models using entropy, Markov random fields and sequential simulation. *In:* Leuangthong, O. & Deutsch, C. V. (eds.), *Geostatistics Banff 2004*, Kluwer Academic, Banff, AB, 215–224.

Descombes, X., Mangin, J.-F., Sigelle, M. & Pechersky, E. 1995. Fine structures preserving Markov model for image processing. *In: Proceedings of the Scandinavian Conference on Image Analysis*, World Scientific, Singapore, 349–356.

Geman, S. & Geman, D. 1984. Stochastic relaxation, Gibbs distribution and the Bayesian restoration of images. *IEEE Transactions on Pattern Analysis and Matching Intelligence*, 6, 721–741.

Geyer, C. J. & Thompson, E. A. 1995. Annealing Markov chain Monte Carlo with applications to ancestral inference. *Journal of the American Statistical Association*, 90, 909–920.

Kindermann, R. & Snell, J. L. 1980. *Markov random fields and their applications*, American Mathematical Society, Providence, RI.

Kjønsberg, H. & Kolbjørnsen, O. 2008. Markov mesh simulations with data conditioning through indicator kriging. *In:* Ortiz, J. M. & Emery, X. (eds.), *Proceedings of the Eighth International Geostatistics Congress 2008*, 1–5 Dec. 2008.

Kolbjørnsen, O., Stien, M., Kjønsberg, H., Fjellvoll, B. & Abrahamsen, P. 2014. Using multiple grids in Markov mesh facies modeling. *Mathematical Geosciences*, 46, 205–225.

Li, S. Z. 2009. *Markov random field modeling in image analysis*, Springer Verlag, Berlin.

Stien, M. & Kolbjørnsen, O. 2011. Facies modeling using a Markov mesh model specification. *Mathematical Geosciences*, 43, 611–624.

Tjelmeland, H. & Austad, H. M. 2012. Exact and approximate recursive calculations for binary Markov random fields defined on graphs. *Journal of Computational and Graphical Statistics*, 21, 758–780.

Tjelmeland, H. & Besag, J. 1998. Markov random fields with higher-order interactions. *Scandinavian Journal of Statistics*, 25, 415–433.

Toftaker, H. & Tjelmeland, H. 2013. Construction of binary multi-grid Markov random field prior models from training images. *Mathematical Geosciences*, 45, 383–409.

Winkler, G. 2003. *Image Analysis, Random Fields and Markov Chain Monte Carlo Methods: A Mathematical Introduction*, Springer Verlag, Berlin.

CHAPTER 5

Nonstationary modeling with training images

In a spatial context, "stationarity of a process" conceptually refers to the assumptions that statistical variation is similar over the entire spatial domain. Loosely stated, this would means that if one looks at a specific portion of the domain, the statistics will be similar to those of any other portion of the domain. More formally, several orders of stationarity can be defined, increasing orders representing a larger degree of complexity in the representation of the phenomenon considered. First-order stationarity means that the probability distribution function (pdf) of the variable considered is preserved by translation:

$$f[Z(\mathbf{x}_1)] = f[Z(\mathbf{x}_2)] = \cdots = f[Z(\mathbf{x}_n)] = f[Z], \qquad \text{(II.5.1)}$$

with $\mathbf{x}$ being the coordinates of a location on the domain. Second-order stationarity implies that the joint distribution of two values only depends on the lag vector separating the two corresponding locations, and not on the absolute location of these values:

$$f[Z(\mathbf{x}_1), Z(\mathbf{x}_2)] = f[Z(\mathbf{x}_1 + \mathbf{h}), Z(\mathbf{x}_2 + \mathbf{h})], \quad \forall \mathbf{h}. \qquad \text{(II.5.2)}$$

Note that these definitions are slightly different from those given in Chilès and Delfiner (1999), who define the second-order nonstationarity as a translation-invariant covariance. By using a joint distribution instead of a covariance, we emphasize that we do not embed assumptions on the form of the relationship between $Z(\mathbf{x}_1)$ and $Z(\mathbf{x}_2)$. Second-order stationarity necessarily involves first-order stationarity, but the opposite is not true. Another classical definition related to stationarity is the intrinsic hypothesis, which considers that the increments between $Z(\mathbf{x}_1)$ and $Z(\mathbf{x}_2)$ are a second-order stationary process (Chilès and Delfiner, 1999). Higher orders of stationarity correspond to the joint distribution of three or more locations.

Nonstationarity is a property inherent to a model, and not a property of the data. In other words, any data set can be modeled with either a stationary or nonstationary model, the choice being essentially a modeling decision. Several types of nonstationarity can be identified. In the simplest case, a first-order nonstationarity is present. This means that the pdf used to stochastically model a given

Multiple-point Geostatistics: Stochastic Modeling with Training Images, First Edition. Gregoire Mariethoz and Jef Caers.
© 2015 John Wiley & Sons, Ltd. Published 2015 by John Wiley & Sons, Ltd.
Companion website: www.wiley.com/go/caers/multiplepointgeostatistics

(a)   (b)   (c)

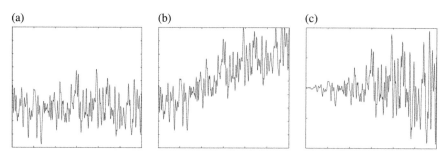

Figure II.5.1 Three cases illustrating nonstationarity in 1D: (a) stationary, (b) nonstationary with varying mean, and (c) nonstationary with variable variance.

phenomenon changes locally (mean and variance). But the relationship between nearby values remains identical. Figure II.5.1 illustrates simple examples of real-izations of a nonstationarity process. In a classical statistical framework, such nonstationarity is dealt with by removing the nonstationary features that are identified, for example by removing a trend or performing local normalization of the data. The modeling is then performed on the resulting stationary process. The nonstationary features are then added on the resulting models (Goovaerts, 1997; Chilès and Delfiner, 1999). This separation into a random variable and a deterministic trend has been discussed in Part I of this volume.

Models that statistically describe spatial continuity necessarily involve multi-ple locations distributed over space. Therefore, all geostatistical methodologies need to rely on some hypothesis of stationarity, otherwise the calculation of statistics would not be feasible. However, it is acknowledged that most physi-cal processes are not stationary. This turns stationarity into not only an unten-able hypothesis but also an impractical one. To accommodate the reality that most real-world models require some nonstationary element, models are often decomposed into a trend component and a stationary stochastic component, as in Figure II.5.1. In many cases, however, the nonstationarity cannot be described by local pdf adjustments, and specific methods need to be used. Such higher-order nonstationarity is especially relevant in the context of multiple-point geostatis-tics, because the complexity of structures modeled with training images often extends to a complexity in the type of nonstationarity. Figure II.5.2 shows an example of such complex nonstationarity that cannot be corrected by histogram transform or trend removal. The types of patterns vary radically from one part of the domain to the other. For example, in Figure II.5.2b, the lower part of the image consists of large white areas, the central part consists of channelized struc-tures, and the top part is isolated dots. The proportion of each category varies across the domain, but the proportion alone is not enough to describe nonsta-tionarity. The spatial interactions between the categories and the connectivity properties also vary in space. Ways to deal with this sort of nonstationarity are reviewed in this chapter.

(a)   (b)

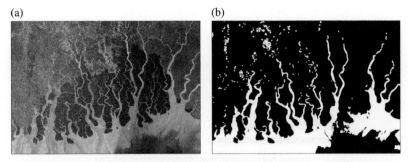

Figure II.5.2 Example of 2D nonstationarity based on a satellite image of the Sundarbans region, Bangladesh. (a) high-order nonstationarity in continuous values, where both the univariate distribution and the type of patterns vary from the bottom to the top of the domain. (b) Similar type of nonstationarity for a categorical variable.

5.1 Modeling nonstationary domains with stationary training images

5.1.1 Zonation

One of the earliest proposals for dealing with nonstationary processes in MPS consists of dividing the modeled domain into a number of stationary zones or regions. This approach was investigated by Strebelle (2002) by using the same TI for all zones but considering different orientations within each zone. The concept was extended, for example, by Wu et al. (2008) by using an entirely separate TI in each zone. Although the approach is simple to implement, one difficulty is that near the edge of each zone, the nodes need to be simulated using neighbors' values from other zones. Such overlapping neighborhoods are needed to maintain spatial continuity across zones and to avoid border effects. However, this does not prevent uncontrollable effects or artifacts from appearing near the transition, especially when the TIs used in both adjacent zones contain incompatible structures. Consider the example depicted in Figure II.5.3. It consists of two zones: the left portion of the domain is simulated using a TI with horizontal connected channels. For the adjacent western zone, the TI consists of isolated patches. The nature of the second TI makes it impossible for the channels to be continuous across the transition between the two zones. Instead, some channels will necessarily terminate abruptly, leading to the simulation of patterns that are in neither TI.

One way to avoid such problems is to smooth the transition between zones. Instead of a sharp transition, one can consider a continuous space where the probability of using one TI is progressively higher while the probability of the other diminishes. More generally, consider K TIs, TI_k, $k = 1, \ldots K$. The nonstationarity needs to be expressed by K probability maps $p_k(\mathbf{x})$, with $\sum_{k=1}^{K} p_k(\mathbf{x}) = 1$, $\forall \mathbf{x}$. The probability maps are normalized such that the probability at each node sums

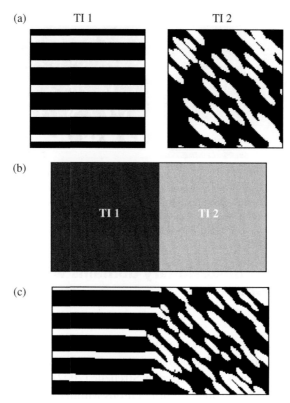

Figure II.5.3 Nonstationary modeling using zones with separate TIs. (a) The two TIs used. (b) The prescribed zone for each TI. (c) One resulting nonstationary realization.

to 1. Then, for the simulation of each node **x**, one TI is chosen according to the probabilities $p_k(\mathbf{x})$. Figure II.5.4 illustrates this approach, with the lower-right panel showing which TI was used for the simulation of each node in the domain. One advantage is that the zonation is no longer deterministic; for different realizations, the selected TI will vary for a given node location. This added flexibility reduces artifacts because the structures in one zone are allowed to continue to a certain degree into the adjacent zones. Artifacts may still appear, especially when the TIs contain incompatible patterns. The resulting incompatibility means that patterns absent from both TIs have to be generated. In the example of Figure II.5.4, such an incompatibility is present, resulting in the generation of artifacts even with the use of a soft transition between zones.

The only way to control which patterns are used at the interface between zones is to explicitly provide such patterns. Transition artifacts can generally be resolved by defining a buffer zone with its own TI explicitly representing the nature of the transition patterns. Another approach is to generate new patterns as an ensemble of some controlled transformation applied to the TI patterns. Such approaches are the subject of the next section.

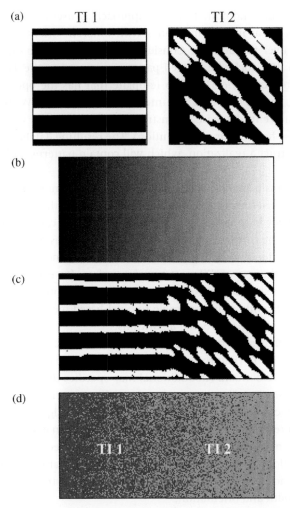

Figure II.5.4 Nonstationary modeling using zones using a continuous transition in the probability of using a TI. (a) Two TIs used. (b) Probability of choosing TI1 (black) or TI2 (white). (c) One resulting nonstationary realization. (d) A posteriori map indicating which TI was used for the simulation of each node.

5.1.2 Data events transformation

A second approach for simulating complex nonstationarity, initially developed for SNESIM, is to use geometrical operations on the simulated data events. Instead of directly lifting data events from the training image, a transformation filter is applied to them, resulting in transformed patterns being simulated. To achieve this, one first constructs a data events catalog using a transformed template, and then simulates values with a nontransformed template (Caers and Zhang, 2004). The most commonly implemented transformations are rotation and affinity (homothety). This approach is useful when only a single stationary TI is available, and zones with different orientation and affinity are desired over the domain. Consider an example where one desires the simulation of a 45° clockwise rotated pattern compared to the TI patterns. To achieve this, one scans the TI with a template that has the opposite transformation applied, in this case a 45° counterclockwise rotation. With this rotated template, a data events database is constructed (a tree, a list, or the transformed template can also be used for a convolution). The resulting database is then used as usual to construct a realization, which will contain patterns with the desired transformation.

Using such transformations, it is possible to simulate a nonstationary domain by imposing different transformations in different zones of the domain, for example 0°, 45°, and 90° in three separate model zones. In this case, three data event databases have to be constructed. Alternatively, continuous transformations can be used with the direct sampling (DS) method. DS does not use databases and therefore allows the use of different transformations for each node instead of zones of fixed transformation. Note that methods based on the transformation of patterns are currently only applicable to pixel-based methods; they have not yet been developed for patch-based algorithms.

Figure II.5.5 illustrates the principle of a zone-specific transformation with a stationary TI. The TI (Figure II.5.5a) consists of essentially horizontal channels over the entire domain. An additional map specifies three different affinity factor zones (Figure II.5.5b). A second map specifies seven categories of angles, with two possible scenarios of angle maps considered (Figure II.5.5c and Figure II.5.5d). Therefore, for each scenario, $3 \times 7 = 21$ pattern databases have to be built, each of them containing the full TI patterns considered at a different angle.

One realization obtained under each angle scenario is displayed in Figure II.5.5e and Figure II.5.5f, showing the expected structures in terms of both channel width and orientation.

5.1.3   Soft probabilities

In some cases, data inform the nonstationarity model component. Examples are seismic data for subsurface reservoirs and remote sensing for Earth surface applications. Although the information content may be weak, it is spatially exhaustive and can therefore provide guidance on general trends, for example facies proportions varying gradually. Such data can often be expressed as soft probabilities, loosely providing information on the likelihood of finding a specific facies at a given simulated node. A straightforward approach to integrate this type of data is the probability aggregation described in Section II.2.8.3.2.

Figure II.5.6 illustrates the use of soft probabilities for modeling nonstationary features in a fan-shaped deltaic reservoir. The TI used is stationary, and the structures are horizontal (Figure II.5.6a). Affinity and rotation transformations, given in Figure II.5.5b and Figure II.5.5c, indicate a general orientation of about

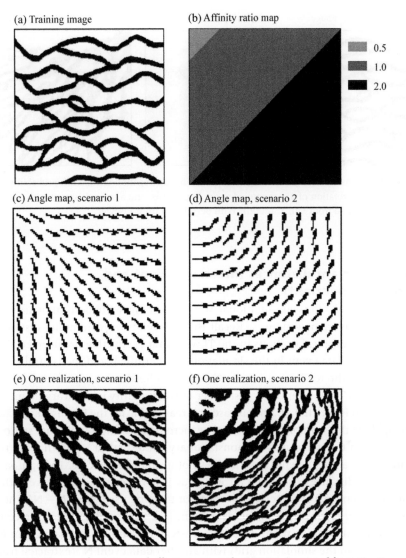

(a) Training image   (b) Affinity ratio map

0.5
1.0
2.0

(c) Angle map, scenario 1   (d) Angle map, scenario 2

(e) One realization, scenario 1   (f) One realization, scenario 2

Figure II.5.5 The use of rotation and affinity zones with SNESIM. Reprinted from Liu, Y. (2006). With permission from Elsevier.

45°. In addition, the soft probabilities, given in Figure II.5.6b, indicate the location of the sand bodies. The soft probability map indicates an absence of channels in the top-right and bottom-left corners of the image.

Probability aggregation is used to combine (1) the local conditional pdfs resulting from the pattern retrievals in the database, and (2) the local pdfs given by the soft probability map. Therefore, each simulated node considers both the spatial continuity given by the TI as well as the soft probability map (Figure II.5.6c).

(a) Training image　　　(b) Soft probability map　　(c) One non-stationary realization

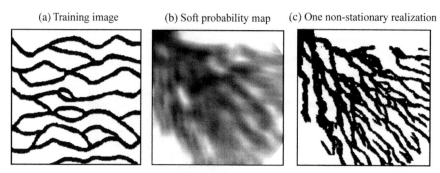

Figure II.5.6 Nonstationarity modeling with probability aggregation using a soft probability map. A uniform rotation of 45° is applied to the TI patterns, and a tau = 1 is used with SNESIM. Reprinted from Liu, Y. (2006). With permission from Elsevier.

5.1.4　Invariant distances (mean invariant and transform invariant)

The DS method offers an additional way of dealing with nonstationarity by considering invariant distances. Recall that DS relies on computing distances between data events. Locations in the TI with data events that are deemed similar to a given data event in the simulation are identified, and the central value is pasted onto the realization (see Section II.3.2.3 for details).

Consider a TI representing a first-order nonstationarity with a varying mean, as depicted in Figure II.5.1b. In this case, it is reasonable for the intrinsic hypothesis to apply to this variable, stating that the increments of the variable are stationary, although the mean may be nonstationary. Note that in Figure II.5.1b, the mean follows a linear trend, but this does not need to be the case. The approach of invariant distances uses such an image as if it were stationary, except that each data event in it is centered on zero prior to being used: $\mathbf{N}' = \mathbf{N} - \overline{\mathbf{N}}$, where $\overline{\mathbf{N}}$ is the mean of all values in the data event (see Section II.2.5.3 for the complete formulation of invariant distances). Instead of comparing a data event $\mathbf{N}_x$ (extracted from the simulation grid) with another data event $\mathbf{N}_y$ (extracted from the TI grid), a mean-invariant distance compares $\mathbf{N}'_x$ with $\mathbf{N}'_y$. As a result, only variations of values within $\mathbf{N}$ matter, not their actual magnitudes. After having located the most similar data event in the TI, the local mean $\overline{\mathbf{N}}_x$ is added before assigning a value to the simulated node.

This approach can also be used when the TI is stationary and nonstationary realizations need to be generated. Consider the case in Figure II.5.7. The TI does not have any local variation in the mean, whereas the data do. Hence, in this case, the data provide the local mean, and any resulting realization will carry the local mean variations present in the data. The same idea can be extended to different types of nonstationarity. When both the mean and standard deviation

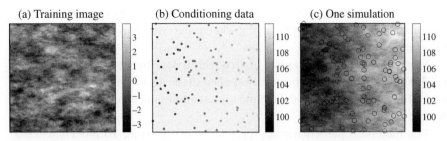

Figure II.5.7 Simulation using a mean-invariant distance. (a) Multi-Gaussian stationary TI. (b) Nonstationary data set (100 points data), with values in a different range than those of the TI. (c) One simulation with a mean-invariant distance. Circles represent the location of the 100 conditioning data (Mariethoz et al., 2010).

are locally variable, a standardization $\mathbf{N}' = (\mathbf{N} - \overline{\mathbf{N}})/\sigma(\mathbf{N})$ can be used similarly, where $\sigma(\mathbf{N})$ is the standard deviation of the local data event.

For modeling more complex forms of nonstationarity, rotation-invariant and affinity-invariant distances can be used. This is done by transforming events $\mathbf{N}_y$ using a range of rotations, affinities, or both. Such a range of transformations can be expressed as a tolerance around a mean value. The transformed data events are denoted as $\mathbf{N}'_y$. Then, the transformation that maximizes the match between data events $\mathbf{N}_x$ and $\mathbf{N}'_y$ is chosen, and the corresponding central value pasted on the simulation grid. An example is provided in Figure II.5.8, with a TI consisting of arrows reflecting the variation of direction of the simulated structures. The tolerance map varies from 0° on the left side of the domain to 180° on the right side. As a result, the patterns can be positioned at any angle, as

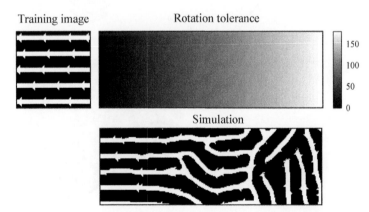

Figure II.5.8 A nonstationary simulation is obtained based on a stationary elementary TI and a rotation tolerance field. The arrows in the channels allow visualizing the rotation of the TI patterns.

long as their neighborhood is compatible with surrounding node values. Their direction becomes increasingly varying when progressing toward the right side of the domain.

From an implementation point of view, this is done by scanning the TI with a random rotation or a random affinity applied to the data event for each scanned node. The data events accepted are the ones that minimize the distance between $\mathbf{N}_x$ and $\mathbf{N}_y'$.

This method is particularly appropriate for constructing 3D models when a suitable 3D TI is difficult to construct or simply not available. In such cases, one may start with a small and simple TI, called an elementary TI. The elementary TI is stationary and does not contain large pattern diversity or complexity. However, the original limited pool of patterns contained in such small TIs is expanded through transform-invariant distances, and the transformation parameters provide a control on how such expansion is achieved. The practical advantage lies in its simplicity: the elementary TI can be constructed with a simple object model, or even based on hand-drawn sketches.

The application of 3D elementary TIs with transform-invariant distances is illustrated in Figure II.5.9, where a 3D orientation field and a 3D tolerance field (not shown) are used in combination with simple layered TIs. Two cases are

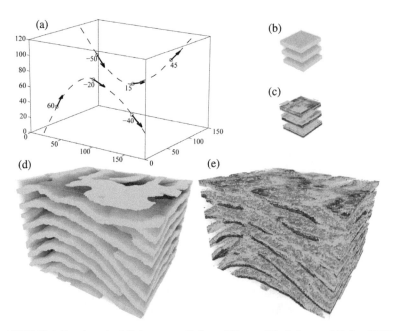

Figure II.5.9 Rotation-invariant distances applied to a 3D case (Mariethoz and Kelly, 2011). (a) Rotation field, interpolated based on discrete values (angles shown in degrees). (b and c) Two possible elementary TIs (categorical and continuous). (d and e) Two corresponding realizations.

provided, one with binary (see Figure II.5.9b) and one with continuous TI (see Figure II.5.9c). The elementary structures warp around the orientation field, resulting in a complex anticline–syncline geological structure (Figure II.5.9d and II.5.9e).

5.2 Modeling nonstationary domains with nonstationary training images

In this section, we discuss the modeling of nonstationarity in the domain by means of nonstationary TIs. The fact that the TI is nonstationary implies that it contains a range of transitional patterns and therefore can offer alternatives to the zonation approach if such an approach results in artifacts at the interface between zones. Nonstationarity TIs usually are more common than stationary ones, in particular when TIs are taken from direct field observations, which are almost inevitably nonstationary, or when it is obtained from process-based models (e.g., Miller et al., 2008; Lopez et al., 2009; Hu et al., 2014).

In pattern databases (Section II.2.4), patterns are typically stored independently of their location in the TI. Therefore, when such patterns are restituted in a simulation, their location-specific character is lost. Figure II.5.10 illustrates this situation: the nonstationary categorical image of the Sundarbans is used to generate one realization with the SNESIM algorithm, with a single tree built based on the entire image. The resulting realization no longer represents the location-specific patterns of the TI.

In order to use nonstationary TIs, one needs a mechanism that explicitly specifies which patterns should go where.

5.2.1 Zonation
The simplest way to use a nonstationary TI is to rely again on zonation, but this time creating zones in the TI and in the simulation domain. The zones of

(a)   (b)

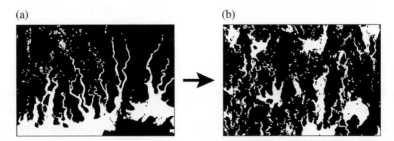

Figure II.5.10 Illustration of using a nonstationary TI (a) without explicit nonstationary modeling. The nonstationary patterns are evenly used on the entire simulation domain, resulting in a stationary simulation (b) with degraded spatial features.

the TI are then treated as if they were separate TIs. The zones in the simulation domain are defined such that the corresponding parts of the TI are used at the desired locations. In the context of the SNESIM algorithm, one can partition the tree-based patterns database into a set of smaller trees, each corresponding to a stationary zone of the TI (Boucher, 2009). This allows increasing the CPU performance and at the same time decreasing the memory load of the SNESIM algorithm.

Generating zones within the TI can be done manually; however, methods from computer vision can be employed to automatically segment an image into stationarity zones. Honarkhah and Caers (2012) propose the use of Gabor filters to compute scores informing the local orientation of the structures as well as the frequency of their occurrence. Using a k-means algorithm, a zonation is then generated based on grouping locations with similar filter scores.

A particular case of nonstationary TIs is a TI derived from observations taken at the same site as the domain being modeled. This can be the case when a deterministic model is already available (either manually built, or the result of a single process model), but one aims to randomize it, thereby generating a model of spatial uncertainty (application Chapter III.2 has several such examples). It is then desirable to generate several realizations that use the existing model as TI, but also use it to infer nonstationary features. The single deterministic model may also be unconditional or only partially conditional to some data; in this case, MPS can then further improve the conditioning. Using a zonation approach, this is done by separating the TI into areas that are considered stationary, and then use the same zonation for the simulation and the local TIs.

One particular implementation that uses this single deterministic model (Hu et al., 2014) employs an algorithm similar to ENESIM (Section II.3.2.3). This method computes the local conditional cumulative distribution function (ccdf) by only scanning the TI locations that are close to the node currently simulated. This is possible because the TI and the simulation grid are in the same coordinate system. The patterns used for simulating a single location are restricted to an area in the vicinity of that location, ensuring that any local nonstationarity present in the TI is reproduced in the simulations. Zhou et al. (2012) employ a similar approach to co-simulate hydraulic properties and state variables in the context of hydrogeological inverse problems (see also Chapter II.9).

5.2.2   Control maps

The use of zones may be too rigid for certain cases, in particular when it is preferable to have a smooth transition between patterns. Control maps are an important tool to alleviate this. Such control maps are essentially a continuous version of the zonation of the TI. Their use with nonstationary TIs was pioneered by Chugunova and Hu (2008). The idea is to introduce one or several control maps (also called "auxiliary variables") to determine which patterns of the TI should go at which locations in the simulation domain. Control maps consist of continuous variables and are defined on the TI as well as on the simulation grid. The control map is not a physical variable, and therefore the units or the range of values in the control map are only conventional; the key point is that they must have similar values in regions where similar patterns are bound to occur.

In the stationarity case, most MPS simulation algorithms consider the local cdf $F[Z(\mathbf{x})|\mathbf{N}_x]$ for the simulation of $Z(\mathbf{x})$, where $\mathbf{N}_x$ is the neighborhood of $\mathbf{x}$. Although the use of control map implementations varies with each simulation algorithm, the general principle is that the following ccdf is used instead:

$$F\left[Z(\mathbf{x})|\mathbf{N}_x, C(\mathbf{x})\right], \qquad\qquad (\text{II.5.3})$$

where $C(\mathbf{x})$ is the value of the control map at the simulated node. The joint probability $F[\mathbf{N}_y, C(\mathbf{y})]$ can be inferred from the TI and its control map. Hence, for every location in the TI, one has a pair of values: one value for the variable simulated, and another value for the control map. In order to use a value $z(\mathbf{y})$ in the simulation, two criteria are considered: (1) its neighborhood must correspond to what is found in the TI, and (2) the value of the control map in the TI $C(\mathbf{y})$ at the location of these patterns must be similar to $C(\mathbf{x})$.

The implementation of Chugunova and Hu (2008) modifies the probability tree structure of SNESIM to store the control map values. In the IMPALA approach (Straubhaar et al., 2011), additional counters are added in the list structure to store the control map values for each pattern. As a result, it is possible to retrieve patterns occurrences conditional to both the neighborhood and to the local control map value. In the DS algorithm, the control map can be used in the multivariate framework as an additional variable.

The control map methods require the specification of the control map exhaustively over the simulation grid, as well as a control map defined over the TI. The requirement of specifying two maps may be problematic for subsurface applications, where it may be difficult to obtain a control map that accurately reflects the TI nonstationarity, and even harder to translate the known nonstationarity of the simulated domain into a continuous variable. However, with applications involving exhaustively known variables obtained by remote sensing, auxiliary variables can be used as control maps (see an application example in Chapter III.3).

In Figure II.5.11, we illustrate the use of control maps for generating a nonstationary realization based on a nonstationary TI. The same Sundarbans TI is used as in Figure II.5.10, but this time an auxiliary variable describes the nonstationarity of the TI. By drawing such a control map, it is assumed that the patterns vary gradually from the bottom to the top of the image; hence, the use of a very simple auxiliary variable (essentially, the y-coordinate of the domain) is appropriate. The pair of TI and auxiliary variable of the TI is denoted as A : A'.

In a first approach, we investigate the case where realizations are desired to exhibit the same nonstationary characteristics as the TI. To accomplish this,

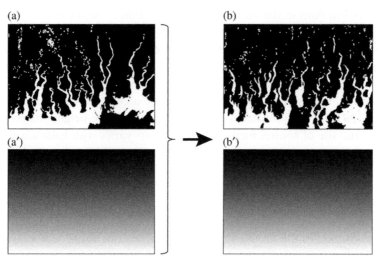

Figure II.5.11 Nonstationarity modeling using control maps. Given a, a′ and b′, the simulation algorithm generates b, which is a nonstationarity simulation presenting the same patterns as in a, and the same relationship with b′ than the relation a : a′.

an auxiliary variable is given, denoted as B′, which is in this case the same as the auxiliary variable of the TI. The simulation algorithm then simulates the main variable – namely, B – such that the spatial relation B : B′ is the same as A : A′. Using the notations of Hertzmann et al. (2001), this relationship between variables can be written as:

$$A : A' :: B : B'. \tag{II.5.4}$$

Note that, in this case, along the x-direction TIs can be considered mildly nonstationary, because in the bottom-right part of the image black areas are present, whereas in the bottom left they are not. Because this element of nonstationarity has not been explicitly incorporated, the simulation presents black over the entire bottom portion of the domain.

The approach is general for all simulation algorithms that use control maps to model nonstationarity (see Chapter II.3 for an overview of the MPS algorithms that use nonstationarity).

Consider now using the same example, except that this time the x-coordinate of the simulation grid nodes is used to describe the nonstationarity of the simulation, whereas the auxiliary variable of the TI remains the y-coordinate. Figure II.5.12 shows the results, where the patterns in the right part of the simulation domain (white auxiliary variable values) correspond to the patterns in the lower part of the TI (where white auxiliary variable values are also present). As a result, patterns of the TI corresponding to similar auxiliary variable values in the simulation are produced at those locations.

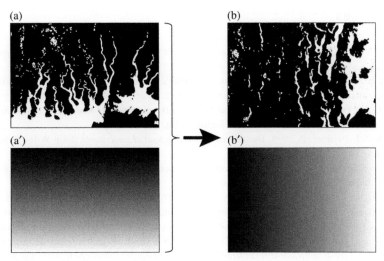

Figure II.5.12 Nonstationarity modeling using control maps (a′ b′), with a different nonstationarity in the simulation (b) and in the TI (a).

The principle of control maps and the concepts underlying Equation (II.5.4) can be extended to more than one auxiliary variable. For example, one could use two (or more) control maps to describe complex types of nonstationarity. Similarly, several ways of dealing with nonstationarity can be combined, such as using zonation combined with control maps to guide the nonstationarity within each zone. Combined applications can be found, for example, in Comunian et al. (2011) or in Pyrcz and Strebelle (2006).

Multiple control maps are particularly relevant when a number of known variables inform nonstationarity within the domain. A typical example is rainfall, which is influenced by a variety of spatial variables, such as temperature, cloud cover, elevation, and distance to the sea. With a more general point of view, we consider K auxiliary variables and the ccdf, written as:

$$F[Z(\mathbf{x})|\mathbf{N}_x, C_1(\mathbf{x}), \ldots, C_K(\mathbf{x})]. \tag{II.5.5}$$

This concept can be further extended by considering that some of the auxiliary variables are unknown or only partially informed, and therefore need to be simulated jointly with the main variable. At this point, the distinction between primary and auxiliary variable vanishes because each variable that is simulated becomes a primary variable. Several variables are then jointly simulated. The topic of joint simulation of variables in the context of MPS will be covered in details in Chapter II.6, but the principle of control maps outlined in this chapter, and in particular Equation (II.5.4), are some of the essential elements that will be used for multivariate MPS simulation.

References

Boucher, A. 2009. Considering complex training images with search tree partitioning. *Computers and Geosciences*, 35, 1151–1158.

Caers, J. & Zhang, T. 2004. Multiple-point geostatistics: A quantitative vehicle for integrating geologic analogs into multiple reservoir models. *In:* Grammer, G. M., Harris, P. M., and Eberli, G. P. (Eds.), *Integration of Outcrop and Modern Analog Data in Reservoir Models, AAPG Memoir 80*, American Association of Petroleum Geologists, Tulsa, OK.

Chilès, J.-P. & Delfiner, P. 1999. *Geostatistics – Modeling Spatial Uncertainty*, John Wiley & Sons, Inc., New York.

Chugunova, T. & Hu, L. 2008. Multiple-point simulations constrained by continuous auxiliary data. *Mathematical Geosciences*, 40, 133–146.

Comunian, A., Renard, P., Straubhaar, J. & Bayer, P. 2011. Three-dimensional high resolution fluvio-glacial aquifer analog: Part 2: Geostatistical modeling. *Journal of Hydrology*, 405, 10–23.

Goovaerts, P. 1997. *Geostatistics for Natural Resources Evaluation*, Oxford University Press, Oxford.

Hertzmann, A., Jacobs, C. E., Oliver, N., Curless, B. & Salesin, D. H. 2001. Image analogies. *In: Proceedings of the ACM SIGGRAPH Conference on Computer Graphics (SIGGRAPH 2001), Los Angeles, CA, 12–17 August*, 327–340.

Honarkhah, M. & Caers, J. 2012. Direct pattern-based simulation of non-stationary geostatistical models. *Mathematical Geosciences*, 44, 651–672.

Hu, L. Y., Liu, Y., Scheepens, C., Shultz, A. W. & Thompson, R. D. 2014. Multiple-point simulation with an existing reservoir model as training image. *Mathematical Geosciences*, 46, 227–240.

Liu, Y. 2006. Using the SNESIM program for multiple-point statistical simulation. *Computers & Geosciences*, 23, 1544–1563.

Lopez, S., Cojan, I., Rivoirard, J. & Galli, A. 2009. Process-based stochastic modelling: Meandering channelized reservoirs. *In: Analogue and Numerical Modelling of Sedimentary Systems: From Understanding to Prediction*, John Wiley & Sons, Inc., Hoboken, NJ.

Mariethoz, G. & Kelly, B. F. J. 2011. Modeling complex geological structures with elementary training images and transform-invariant distances. *Water Resources Research*, 47.

Mariethoz, G., Renard, P. & Straubhaar, J. 2010. The direct sampling method to perform multiple-point geostatistical simulations. *Water Resources Research*, 46.

Miller, J., Sun, H., Stewart, C., Genty, D., Li, D. & Lyttle, C. 2008. Direct modeling of reservoirs through forward process based models: Can we get there? IPTC Paper 12729, International Petroleum Technology Conference, Kuala Lumpur, Malaysia.

Pyrcz, M. & Strebelle, S. 2006. Event-based geostatistical modeling of deepwater systems. 26th Annual Gulf Coast Section SEPM Conference, Houston, TX.

Straubhaar, J., Renard, P., Mariethoz, G., Froidevaux, R. & Besson, O. 2011. An improved parallel multiple-point algorithm using a list approach. *Mathematical Geosciences*, 43, 305–328.

Strebelle, S. 2002. Conditional simulation of complex geological structures using multiple-point statistics. *Mathematical Geology*, 34, 1–22.

Wu, J., Boucher, A. & Zhang, T. 2008. A SGeMS code for pattern simulation of continuous and categorical variables: FILTERSIM. *Computers & Geosciences*, 34, 1863–1876.

Zhou, H., Gomez-Hernandez, J. & Li, L. 2012. A pattern search based inverse method. *Water Resources Research*, 48, W03505.

CHAPTER 6

Multivariate modeling with training images

Many applications of geostatistics involve the simulation of several variables on the same spatial domain. In this chapter, we explore the joint simulation of variables using multiple-point geostatistics. The basis of such joint MPS simulation is a multivariate training image: a set of collocated variables each representing the spatial structure being modeled (within each variable), and representing physically realistic spatial dependences between variables. Alternatively, similar information coming from very dense data sets can also be used (for details, see the case study on remote sensing in Chapter III.3). If such multivariate TIs or training data are not available, the methods discussed in this chapter are not applicable.

In order to illustrate the challenges involved in multivariate simulation, consider a tangible example of aquifer characterization in the presence of multiple sources of data. Typical data types could include electrical resistivity tomography (spatially distributed data), hydraulic conductivity measured at discrete locations, and local head (pressure) measurements. Hydraulic conductivity and head data are typically sparse. Additional information can potentially be given by exhaustively known covariates, for example vegetation, or partially known covariates, such as locally observed geological attributes.

Sometimes, the relationship between variables can be modeled with a parametric function. Oftentimes, however, the information on the relationship between variables cannot be fully represented by simple parametric relationships. This is where multivariate training images can be used to convey such information. Such use consists of providing an example of the relationship, and then generating simulations that replicate this example. The use of multivariate MPS is most valuable in the following cases:

- When the multivariate relationship can be informed by external information that is not in the local data. This external information may come from analog sites, physics-based models, or measurements taken at previous time steps (in the case of temporally varying processes).
- When the relationship between variables and/or when the spatial variability of each variable considered is too complex to be represented by cross-variogram-based approaches.

Multiple-point Geostatistics: Stochastic Modeling with Training Images, First Edition. Gregoire Mariethoz and Jef Caers.
© 2015 John Wiley & Sons, Ltd. Published 2015 by John Wiley & Sons, Ltd.
Companion website: www.wiley.com/go/caers/multiplepointgeostatistics

It may be possible to build an example relationship based on physics. In the case of aquifer characterization, one could look at a similar site where the hydraulic properties have been densely mapped. If the available variable relates to a geophysical attribute, a forward geophysical model can be applied to obtain the geophysical resistivity map corresponding to the analog aquifer. The pair flow properties–resistivity then constitutes a physically consistent multivariate TI (see Section II.7.8 for more details on multivariate TI construction).

The general concepts used in multivariate MPS simulation are very different from the techniques that have been proposed in the context of variogram-based geostatistics for co-simulations. In this classical framework, two or more variables are simulated jointly by respecting certain analytically tractable relationships between them. This is the case for example of co-kriging or co-simulation. As is commonly the case with statistics-based methods, a two-stage procedure is involved in the simulation procedure:

- A first step consists of the inference of a model for the joint relationship between the variables considered. Linear relationships (or linear relationships of transformed variables) are typically considered. The relationship between the spatial variability of the variables is captured using cross-variograms.
- A second step generates realizations of the variables considered using this multivariate model. The spatial variability of each variable corresponds to the modeled variograms, and the dependence between each pair of variables corresponds to modeled cross-variograms.

In many practical applications, the only data available to infer such a joint relationship (first step) are sparse locations where several variables are observed simultaneously. Although cross-variograms have been widely and successfully used, they have faced two types of limitations. On the one hand, when the data quantity or quality is poor, it can be difficult to estimate the parameters of the statistical model. The experimental variograms and cross-variograms may not be representative, and adjusting variogram models may be cumbersome (see also Part I). On the other hand, when large amounts of data are present, they reveal a complexity such that assumptions of variogram-based methods no longer hold. For example, it may be found that the variables are not linearly correlated, and that their spatial dependence cannot be represented using covariances.

This being said, the multivariate MPS also has its limitations, the main one being the availability of multivariate TIs. Therefore, in our view, multivariate MPS and multivariate variogram-based geostatistics are complementary tools that can be alternatively used in different cases, depending on the application and the available data.

6.1 Multivariate and multiple-point relationships

Before presenting in detail how multivariate MPS is implemented, consider first the ultimate aim of such methods. Even in the multivariate case, the requirements regarding the spatial variability of each single variable still apply. Each simulated variable needs to reflect patterns of the TI corresponding to that variable. This calls for considering the joint relationships over all nodes in each of the M univariate neighborhoods in the TI, and applying the same interactions in the simulation:

$$\text{Prob}\left[Z_k(\mathbf{x})|Z_k(\mathbf{x}+\mathbf{h}_1^1),...,Z_k(\mathbf{x}+\mathbf{h}_{n_k}^k)\right]$$
$$\cong \text{Prob}\left[Z_k(\mathbf{y})|Z_k(\mathbf{y}+\mathbf{h}_1^1),...,Z_k(\mathbf{y}+\mathbf{h}_{n_k}^k)\right], \quad k=1...M \quad (\text{II.6.1})$$

where n_k is the number of neighbors defined for variable k; $\mathbf{x}$ denotes a location in the simulation grid; and $\mathbf{y}$ is a location in the TI. Note that k is used for the variable currently simulated, whereas M denotes the total number of variables. The lag vectors $\mathbf{h}^k$ can be different for each variable. Using the notations introduced in Section II.2.3, Equation (II.6.1) can be simplified into:

$$\text{Prob}_{SG}\left[Z_k|\mathbf{N}_k\right] \cong \text{Prob}_{TI}\left[Z_k|\mathbf{N}_k\right], \quad k=1...M. \quad (\text{II.6.2})$$

where SG denotes the simulation grid and TI the TI grid. At this stage, no relationships between variables are considered, and using Equation (II.6.2) therefore considers each variable as spatially independent, as if separate univariate realizations were generated.

In the general case, interactions exist between neighboring nodes as well as between variables. Including a complete multiple-point and multivariate relationship requires extending Equation (II.6.1) to the case of a multivariate neighborhood. We consider the relationship between a node and multiple neighboring locations on the same and other variables. This relationship is written:

$$\text{Prob}\left[Z_{k_{central}}(\mathbf{x})|\underbrace{Z_1(\mathbf{x}+h_1^1),...,Z_1(\mathbf{x}+h_{n_k}^k)}_{\text{neighborhood of variable 1}},...,\underbrace{Z_M(\mathbf{x}+h_1),...,Z_M(\mathbf{x}+h_{n_M}^M)}_{\text{neighborhood of variable }M}\right]$$

$$\cong \text{Prob}\left[Z_{k_{central}}(\mathbf{y})|\underbrace{Z_1(\mathbf{y}+h_1^1),...,Z_1(\mathbf{y}+h_{n_k}^k)}_{\text{neighborhood of variable 1}},...,\underbrace{Z_M(\mathbf{y}+h_1),...,Z_M(\mathbf{y}+h_{n_M}^M)}_{\text{neighborhood of variable }M}\right],$$

$$k=1...M \quad (\text{II.6.3})$$

conveniently written as:

$$\text{Prob}_{SG}\left[Z_{k_{central}}|\mathbf{N}_1,...,\mathbf{N}_M\right] \cong \text{Prob}_{TI}\left[Z_{k_{central}}|\mathbf{N}_1,...,\mathbf{N}_M\right], \quad k=1...M. \quad (\text{II.6.4})$$

Note that, as defined in Section II.2.3, the variable $k_{central}$ is the variable that carries the central node of the data event considered. From Equation (II.6.4), it appears that there is no hierarchy between the different variables considered because the joint probability distribution function (pdf) of all neighborhoods $\mathbf{N}_k$ is considered. In other words, when multiple variables are jointly simulated, all variables can be simulated simultaneously with a single random path running through all variables of each grid node.

An important observation is that the conditioning event is nothing else than the joint occurrence of the individual neighborhoods for all variables considered. Recalling Equations (II.2.6) and (II.2.7), we obtain the conditional cdf for the value at $\mathbf{x}$ in the categorical case:

$$P\left[Z_{k_{central}}(\mathbf{x}) = c|\mathbf{N}\right] = \frac{\#\left(Z_{TI}^{k_{central}} = c|\mathbf{N}\right)}{\sum\limits_{b=1}^{C} \#\left(Z_{TI}^{k_{central}} = b|\mathbf{N}\right)}, \quad \mathbf{N} = \bigcup_{k=1}^{M}\mathbf{N}_k \qquad \text{(II.6.5)}$$

Similarly, for the continuous case:

$$F\left[Z(\mathbf{x}) \leq z|\mathbf{N}\right] \cong F\left[Z_{TI}(\mathbf{y}^\dagger) \leq z\right], \text{ with } \mathbf{y}^\dagger : \left\{\left[TI(\mathbf{y}^\dagger)^*\mathbf{N}\right] < t\right\}, \quad \mathbf{N} = \bigcup_{k=1}^{M}\mathbf{N}_k.$$
$$\text{(II.6.6)}$$

where the star operator denotes a convolution operator (an image filter, as described in Section II.2.4.1.2). Because convolutions can be performed for both continuous and categorical variables, Equation (II.6.6) can also be applied to the categorical case, leading to:

$$P\left[Z(\mathbf{x}) = c|\mathbf{N}\right] = \frac{\#\left[Z_{TI}(\mathbf{y}^\dagger) = c|\mathbf{N}\right]}{\sum\limits_{b=1}^{C} \#\left[Z_{TI}(\mathbf{y}^\dagger) = b|\mathbf{N}\right]} \text{ with } \mathbf{y}^\dagger : \left\{\left[TI(\mathbf{y}^\dagger)^*\mathbf{N}\right] < t\right\}, \quad \mathbf{N} = \bigcup_{k=1}^{M}\mathbf{N}_k$$
$$\text{(II.6.7)}$$

These formulations are the basis for implementing multivariate MPS simulation.

6.2   Multivariate conditional simulation: Implementation issues

At least in theory, multivariate conditional simulation can be achieved once the conditional distributions of Equation (II.6.5) have been inferred. One option is to store such conditional probabilities in a search tree as is done for the univariate categorical, as proposed in the original SNESIM implementation (Strebelle, 2002). However, such an approach will only work for relatively simple cases and perhaps no more than two variables, because the search tree grows exponentially in size when the number of categories or the number of variables considered increases. Chugunova and Hu (2008) and Straubhaar et al. (2011) propose methods that change the organization of the tree or the list such that multivariate data events can be stored without this exponential increase. Even then, their extension only covers conditional probability models accounting for collocated relationships (a single value for the collocated node on other variables).

In the case where multiple-point multivariate data events are required to fully capture intricate relationships between variables, storage in trees, lists, or hash tables is no longer practically feasible. For each data event related to each variable, counters would need to be established for all possible combinations of data events related to the other variables. Another difficulty is that some variables may be continuous and others may be categorical. Equation (II.6.5) only works for categorical variables, meaning that continuous variables will need to be categorized using thresholds, thereby reducing the available information, unless a large number of thresholds are considered, thereby resulting in large storage.

These difficulties in dealing with complete multivariate data events rule out methods that store patterns in a database. This leaves only the methods using raw storage (Section II.2.4.1), relying instead on convolutions to determine local conditional probabilities. Equations (II.6.6) and (II.6.7) are based on convolutions and are precisely applicable for both continuous and categorical variables. Among the MPS methods outlined in Chapter II.3, the ones that have been used for multivariate simulation are the cross-correlation simulation (CCSIM), image quilting (IQ), as well as direct sampling (DS). These are the methods that are based on convolutions. Simulated annealing (SA) has the potential to consider multivariate simulation by means of an objective function formulation, but such implementation has not yet been developed.

The multivariate versions of the CCSIM, IQ, and DS algorithms are essentially the same as for the univariate case; the difference lies in the use of multivariate neighborhoods and multivariate distances to perform convolutions. For each univariate neighborhood, a separate convolution is performed. Any norm can be used for each of the variables, whether it is continuous or categorical. The convolution result for each variable is then normalized in the interval [0 1]. Using Equation II.2.32, a single convolution map is obtained, which can be used as a basis for determining pattern frequencies as in the univariate case.

Multivariate convolutions are illustrated in Figure II.6.1, using the same example as in Section II.2.4.1.2 for univariate convolutions, but now using the different color bands as covariates. Four variables are considered, consisting of red, green, and blue (RGB) color bands of the image, as well as one categorical variable delineating areas of high and low color intensity (the color intensity is the norm of the RGB vector). A multivariate data event $\mathbf{N} = [\mathbf{N}_1 \ \mathbf{N}_2 \ \mathbf{N}_3 \ \mathbf{N}_4]$ is applied at each location of the multivariate image. The convolutions for the first three variables use a Euclidean distance, and the last one uses the number

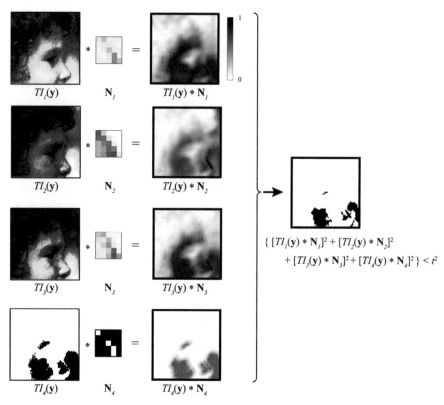

Figure II.6.1 Illustration of a multivariate convolution involving different types of variables. Variable 1: red band; variable 2: green band; variable 3: blue band; and variable 4: delineation of two categories of color intensity values (high intensity in black and low intensity in white).

of mismatching nodes distance (Equation II.2.21). These four convolutions are then assembled into a single error map, which is the result of the multivariate convolution:

$$\left\{ \frac{1}{4}\left[TI_1(\mathbf{y}) * \mathbf{N}_1 \right]^2 + \frac{1}{4}\left[TI_2(\mathbf{y}) * \mathbf{N}_2 \right]^2 + \frac{1}{4}\left[TI_3(\mathbf{y}) * \mathbf{N}_3 \right]^2 + \frac{1}{4}\left[TI_4(\mathbf{y}) * \mathbf{N}_4 \right]^2 \right\}$$

$$(II.6.8)$$

In this case, each variable has an equal weight of 1/4. Applying a threshold on this error map yields the matching set $\mathbf{y}^\dagger$. Sampling a location from $\mathbf{y}^\dagger$ is then a good approximation of a sampling from the conditional pdf $F\left[Z(\mathbf{x}) \leq z|\mathbf{N} \right]$, in the same fashion as for the univariate case (see Equation II.2.10).

Note that the conditional probabilities are not explicitly formulated or estimated (as would be done in an MRF model; Chapter II.4). These conditional probabilities are very high-dimensional and would require large TIs for inference. Indeed, considering multiple-point multivariate relationships would require joint distributions of dimension $\sum_{k=1}^{K} n_k$. In many cases, it is impossible to comprehensively inform such high-dimensional distributions, even when using very large TIs. Methods relying on convolutions avoid this bottleneck by directly retrieving a sample from the TI instead of estimating or modeling conditional or joint distributions.

6.3   An example

Section II.6.2 provided general concepts on multivariate MPS. Here, we provide a simple illustrative example on a synthetic case. A comprehensive case study involving multivariate modeling is presented in Chapter III.3.

In practical multivariate modeling, it is often the case that the data available for each variable are not collocated. One variable may be sampled at some locations, whereas another variable is sampled at different locations. In some areas, several variables may be known simultaneously. Some variables may not be informed (e.g., contaminant concentration at an unsampled location), and other variables may be completely informed (e.g., topography).

Our synthetic example is based on the reference model shown in Figure II.6.2. The grid size is 2D with 190×190 nodes, and three variables are considered. Variable 1 is obtained by taking the absolute value of a realization sampled from a multi-Gaussian model with zero mean and an isotropic exponential variogram of range 100 nodes. The resulting image has an asymmetric histogram. A histogram transformation (normal-score transform) is then used to restore the standard Gaussian histogram. Because of the absolute value operator applied, the values previously near the mode of the histogram have become the lowest values. Because values of a multi-Gaussian model around the median are usually highly connected and because of the nature of the transformation, low values are now connected. The resulting complex patterns generated become difficult to describe with covariance models. One such realization is shown in Figure II.6.2a (note that the values are inversed in Figure II.6.2a such that high values are connected instead of low ones).

Variable 2 is defined as a smoothing of variable 1 with a moving average using a window of 17×17 pixels, which is then shifted by a vector [10 10] (Figure II.6.2b). The shift incurs a dependency whereby co-located values carry less information than values at a distance. This is visible when comparing Figure II.6.2a and II.6.2b (where there is a clear spatial dependence between variables 1 and 2) with Figure II.6.3a (where these variables are only poorly correlated when compared on a pixel-wise basis).

Variable 3 is categorical and very broadly delineates areas of high and low values in variable 1. It is obtained by first applying a moving average on variable 1 using a large window of 21×21 pixels, and then applying a threshold on this moving average (Figure II.6.2c). Because variable 3 is binary, the scatterplots of

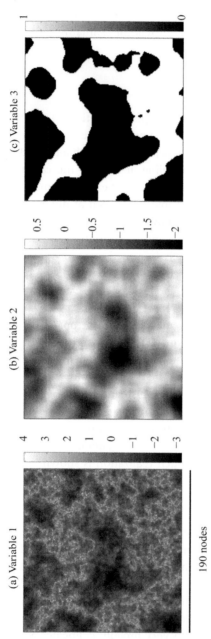

Figure II.6.2 Trivariate reference case. (a) Variable 1: a transformed multi-Gaussian field (Zinn and Harvey, 2003). (b) Variable 2: moving average and shifting applied to variable 1. (c) Variable 3: smoothing, then thresholding of variable 1.

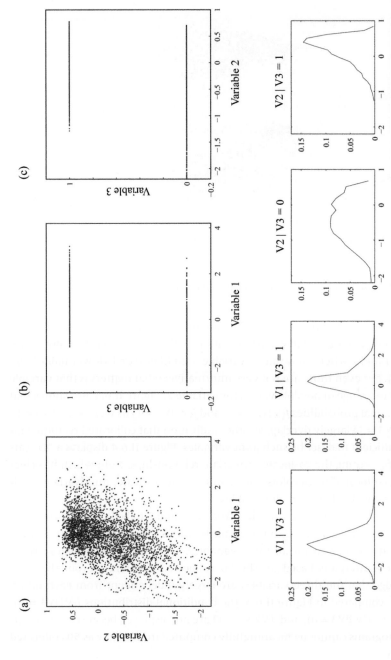

Figure II.6.3 (a–c) Scatterplots of all co-located values, with variables considered two by two. Lower half of the figure: conditional histograms illustrating the co-located relationship with binary variable 3.

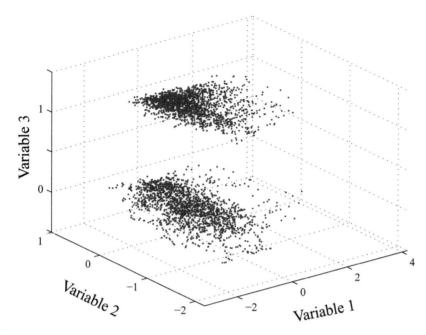

Figure II.6.4 3D scatterplot of all three variables in the reference.

variable 3 against the other two variables are not very informative (Figure II.6.3b and II.6.3c). A more suitable representation is the conditional histograms for variables 1 and 2 when the value of variable 3 is either 0 or 1 (lower half of Figure II.6.3), but even those are not very informative: what matters is that variable 3 restricts the range of possible values for variable 2. Figure II.6.3 also shows that the conditional probabilities $f(V1|V3 = 0)$ and $f(V1|V3 = 1)$ as well as $f(V2|V3 = 0)$ and $f(V2|V3 = 1)$ largely overlap, another indication that collocated relationships are not sufficient to describe such joint variables. Figure II.6.4 displays a 3D scatterplot representing the co-located trivariate relationships, from which it is clear that such co-located scatterplots hardly inform the intricate, multivariate, multipoint relationship between all variables.

In this synthetic case, we consider variables 1 and 2 to be known at 100 sample locations, and variable 3 at 1000 sample locations (Figure II.6.5). Fifty of these locations are collocated for all three variables. The lower part of Figure II.6.5 shows the scatterplots based on the 50 collocated data only. From this, one would judge that the three variables are largely independent from each other. Note that, compared to Figure II.6.3, the conditional distributions $f(V1|V3 = 0)$, $f(V1|V3 = 1)$, $f(V2|V3 = 0)$, and $f(V2|V3 = 1)$ are not displayed because such conditional histograms cannot be meaningfully computed from as few as 50 collocated data points.

We assume that a multivariate TI is available, accurately representing variability and relationships; see Figure II.6.6. This TI was obtained using the same

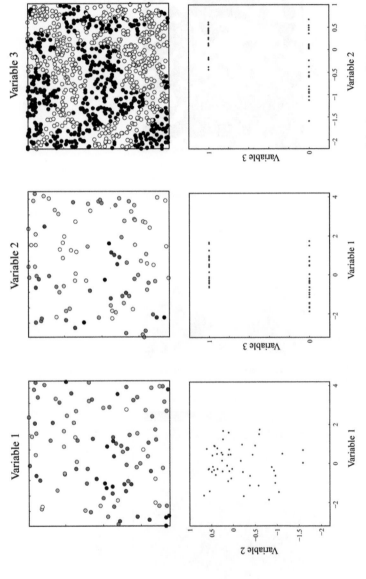

Figure II.6.5 The data available and the scatterplots based on 50 collocated sample locations. Lower part of the figure: conditional histograms to better illustrate the relationship with binary variable 3.

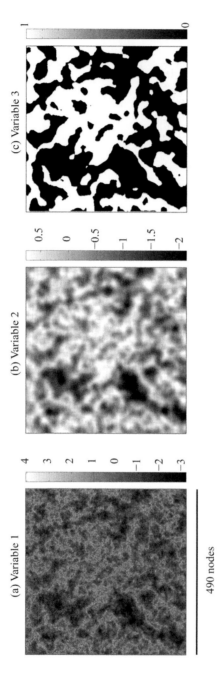

Figure II.6.6 (a–c) Multivariate training images.

procedure as for the reference, except that the TI domain is of the size 490×490 nodes. Figure II.6.7 and Figure II.6.8 show the co-located scatterplots for the TI, displaying similar relationships as for the reference case.

With the TI defined, all ingredients are present to perform multivariate MPS simulation. DS is used because its pixel-based nature easily allows conditioning to dense point data. The neighborhood for each variable consists of $n_1 = n_2 = n_3 = 15$ nodes. In comparison with applications of DS to univariate problems elsewhere in this book, the size of the neighborhood for each variable is rather small (15 nodes), but this number is sufficient because for all three variables, the neighborhood is jointly informed by 45 values – enough to represent complex spatial patterns.

The Euclidean distance (L2 norm) is used for variables 1 and 2 (Equation II.2.30), and the number of mismatching nodes is used for variable 3 (Equation II.2.31). The distances for the different variables are combined by assigning an equal weight of 1/3 to each variable (Equation II.2.33). A distance threshold of $t = 0.0001$ is used in DS. A fully random simulation path is adopted, which means that each variable at each location is visited in a random order. As a consequence, the simulated triplets $[v_1, v_2, v_3]$ at any location may not correspond to any triplet values in the TI, which is desired to generate spatial uncertainty.

As a result of this fully random path, all variables are simulated jointly without hierarchy or order. An alternative to this full random path would be to paste an entire vector $[v_1, v_2, v_3]$ from the TI instead of a single value at a time. This would accelerate the simulation, but it would only allow triplets of values $[v_1, v_2, v_3]$ that are already present in the TI.

Figure II.6.9 shows a comparison between one conditional realization and the reference. A good similarity is observed in terms of local features (given by the conditioning data), spatial continuity (given by the TI), as well as relationships between variables (also given by the joint TIs). The resulting co-located scatterplots are shown in Figure II.6.10 and Figure II.6.11, and they correspond to what is observed in the reference and in the TI. Visually, one observes that the diagonal shift of [10 10] nodes between variable 1 and variable 2 is respected (e.g., the dark spot in the center of Figure II.6.9a and Figure II.6.9b is displaced by this amount). The areas of high values in variable 1 are also correctly delineated by variable 3, and this is visible in the ensemble average.

For this case, we used a TI built using the same processes as in the reference. In this regard, our example represents a rather ideal case. These methods are suitable for cases where a multivariate TI is available and modelers have high confidence in such multivariate TIs, for example in remote sensing applications where large multivariate data sets are typically available, or through modeling of physical processes that relate variables between each other (see Chapter III.3 for a real case application). However, one may encounter difficulties in application domains where such complete and high-confidence data are not available.

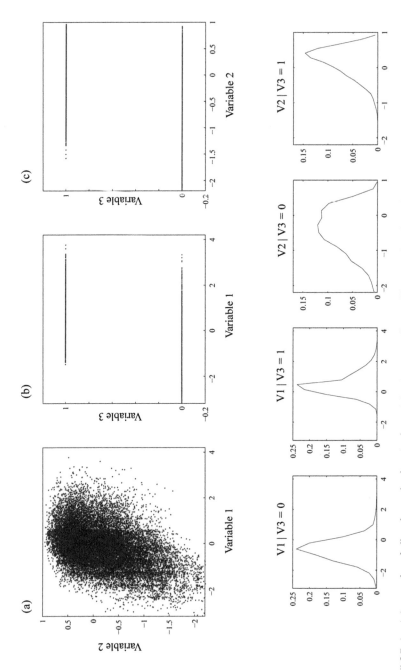

Figure II.6.7 (a–c) Scatterplot of all co-located values in the training image, with variables considered two by two. Lower half of the figure: conditional histograms to better illustrate the relationship with binary variable 3.

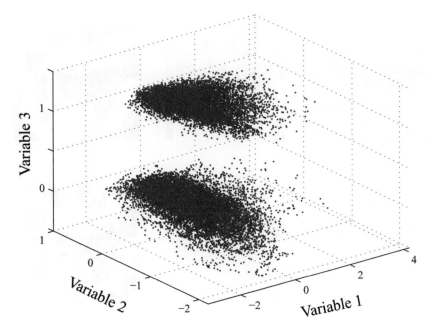

Figure II.6.8 3D scatterplot of all three variables in the training image.

This is, for example, the case with subsurface 3D modeling where multiple exhaustive measurements are rather the exception. If the TI is poorly chosen or constructed, mimicking incorrect multivariate relationships can lead to significant biases. However, these biases have to be put in perspective with the biases resulting from other, more traditional methods that simplify complex relationships (e.g., assumptions of linear correlation) or assume extreme values as being disconnected. Ways of constructing physically consistent multivariate TIs are further discussed in Section II.7.3.

In our example, the amount of conditioning data varies greatly between variables, with variable 3 being more densely sampled than variables 2 and 3. In some applications, the difference of information content between variables can be even larger, with some variables that can be exhaustively informed by remote sensing, and other variables that may not be informed at all. In such cases, the input data are similar as in traditional methods based on external drift (i.e., an auxiliary variable that is known and an empty grid for the main variable). The difference is that with multivariate MPS, the exhaustively known variable is not considered as only affecting the mean, but may influence what type of patterns are locally present. In the example above, the regions where variable 3 is white correspond to connected patterns in variable 1. The black regions in variable 3 correspond to disconnected patterns in variable 1. Therefore, not only the local means are affected, but in fact the entire spatial structure is.

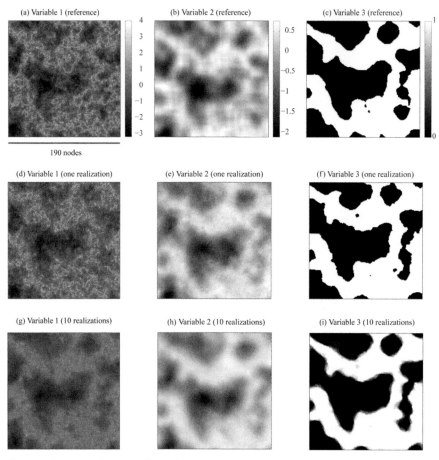

Figure II.6.9 (a–c) Reference (identical to Figure II.6.2), shown here for comparison. (d–f) One multivariate conditional realization obtained using direct sampling. (g–i) The average of 10 conditional realizations.

6.4 Multivariate simulation as a filtering problem

To conclude this chapter, we briefly introduce notions within the field of texture synthesis that apply to the context of multivariate modeling. In texture synthesis, one often considers several variables as filtered versions of each other. One of the variables can be known, forcing certain types of structures to occur at specific locations in the other variable (Efros and Freeman, 2001), which is sometimes called "texture transfer."

Hertzmann et al. (2001) develop this idea as a framework named "image analogies" and first define the notation A : A' :: B : B' (already introduced previously in this book: see Equation II.5.4). It clearly states the problem, where given a bivariate TI, the relationship between both variables A and A' (whatever

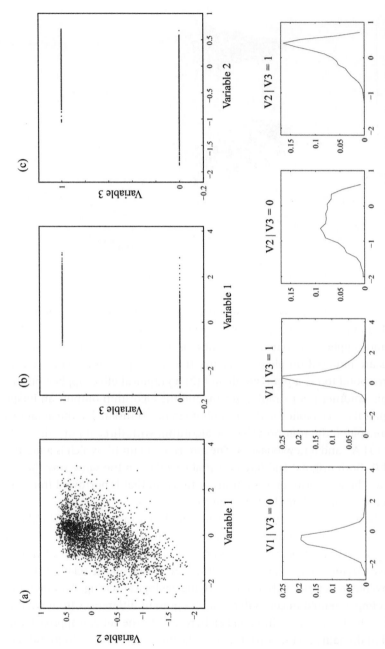

Figure II.6.10 (a–c) Scatterplot of all co-located values in one realization. Lower half of the figure: conditional histograms to better illustrate the relationship with binary variable 3.

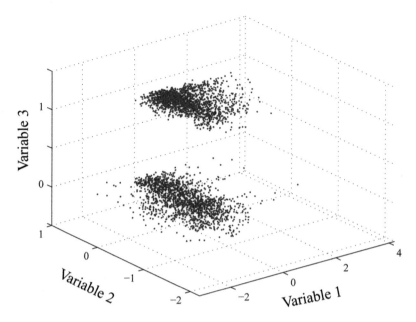

Figure II.6.11 3D scatterplot of all three variables in the realization.

its nature and complexity) is expressed as A : A'. The aim is that the simulated variables B and B' follow the same relationship noted B : B'.

A common application in texture synthesis is by-example filtering where special effects are imposed on pictures (Figure II.6.12). A pair of images is given, which correspond to (1) a photograph and (2) a graphical effect applied on this photograph. Another photograph is given, and the algorithm outputs an image of the graphical effect applied to the new photograph. Translated in the notations of image analogies, the user provides the algorithm with three inputs: the TI A, the filtered TI A', and a new image B. The aim is to obtain B', which is a filtered version of B. Both original and filtered variables relate in the same way that A' relates to A. The graphical filter is a transfer function that is borrowed from the training set and imposed on the target image.

For applications in geostatistics, the concept of by-example filters should be turned around: the user provides as inputs the TIs A and A' and the filtered output B'. Image analogies can then be used to find B. For example, B' can be an auxiliary variable such as a geophysical survey. Based on the relationship A : A', it is possible to determine B, the underlying geological setting. Although the algorithms implemented in computer graphics present differences from what is presented in this book (Mariethoz and Lefebvre, 2014), the general approach can be applied in domains that involve the identification of a variable from indirect measurements, such as geophysics, remote sensing, or flow.

Figure II.6.13 illustrates a potential application of image analogies to the identification of geological features based on georadar surveys. A and A' consist of the

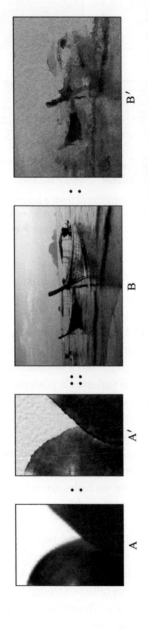

Figure II.6.12 Illustration of the image analogy notation with a by-example filtering application. Modified from Hertzmann, A., C. E. Jacobs, et al. (2001).

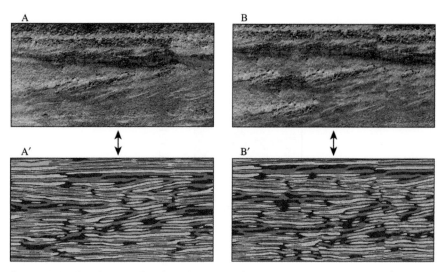

Figure II.6.13 Identification of geological structures from a georadar survey. A and A′: training image (Bayer et al., 2011). B′: another georadar survey taken in a similar geological environment. B: simulated photograph honoring the relationship B : B′.

photograph of an outcrop at the Herten site in Germany (Bayer et al., 2010) and a georadar image taken at the same location. Given another georadar survey taken on the same site but at a different location (B′), it is possible to generate conditional realizations such as (B) of what a photograph would look like at the location of the second georadar survey if the site was excavated at this location. Note that B is not an actual photograph, but a simulation of what the photograph could be. The dependence between variables is preserved as shown on the simulated photograph: the simulated outcrop photograph presents spatial features at locations that are coherent with those of the georadar image B′.

Figure II.6.14 shows the scatterplots of A versus A′ and B versus B′. The fact that both scatterplots are similar is an additional indication that the filtering relationship A : A′ :: B : B′ is respected. These scatterplots only represent the co-located relationship between the variables considered and therefore are not representative of the full multivariate multiple-point relationship expressed in Equation (II.6.4).

The same method can be applied to model head or contaminant concentration in an aquifer. The variables considered are then a combination of static and dynamic attributes, for example hydraulic conductivity and head. In this case, however, the head variable will generally be strongly nonstationary, even if the aquifer flow properties are stationary. Therefore, localized TIs need to be used (Section II.5.2.1) that have the same size and boundary condition as the simulation domain. This approach has been used by Zhou et al. (2012) to solve hydrogeological inverse problems.

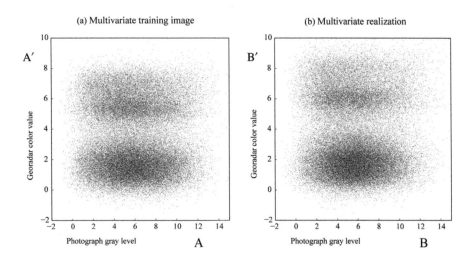

Figure II.6.14 Scatterplots of (a) the multivariate training images and (b) multivariate realization of Figure II.6.13.

References

Bayer, P., De Paly, M. & Bürger, C. 2010. Optimization of high-reliability based hydrological design problems by robust automatic sampling of critical model realizations. *Water Resources Research*, 46, W05504.

Bayer, P., Huggenberger, P., Renard, P. & Comunian, A. 2011. Three-dimensional high resolution fluvio-glacial aquifer analog: Part 1: Field study. *Journal of hydrology*, 405, 1–9.

Chugunova, T. & Hu, L. 2008. Multiple-point simulations constrained by continuous auxiliary data. *Mathematical Geosciences*, 40, 133–146.

Efros, A. A. & Freeman, W. T. 2001. Image quilting for texture synthesis and transfer. *In: Proceedings of the ACM SIGGRAPH Conference on Computer Graphics*, 341–346.

Hertzmann, A., Jacobs, C. E., Oliver, N., Curless, B. & Salesin, D. H. 2001. Image analogies. *In: Proceedings of the ACM SIGGRAPH Conference on Computer Graphics (SIGGRAPH 2001)*, Los Angeles, CA, 12–17 August 2001, 327–340.

Mariethoz, G. & Lefebvre, S. 2014. Bridges between multiple-point geostatistics and texture synthesis. *Computers & Geosciences*, 66, 66–80.

Straubhaar, J., Renard, P., Mariethoz, G., Froidevaux, R. & Besson, O. 2011. An improved parallel multiple-point algorithm using a list approach. *Mathematical Geosciences*, 43, 305–328.

Strebelle, S. 2002. Conditional simulation of complex geological structures using multiple-point statistics. *Mathematical Geology*, 34, 1–22.

Zhou, H., Gomez-Hernandez, J. & Li, L. 2012. A pattern search based inverse method. *Water Resources Research*, 48, W03505.

Zinn, B. & Harvey, C. 2003. When good statistical models of aquifer heterogeneity go bad: A comparison of flow, dispersion, and mass transfer in connected and multivariate Gaussian hydraulic conductivity fields. *Water Resources Research*, 39, WR001146.

CHAPTER 7

Training image construction

As illustrated in previous chapters of this book, one of the main challenges in using multiple-point geostatistics is the availability of a training image. This is a general problem that has been partially discussed in Chapter II.6 for the case of multivariate TIs. At this time, there is no universal public database where users can browse for large numbers of TIs. On the website accompanying this book, http://www.trainingimages.org, interested readers will find electronic versions of most TIs used in this book. However, as with other databases, this only represents a very limited number of TIs. Therefore, the solution often adopted is to custom-build TIs for each application, or to use specific databases that have been developed for limited application domains. A number of tools are available to this end, regardless of whether the TIs are constructed based on physical principles or statistical methods or by using empirical rules. The review of these different methods is the topic of this chapter.

7.1 Choosing for training images

Earth scientists, geologists and engineers need images of natural phenomena. These images are crucial tools for the interpretation, representation, and modeling of processes. The term "image" is defined here in a very broad sense as an exhaustive representation of a process deployed in space and time. Geostatistical simulation methods generate images, termed "realizations," which mimic real-world processes. In a traditional statistical framework, this would be accomplished by isolating the statistical properties associated with the outcome of natural processes, and then using these statistics as constraints for the generation of images. However, natural processes may consist of a complex combination of randomness and deterministic physical laws. As such, real processes may present complex idiosyncrasies. Representing a natural phenomenon by the statistics of its outcomes is a simplification of this complexity, which is often described with terms such as "non-stationarity," "non-Gaussianity," and "non-heteroscedasticity." The negative characterization of these terms is an indication that traditional statistical descriptions often do not apply to these cases.

Multiple-point Geostatistics: Stochastic Modeling with Training Images, First Edition. Gregoire Mariethoz and Jef Caers.
© 2015 John Wiley & Sons, Ltd. Published 2015 by John Wiley & Sons, Ltd.
Companion website: www.wiley.com/go/caers/multiplepointgeostatistics

As a consequence, statistics are often not satisfactory to describe natural processes because they require repetition. An example presents itself when extracting multiple-points statistics from a TI. As soon as the image presents some degree of complexity, most of the multiple-points statistics have only one single replicate. Although the TI may be considered a reasonable depiction of the natural system under study, it is clearly not statistically representative.

This discussion suggests that there is a trade-off between statistical representation and empirical realism. In natural sciences, where subjective interpretations and concepts such as realism are key, the use of complex and perhaps unique TIs may be required, even if it comes at the price of reduced statistical representativity. The overwhelming complexity of natural processes and the interpretative character of the available data mostly produce cases that are unique in nature. On the other hand, when simplifications or generalizations are acceptable and when data are present to support a statistical analysis, traditional statistical methods may be more applicable. The utility of such statistical analysis is clear. However, although reproducing statistical quantities is one of the criteria that may eventually lead to obtaining realistic models of natural systems, it is not a guarantee that the physical processes are correctly represented.

Using TIs offers a framework to include subjectivity in Earth science models. However, subjectivity is a double-edged sword because now modeling assumptions are explicitly laid out in the TI. Hence, choosing a TI or, even better, a set of TIs (see Chapter II.9) is often the most critical modeling decision in the type of modeling described in this book. MPS forces modelers to be explicit in their modeling choices and not duck the question of the exact nature of spatial variability or leave it to parametric models that are not always very well understood by practitioners. The resulting realizations should not be a surprise to the modeler of MPS, but rather should be the result of a set of conscious and explicit modeling and algorithmic choices.

7.2 Object-based methods

Object-based methods (also called Boolean methods) are a family of stochastic simulation methods that has been heavily used in geological modeling (Haldorsen and Chang, 1986; Lantuéjoul, 2002), but also has applications in other fields such as rainfall simulation (Zhang and Switzer, 2007). The principle of object-based methods is to define 2D or 3D shapes that are used to populate a given model using a set of rules that define the probability of occurrence of objects as well as the interactions between these objects. Conditioning to point data and to nonstationarity in the proportions of objects can be achieved by using a birth and death algorithm (Allard et al., 2006).

In some cases, when large amounts of data are present, it can be computationally demanding to condition object-based models. Moreover, the predefined

Figure II.7.1 A simple object-based model of channelized structures. Image obtained with the TiGenerator software (Maharaja, 2008). Grid size: 250×250×100 nodes.

shapes used can sometimes lack flexibility when complex structures are modeled or when data are plenty. For example, representing a meandering channel as a sinusoid can be an oversimplification (even if the parameters of the sinusoid are randomized within certain bounds). These issues can be addressed by MPS by using a two-step approach consisting of (1) generating an initial unconditional model with an object-based simulation method, and (2) using this model as a TI for conditional MPS simulations. The MPS simulation guarantees local data conditioning, and at the same time offers increased flexibility in terms of the shape of the bodies simulated. For example, a TI consisting of analytically defined sinusoids (which each have a defined and fixed amplitude and wavelength) will produce a simulation with curves that are not constrained by fixed characteristics. This can be desirable because the shape of natural channels is known to be more complex than a simple sinusoid. By adding a level of randomness, MPS simulation impairs additional flexibility.

Figure II.7.1 and Figure II.7.2 show examples of object-based simulations of 3D channelized systems. Although the model in Figure II.7.1 uses a simple representation of the structures (i.e., sinusoids of constant width and depth), Figure II.7.2 depicts the application of more complex rules, including levees along the channels (in orange) and variation in the object parameters.

7.3   Process-based models

The main difference between the models depicted in Figure II.7.1 and Figure II.7.2 is that the model of Figure II.7.2 is constrained by physically inspired

Figure II.7.2 An object-based model considering variable channel width and depth, randomized sinuosity, and different facies for the various architectural elements (main channel and levees). Image obtained with the FLUVSIM software (Deutsch and Tran, 2002). Grid size: 250×250×100 nodes.

rules, and therefore is more likely to correspond to processes occurring in the real world. One can take the idea of physical reality further and use geological process–based models, which numerically forward-simulate the deposition and erosion of sediments to obtain a 3D volume representing a geological formation with a very high degree of detail.

A good example of such forward geological modeling is given by Koltermann and Gorelick (1992), who create a 3D model that simulates 600,000 years of sedimentary processes in an alluvial fan in northern California. The simulation accounts for the dynamics of river flooding, sedimentation, subsidence, and land movement that resulted from faulting and sea level changes. The domain is discretized in time and space, and the partial differential equations describing these various processes are solved for each time step. Critically, the model takes as inputs water and sediment time series for the entire 600,000-year period (Figure II.7.3). Although the initial conditions of the model are known, its forward evolution is largely out of the control of the modeler. As a result, the model presents physically plausible geological structures, but it is difficult to constrain such models locally to data. In the example of Figure II.7.3, the boundary conditions of the deposition model control the presence of large-scale structures which therefore occur at observed locations. However, it is much more difficult to condition such models to data at a finer scale, such as facies observations on the surface of the model, because the values of the surface nodes also depend of the model outcomes on all layers located at depth.

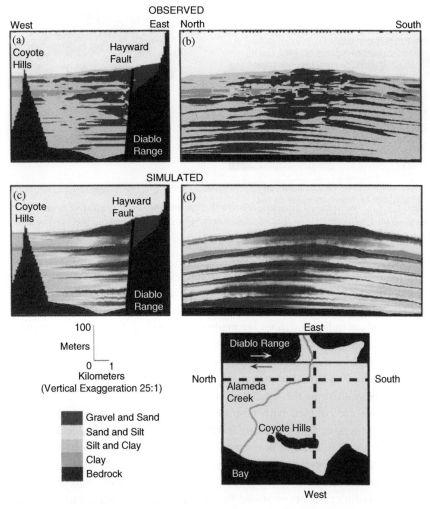

Figure II.7.3 Process-based reconstruction of the Alameda Creek alluvial fan. Reprinted from Paleoclimatic signature in terrestrial flood deposits. Science 256(5065): 1775–1782. With permission from AAAS.

In the domain of process-based models, a large body of work addresses the physics of meandering rivers, which can be used to forward-model the evolution of alluvial reservoirs (e.g., inland basins and deltas). For example, Ikeda et al. (1981) describe the evolution of a meander taking into account the near-bank excess velocity and the planform curvature. Seminara (2006) included more complex bank erosion processes. These models can reproduce many features of real channels, but they require calibration of empirical parameters, which can be difficult. As a result, very accurate process-based models are available for obtaining TIs for this sort of geological environment (Gross and Small, 1998; Lancaster and Bras, 2002; Pyrcz et al., 2009).

(a)
(b)

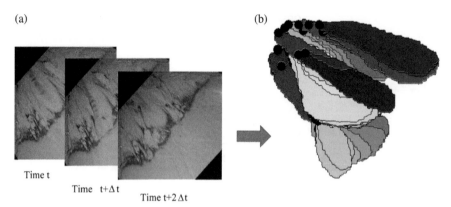

Time t

Time t+Δt

Time t+2Δt

Figure II.7.4 (a) Process-based modeling from tank experiments (data courtesy of Saint Anthony Falls Laboratory, University of Minnesota). (b) Interpretation from the overhead photos (data courtesy of Siyao Xu).

Figure II.7.4 shows an example of a process-based model extracted from a tank experiment, with the presence of channels and lobes. Overheard cameras take snapshots over time. Lasers scan the generated topography. From these data, interpretations of individual lobes are made that provide information on stacking patterns and changes in geometry over time.

Computer-simulated process models solve equations that govern the physical processes such as turbidite deposition and erosion through time, dependent on fluid flow velocity in the overlying water (Michael et al., 2010). The fluid flow equations are coupled with empirical equations describing erosional and depositional processes at the bedding level (Sun et al., 2001). Solutions require the specification of initial and boundary conditions, consisting of the bed grain size distribution, suspended sediment composition and concentration, and specification of temporally and spatially variable sediment sources, with given flow velocity and sediment concentration. The simulation of one model may require days to weeks depending on the size and required details.

7.4 Process-mimicking models

Process-based models produce realistic geometries, but often accurate representation of the physics requires extreme spatial and temporal discretization, which results in unfeasible computational times. Because of their considerable CPU demand, "pure" process models are often supplanted by process-mimicking models. These models are forward models as well, but no physical equations are solved. Several methodologies, algorithms, and forms of software have been proposed for both clastic and carbonate environments (Hill and Griffiths, 2009; McHargue et al., 2011). Some are termed event-based models (Pyrcz et al., 2009) or surface-based models (Bertoncello et al., 2013; Michael et al., 2010). Such models are of particular relevance when the spatial structures have been created

Figure II.7.5 A process-mimicking model of an alluvial reservoir using the FLUMY simulation method (Lopez et al., 2008). The spatial organization of the structural elements is based on physical equations. Grid size: 500×500×200 nodes.

by a geological process for which the individual depositional or erosional events are easily identified, resulting in a complex layering structure. Other process-mimicking algorithms are pixel based, such as cellular automata (Coulthard and Van De Wiel, 2006).

Process-mimicking models do not aim at being physically realistic but remain realistic in terms of the geometries generated by such processes. In this way, they aim to mimic the end result created by process models, albeit at a fraction of the computational time. Process-mimicking models suffer from the same conditioning problems as process models; however, because of their favorable CPU requirements, iterative schemes of conditioning can be employed for conditioning to some limited well or seismic data (Bertoncello et al., 2013). Alternatively, forward models generated (with some minimal conditioning) can be used as TIs for further rapid generation of conditional reservoir models.

A simulated model of a fluvial meandering system using a process-mimicking model is shown in Figure II.7.5 (Lopez, 2003), with improved realism compared to object-based models. Similarly as for the example of Figure II.7.3, this particular model offers some of the conditioning capabilities of the large-scale features by setting a higher bank erodability at the locations where sand bodies have been measured, therefore increasing the chance of a channel evolving in such areas (Lopez et al., 2008). This method, however, does not allow conditioning to dense point data (wells), but here again the results of such process-based models are ideal candidates to be used as 3D TIs.

Accurate conditioning of process-based models to dense data, such as well logs and seismic in the case of geological models, has not been achieved in any realistic modeling context to date. The question is often raised on how to use the output of process models as TIs for geostatistical modeling. The answer to this question is not trivial because the process model, by its very nature, is nonstationary because

Process-based simulation

Process-mimicking simulation

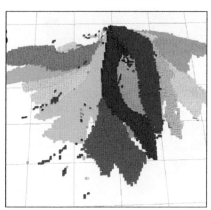 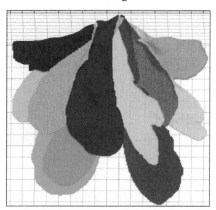

Weeks to simulate

Seconds to simulate

Figure II.7.6 Process-based (data courtesy of Exxon) versus process-mimicking models (Michael et al., 2010).

it represents a real depositional system (Comunian et al., 2013). Two views on this have been formulated:

1 *The localized TI*: one can use the process-based model directly as a nonstationary TI in geostatistical modeling, employing the methods that are available for dealing with nonstationarity (Chapter II.5). This often means that the process model represents a localized TI, having the same dimensions and grid structure as the modeling domain. This would only be meaningful if the process model was already somewhat constrained to the particular reservoir setting and structures. However, instead of focusing on well conditioning, one could tune process models to reflect some major reservoir structures.

2 *Hybridization*: the second approach is to lift meaningful statistics from the process model. After all, one is not interested in how the model was created but in the end result of the process, namely, the geometrical structures that were created as well as their spatial variation. These statistics can, for example, describe the thickness of events generated (lobes and channels in Figure II.7.6), provide rules on how these lobes and channels stack vertically or on how they erode each other, or provide localized TIs for certain architectural elements. These rules and statistics can then be used to parameterize a hybrid object-based model where the objects are themselves generated using MPS.

7.5 Elementary training images

MPS is difficult to apply in cases when one does not have enough information to clearly decide which TI to use. The choice of a TI is often a discrete decision (either image A or B), which sometimes lacks nuance. It is therefore desirable in certain cases to parameterize the spatial structure in a similar way as with multi-Gaussian models that allow fitting the parameters of the semivariogram. This is generally not possible because the TIs normally used are too large and complex to be parameterized simply. Some (e.g. Emery and Lantuejoul, 2014) recommend to use very large TIs that contain as much diversity as possible to encompass all possible patterns to be simulated.

A different approach is to consider simple elementary TIs as an expression of prior spatial continuity. During the simulation, the diversity of patterns is enriched by applying random transformations (see using transform-invariant distances in Section II.2.5.3). This method enables the generation of complex models whose spatial structure can be parameterized by adjusting the statistics of the random transformations. It addresses the problem of building TIs in 3D, because building elementary images is easy even in 3D (see Figure II.7.7 and Figure II.7.8). The TI no longer represents the spatial texture that is desired in the simulation but rather constitutes a basic element that forms an initial basis for defining complex structures obtained by the application of transformations of these elementary patterns.

In the case of Figure II.7.7 and Figure II.7.8, the TI only conveys the concept of disconnected bodies that are separated by thin gaps, the gaps being connected with each other. The simulation is obtained with the direct-sampling method using rotation-invariant distances. Such a setting can be used to generate samples of porous media such as sandstone (Figure II.7.7) or fractured rock (Figure II.7.8), depending on the transformation parameters used.

Figure II.7.7 Left: elementary training image (size: 50×50×50 nodes). Right: one realization using rotation-invariant distances (size: 180×150×120 nodes). Rotation-invariant distances are considering angles of +/− 90° in all directions.

Figure II.7.8 Left: elementary training image (size: 50×50×50 nodes). Right: one realization using rotation-invariant distances (size: 180×150×120 nodes). Rotation-invariant distances are considering angles of +/− 20° in all directions.

7.6 From 2D to 3D

For many geological applications of MPS, 3D TIs are needed. However, most measurements that can be used as TIs are 2D. These include, for example outcrops or geophysical transects. Typical textbook conceptualizations of geological structures are represented as 2D sketches that should ideally be usable as TIs. As a consequence, it is important to be able to use 2D TIs and somehow combine them to generate 3D simulations. It is clear that this is a difficult problem that does not, in general, have a well-defined solution, because of the lack of information on how to join the known sections. A notable exception is when the 2D TIs are the faces of a quarry front that provide cross-sections that are in the same coordinate system and that intersect each other.

A first approach is to use probability aggregation (see Section II.2.8.3.2) to combine the information coming from several orthogonal 2D TIs (Caers, 2006). The classical approach to 3D MPS simulation would call for determining a local ccdf for each node to simulate, based on its 3D neighborhood. Here, a given 3D neighborhood is separated in several 2D neighborhoods oriented in the direction of each 2D TI. A separate ccdf is computed for each of these subneighborhoods, based on the corresponding TI. In a final step, all resulting 2D ccdfs are combined (aggregated) using one of the formulas described in Section II.2.8.3.2, from which an outcome can be drawn and used for the simulated node.

Another method consists in assembling together compatible orthogonal 2D data events to obtain 3D data events. Two orthogonal data events are considered compatible if all the nodes that they have in common have the same value (see

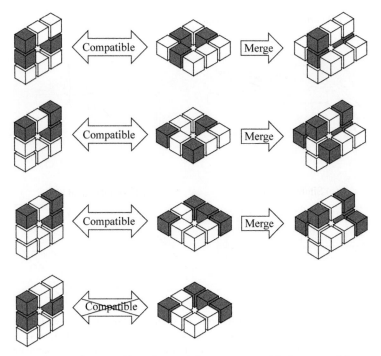

Figure II.7.9 Merging of compatible orthogonal data events. Modified/Reprinted from Comunian, A., P. Renard, et al. (2012). With permission from Elsevier.

Figure II.7.9). Based on these reconstructed data events, a pattern database can be constructed and used for simulation (Comunian et al., 2012).

Comunian et al. (2012) compared both methods of probability aggregation and merging of orthogonal data events. They found that the latter method generally gave better results, based on tests carried out on porous media samples that are exhaustively known in 3D.

Kessler et al. (2013) proposed another method specifically tailored for the case where the available cross-sections are in the same coordinate system. It obtains 3D simulations by sequentially generating 2D simulations perpendicular to each other. Each new section is conditioned to the locations of the intersections with previously simulated sections, as well as conditioning data if those are present. The order and location of the simulated cross-sections are chosen such that parallel sections are initially as distant as possible from each other, and at the same time they occur at locations where conditioning data are available. The methodology is illustrated in Figure II.7.10.

7.7 Training data

It is common practice in geological applications to use object-based or process-based models to obtain TIs. Although such a practice is justified, it does not allow

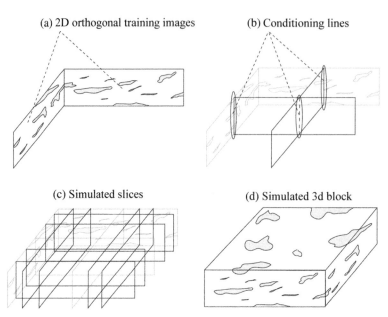

(a) 2D orthogonal training images

(b) Conditioning lines

(c) Simulated slices

(d) Simulated 3d block

Figure II.7.10 Method of perpendicular sections. (a) Orthogonal training images. (b) The first sections simulated are those that intersect at the location of conditioning data (wells). (c) The remaining sections are generated sequentially until (d) the entire 3D volume is simulated (modified from Kessler et al., 2013).

using the full potential of MPS as a way to mimic processes observed in real-world analogs. MPS simulation can then be seen as a way of conditioning existing object-based or process-based models. However, if a complete analog case is available, it can provide firsthand information on the type of spatial patterns that are actually present.

In some applications, such as Earth surface characterization using remote sensing data, very large amounts of spatial information are available. This can also be the case in geological applications, such as mining where the density of drill holes is very high or when outcrops allow for an exhaustive mapping of the geological structures. In such cases, no TI is needed, and available data can be directly used to populate frequencies in a pattern database or to be used for convolutions.

Although the idea to use training data instead of TIs is appealing, some technical difficulties have to be addressed. The main one is that training data are generally not located along regular patterns. However, templates are made of nodes located at fixed positions relatively to a central node. The data, however, can have any geometry and can change for each simulated node. Therefore, using template-based methods will inevitably result in some patterns being incomplete. Because incomplete patterns cannot be used to inform frequencies in a pattern database, only the subset of the training data that can fit in the template can be

(a) Partial image   (b) Reconstruction

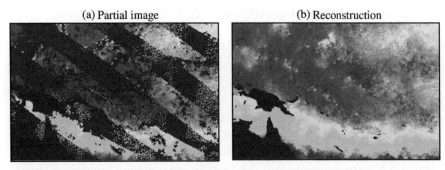

Figure II.7.11 Reconstruction of a partial image based on the principle of training data. (a) Infrared satellite data of the Pacific Ocean presenting gaps (source: National Oceanic and Atmospheric Administration (NOAA)). (b) One reconstruction simulation based on (a); here, the data are used as both training data and conditioning data. Dark blue represents gaps, and darker blue represents continents.

considered. Other issues are related to the use of multiple grids, where the relocation of data points to coarse multiple grids can be challenging with very dense data sets. Despite these shortcomings, successful use of training data is possible with MPS simulation techniques based on the storage of patterns (see, e.g., Wu et al., 2008).

On the other hand, simulation methods based on convolutions do not store patterns and are not bound to the use of templates. They are therefore particularly suited to applications where training data are used instead of TIs. Figure II.7.11a shows an example of remote sensing infrared-based temperature measurements acquired over the Pacific Ocean. The measurements present gaps due to the orbital characteristics of the satellite, and also because of cloud occlusions. As a result, the gaps have complex shapes, including large contiguous informed areas as well as isolated informed pixels. These training data are used to perform one simulation using the direct sampling method (Figure II.7.11b). Here, the training data are used to infer the spatial structure (i.e., the informed nodes are scanned to search for matching data events), but the same data are also used for conditioning. The result is that rather than a simple simulation exercise, a reconstruction is performed. Some data are present in all regions of the image, and this is enough to inform the global nonstationary trend in the values (warmer values toward the top of the image, corresponding to the Equator).

This approach is entirely data driven. An advantage is that the spatial information contained in the data is based on the true field, which is a more direct source of information than a TI derived from interpretation. The danger of using training data is that, on many occasions, the data can present errors or biases, which are then reproduced in the simulation. Training data should therefore be used with care when the data are subject to significant uncertainty.

7.8 Construction of multivariate training images

Obtaining multivariate TIs can arguably be one of the most difficult and challenging cases discussed in this chapter. There are situations where the multivariate training data are readily available. For example, a multivariate variant of Figure II.7.11 is an ideal case for multivariate MPS applications.

The most general way of creating multivariate TIs is to use a physical model. When process-based models can be used, those often produce additional variables that can be used to create multivariate TIs. If we consider again the example of Figure II.7.5, the equations that describe the generation of lithological structures also consider the grain size of the deposited sediments. Moreover, because these equations are developed in time, it is straightforward to map the age of the sediments. Figure II.7.12 represents these three variables, which present correct spatial relationships based on the physical laws used in the process-based models. Moreover, additional variables could be derived from the existing ones. For example, based on grain size and using sediment age as a proxy for compaction, one could infer sediment porosity. Then, based on empirical laws such as the Hagen–Poiseuille law, one could in turn derive the permeability of the sediments.

Another possibility is to start with a univariate TI and derive additional variables based on it using a physics-based forward model. Figure II.7.13 illustrates a multivariate TI composed of two variables: the first one (Figure II.7.13a) is a permeability field that can be obtained using any of the TI generation methods described here. The other variables (Figure II.7.13b–f) are obtained by simulating flow and transport in a domain having a spatial distribution of hydraulic properties determined by the first variable, and recording the contaminant distribution at different time stamps. Multivariate relationships then allow describing the interactions between patterns of connected high values and the fingering that occurs in the patterns of the contaminant concentration. A challenge is that the TI is then highly nonstationary, but this can be handled by using additional variables or zonation, as described in Chapter II.5. Such multivariate TIs describing flow processes have been used in the context of inverse problems where the interaction between both variables is highly nonlinear and spatially dependant (Zhou et al., 2012). This is further discussed in Section II.6.4.

7.9 Training image databases

Boolean models, whether stationary or nonstationary, are still the prime tool for creating TIs. TI libraries have been developed, such as those based on Boolean algorithms or pseudo-genetic or process-mimicking techniques (Zhang et al., 2011). The libraries consist of several TIs generated by varying the input parameters of the Boolean codes (Pyrcz et al., 2008; Colombera et al., 2012), and they

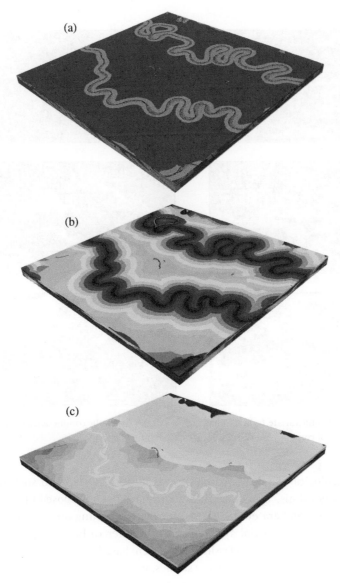

Figure II.7.12 (a) Lithological facies. (b) Sediment grain size. (c) Sediment age since the time of deposition.

have generally been developed for alluvial sedimentary environments. However, such libraries are still more geostatistical than geological in nature. They do not address the questions of what parameter values should be chosen, where that information originates from, and how this selection process proceeds. Geologists reason by means of analogs and through the understanding of processes, not through input parameters of a Boolean code. A more in-depth proposal

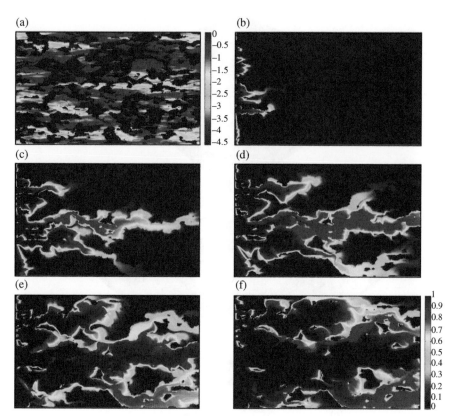

Figure II.7.13 A multivariate training image made of (a) a hydraulic conductivity field; and (b–f) snapshots of contaminant distribution at different time stamps.

is to link geological databases based on classification systems to such library-generating algorithms. Two database systems have been proposed in the recent past: FAKTS (Colombera et al., 2012) and CARBDB (Jung and Aigner, 2012). The former addresses fluvial systems, and the latter carbonate systems. Both relational databases employ a classification system based on geological reasoning. The FAKTS database classifies existing studies, whether literature derived or field derived from modern or ancient river systems, according to controlling factors (e.g., climate) and context-descriptive characteristics (e.g., river patterns). The database can therefore be queried on both architectural features and boundary conditions to provide the analogs for the studies' reservoirs. Modeling parameters can then be extracted and input to Boolean modeling codes for TI generation.

CARBDB is a relational database specific to carbonate rocks where controlling factors may be more complex than for sedimentary systems. Yet such rocks contain most of the world's oil reserves. CARBDB introduces a classification system as well as a terminology that allow better linking with quantitative reservoir modeling. The classification system directly identifies the dimensions, shape, and internal structures of carbonate bodies. The following factors are used in the classification of both modern and ancient systems: depo-time (geological age), depo-system (platform), depo-zone (hosting zone), depo-shape (geometry of bodies), depo-element (building blocks of bodies), and depo-facies (lithofacies). Most important to reservoir modeling with TIs are the notions of bodies and elements. Interpretation of the larger-scale elements (time, platform, and zone) can help in narrowing the potential variation and dimension of subseismic bodies. This system was tested in modeling a reservoir analog (Muschelkalk, Germany), where actual wells have been drilled and the outcropping of rock may help validate such a modeling approach.

References

Allard, D., Froidevaux, R. & Biver, P. 2006. Conditional simulation of multi-type non stationary markov object models respecting specified proportions. *Mathematical Geology*, 38, 959–986.

Bertoncello, A., Sun, T., Li, H., Mariethoz, G. & Caers, J. 2013. Conditioning surface-based geological models to well and thickness data. *Mathematical Geosciences*, 45, 873–893.

Caers, J. A 2006. General algorithm for building 3D spatial laws from lower dimensional structural information. Paper presented at the 19th Stanford Center for Reservoir Forecasting Meeting, Stanford University, Stanford, CA, May 8–9.

Colombera, L., Felletti, F., Mountney, N. P. & McCaffrey, W. D. 2012. A database approach for constraining stochastic simulations of the sedimentary heterogeneity of fluvial reservoirs. *AAPG Bulletin*, 96, 2143–2166.

Comunian, A., Jha, S. K., Giambastiani, B. M. S., Mariethoz, G. & Kelly, B. F. J. 2013. Training images from process-imitating methods – an application to the lower namoi aquifer, Murray-Darling Basin, Australia. *Mathematical Geosciences*, 1–20.

Comunian, A., Renard, P. & Straubhaar, J. 2012. 3D multiple-point statistics simulation using 2D training images. *Computers and Geosciences*, 40, 49–65.

Coulthard, T. J. & Van de Wiel, M. J. 2006. A cellular model of river meandering. *Earth Surface Processes and Landforms*, 31, 123–132.

Deutsch, C. V. & Tran, T. T. 2002. FLUVSIM: A program for object-based stochastic modeling of fluvial depositional systems. *Computers and Geosciences*, 28, 525–535.

Emery, J. & Lantuejoul, C. 2014. Can a training image be a substitute for a random field model? *Mathematical Geosciences*, 46, 133–147.

Gross, L. J. & Small, M. J. 1998. River and floodplain process simulation for subsurface characterization. *Water Resources Research*, 34, 2365–2376.

Haldorsen, H. H. & Chang, D. M. 1986. Notes on stochastic shales from outcrop to simulation models. *In:* Lake, L. W. & Carrol, H. B. (eds.), *Reservoir Characterization*, Academic Press, New York.

Hill, E. J. & Griffiths, C. M. 2009. Describing and generating facies models for reservoir characterisation: 2D map view. *Marine and Petroleum Geology*, 26, 1554–1563.

Ikeda, S., Parker, G. & Sawai, K. 1981. Bend theory of river meanders. – 1. Linear development. *Journal of Fluid Mechanics*, 112, 363–377.

Jung, A. & Aigner, T. 2012. Carbonate geobodies: Hierarchical classification and database – a new workflow for 3D reservoir modelling. *Journal of Petroleum Geology*, 35(1), 49–65.

Kessler, T. C., Comunian, A., Oriani, F., Renard, P., Nilsson, B., Klint, K. E. & Bjerg, P. L. 2013. Modeling fine-scale geological heterogeneity – examples of sand lenses in tills. *Groundwater*, 51, 692–705.

Koltermann, C. E. & Gorelick, S. M. 1992. Paleoclimatic signature in terrestrial flood deposits. *Science*, 256, 1775–1782.

Lancaster, S. T. & Bras, R. L. 2002. A simple model of river meandering and its comparison to natural channels. *Hydrological Processes*, 16, 1–26.

Lantuejoul, C. 2002. *Geostatistical Simulation: Models and Algorithms*, Springer, Berlin.

Lopez, S. 2003. *Modélisation de Réservoirs Chenalisés Méandriformes, Approche Génétique et Stochastique*, Paris School of Mines, Paris.

Lopez, S., Cojan, I., Rivoirard, J. & Galli, A. 2008. Process-based stochastic modelling: Meandering channelized reservoirs. *Special Publications of the International Association of Sedimentologists*, 144, 139–144.

Maharaja, A. 2008. TiGenerator: Object-based training image generator. *Computers and Geosciences*, 34, 1753–1761.

Mchargue, T., Pyrcz, M. J., Sullivan, M. D., Clark, J. D., Fildani, A., Romans, B. W., Covault, J. A., Levy, M., Posamentier, H. W. & Drinkwater, N. J. 2011. Architecture of turbidite channel systems on the continental slope: Patterns and predictions. *Marine and Petroleum Geology*, 28, 728–743.

Michael, H., Boucher, A., Sun, T., Caers, J. & Gorelick, S. 2010. Combining geologic-process models and geostatistics for conditional simulation of 3-D subsurface heterogeneity. *Water Resources Research*, 46.

Pyrcz, M. J., Boisvert, J. B. & Deutsch, C. V. 2008. A library of training images for fluvial and deepwater reservoirs and associated code. *Computers and Geosciences*, 34, 542–560.

Pyrcz, M., Boisvert, J. & Deutsch, C. 2009. ALLUVSIM: A program for event-based stochastic modeling of fluvial depositional systems. *Computers & Geosciences*, 35, 1671–1685.

Seminara, G. 2006. Meanders. *Journal of Fluid Mechanics*, 554, 271–297.

Sun, T., Meakin, P. & Jøssang, T. 2001. A computer model for meandering rivers with multiple bed load sediment sizes 2. Computer simulations. *Water Resources Research*, 37, 2243–2258.

Wu, J., Boucher, A. & Zhang, T. 2008. A SGeMS code for pattern simulation of continuous and categorical variables: FILTERSIM. *Computers & Geosciences*, 34, 1863–1876.

Zhang, T., Li, T., Reeder, S., Yue, G. & Thachaparabil, M. 2011. An integrated multi-point statistics modeling workflow driven by quantification of comprehensive analogue database. *In*: *73rd EAGE Conference & Exhibition*, May 23, 1054–1058.

Zhang, Z. & Switzer, P. 2007. Stochastic space-time regional rainfall modeling adapted to historical rain gauge data. *Water Resources Research*, 43.

Zhou, H., Gómez-Hernández, J. J. & Li, L. 2012. A pattern-search-based inverse method. *Water Resources Research*, 48.

CHAPTER 8

Validation and quality control

8.1 Introduction

In this chapter, we discuss the issue of model-to-data validation and quality control of the generated MPS realizations. Although these are two different topics, they share some common methodologies. In traditional covariance-based geostatistics, model parameters are often directly inferred from data; hence, the risk of specifying models inconsistent with such data is virtually nonexistent. This is no longer the case when spatial continuity is borrowed from training images. Such training images may still be inferred from data, but such inference is not necessarily rigorous statistically; it is likely to be more broadly based on expert interpretation of the data. As a consequence, the training image may contain details or aspects that are inconsistent with the data (e.g., hard, soft, and auxiliary), and such inconsistency may not always be obvious either because of the complexity of the data or training image or because of the presence of dense (in particular, hard) data.

Secondly, in practice, the geostatistical realizations generated often need to undergo a posterior quality control. In the traditional framework, reproduction of marginal distribution and semivariograms are checked against the inputs of the simulation algorithm. Such statistics need not be reproduced exactly, while fluctuations can be expected due to the limited size of the model domain (see the discussion on ergodicity in Part I). Most algorithms have various tuning parameters that may have intractable effects. For example, sequential Gaussian simulation relies on a search neighborhood; Markov random field models require Markov chain Monte Carlo sampling, which relies on some tuning in the convergence of the generated chains. Virtually all methods are impacted by the choices of tuning parameters, whether they are derived from theory or are more algorithmically inspired.

The questions and methods we address in this chapter relate to the a priori (data-to-model validation) and a posteriori (quality control) checks needed when dealing with training images as the source of spatial continuity. The terms "prior" and "posterior" here refer to what is done before and after executing the stochastic simulation algorithm (hitting the "run" button). The involvement of multiple-point statistics as well as training images (a nonparametric representation) makes

Multiple-point Geostatistics: Stochastic Modeling with Training Images, First Edition. Gregoire Mariethoz and Jef Caers.
© 2015 John Wiley & Sons, Ltd. Published 2015 by John Wiley & Sons, Ltd.
Companion website: www.wiley.com/go/caers/multiplepointgeostatistics

such validation a bit more involved than simply calculating first- and second-order statistics, in particular:

> *Data-to-model validation*: because of the interpretative nature of selecting a set of representative training images, the relationship between data and training image is defined, at least in part, based on subjective criteria. This poses few problems if data are few or rather uninformative, but it may cause inconsistencies if large amounts of data bring reliable information that may be in contradiction with a given training image. The training image case is perhaps reflective of Boolean modeling, where certain parameters are lifted from data and others are lifted from analog information. This may result in conflicts and nonconvergence when sampling from such Boolean models (Caers, 2005). Nevertheless, there is a need to apply tools for aiding the expert validating the selection of training images against the actual field data.

> *Quality control*: in Gaussian theory, the a posteriori distribution is well known and defined. Several methodologies are available to check whether outputs from a particular stochastic simulation method are sampled correctly, based on probability theory (e.g., Mardia, 1970; Deutsch and Journel, 1992). Among those tests, the simplest would involve checking the reproduction of univariate statistics (histogram) and semivariograms. With MPS, different tools are needed to validate models against higher-order criteria. This requires developing tools for checking (1) whether conditioning is properly achieved, and (2) how statistics of the training images compare with statistics of the realizations.

In this chapter, we present several methods that attempt to address these two issues. Development of such methods is an ongoing area of research because most of the focus in the last decade has been on developing algorithms and methods for generating training images. We also do not develop any formal hypothesis testing, which would be the more statistical approach to this problem. To get to any such hypothesis test, many additional subjective decisions are required (e.g., the choice of templates, the nature of the search radius, and assumptions of independence) for which no formal objective test can be designed (Journel, 1999). In this context, we consider these methods more as informal checks rather than rigorous hypothesis tests.

8.2 Training image – data validation

8.2.1 Validation against point data
8.2.1.1 Wells

Data for spatial modeling may be numerous in type and origin depending on the application. In the case studies (Part III of this book), we provide more details on how such validation works for these particular areas of interest. Here, we roughly classify data into two major groups: point data and soft data. The validation issues related to dynamic data (typically, time-varying) requiring the explicit forward modeling of such data are treated in the chapter on inverse modeling (Chapter II.9; and see also Chapter III.1).

In certain applications, such as for subsurface problems, the lack of data as well as the availability of specific analog information and expertise are often important reasons for using training images. Semivariograms are often meaningful only in the vertical direction, where sampling is most dense along boreholes. Any consistency check could therefore focus mostly on vertical variations: should wells be drilled vertically, or along the well-bore path, in the case of the now more commonly deviated well paths?

A statistical method for analyzing the distribution of categorical variables along a path is the distribution of runs (Mood, 1940; Boisvert et al., 2007). The concept is to count the number of consecutive occurrences of a particular category along a given direction, in this case the vertical direction. A counting distribution is then obtained starting from 1 (an isolated category) to some specified maximum. Because runs are calculated along wells, an assumption underlying this calculation is that the wells are placed in such a way that they are representative of all areas of the domain (i.e., the sampling is not biased). The run analysis results, for each category, in a counting distribution for the well data and another counting distribution for the training image. These frequency distributions can then be statistically compared using any of the methods described in Section II.2.5.5.

An alternative is to use a multiple-point histogram (MPH; see Section II.2.4.4) where the histogram is constructed using a 1D vertical template. Now, the counting captures the number of times each configuration of a binary indicator variable occurs within the given 1D template. This can be repeated for each binary indicator variable representing a category.

These two techniques work well for categorical variables and relatively small templates. In cases of larger templates or continuous variables, the same method can be extended using a cluster-based histogram of patterns (CHP, see Section II.2.4.5).

8.2.1.2   Dense point data

If point data become denser and more regularly gridded, then the explicit calculation of higher-order statistics becomes possible. For such dense data, the calculation of higher-order moments or cumulants is feasible (see Section II.2.4.6). These data cumulants can then be compared with cumulants extracted from the training image.

An alternative is to use a brute force approach that isolates patterns of points in the data and determines the frequency of these patterns in the training image. If the frequency of the data patterns is high in the training image, then this can be interpreted as a high degree of consistency between data and training image. It is easy to determine experimental frequencies in the training image by exhaustive scanning. The computational cost of such a scan grows with the number of data patterns.

An interesting aspect of this approach is that patterns of different sizes can be used for comparison. If patterns consist of a single node, then a comparison of univariate statistics (histogram) between the data and training image is made. Patterns of two nodes would compare bivariate statistics and so on. In practice, the comparison is carried out for all orders, and one can determine for example if the high-order statistics are incompatible even though the univariate proportions of data and training image are matching (Perez et al., 2014).

8.2.2 Dealing with nonstationarity

The early view (Caers and Zhang, 2004) was that training images need to be stationary and that only stationary training images are "statistically valid" (Mirowski et al., 2009). This makes sense when considering methods and algorithms that rely on such stationary assumptions. The SNESIM algorithm (Section II.3.2.2) is one such example: it relies on scanning the entire image in order to derive frequency information of higher order data events. Therefore, it makes sense to first determine whether such training images can be used in stationary modeling. One such quantitative method, based on a directional and scaling stationarity over the domain, was developed in Mirowski et al. (2009).

Most real field applications require nonstationary modeling approaches, which have been presented in Chapter II.5. Point data may exhibit clear trends, for example wells in the vertical. On the other hand, training images may contain statistics that are stationary. These stationary statistics are often transformed into nonstationary models through various methodologies such as the use of regions, orientation, and affinity transformations. It would therefore make little sense to compare stationary training image statistics with statistics coming from nonstationary data or realizations.

When dealing with nonstationarity in either data or training images, we need to distinguish several cases:

- The data exhibit trends, but the training image is stationary;
- The data exhibit a trend, and the training image is nonstationary with the same trend; and
- The data exhibit a trend, but the training image is nonstationary with a different trend.

The "trend" concept can be quite broad and should not just be interpreted in terms of proportions, marginal distribution, orientation, scaling, or some external drift; it could be specified on any feature of the training image patterns or data. For example, one type of pattern may become more frequent than another, but with both types of patterns containing the same values. In such a case, the patterns would gradually differ, but the marginal distribution would remain identical. Given the variety of possible types of nonstationarity, the question of validation should therefore be divided into a question of (1) validation of the trends and (2) validation of the stationary statistics.

To address this more general case of validating both trend and training image, one cannot avoid generating a few realizations. Indeed, generating a number of unconditional realizations, which reflect the nonstationary modeling component as well as the training image patterns, but not any local point or other conditioning data, provides a sample set from the population of unconditional models: explicit samples of the nonstationary prior. The conditioning data can then be forward modeled on each of these unconditional realizations. In the case of point data, this simply entails reading off the values in the unconditional realizations at the point data locations. This creates sets of point data representing a limited sample of the entire population of possible point data sets. Consistency between data and trend–training image would entail that the actual point data from the field is part of this population represented by its limited set.

Consider as illustration the example in Figure II.8.1, where three alternative binary training images and two alternative trend models for the vertical proportion of the blue category are provided. The data comprise a single vertical well. The question is now what combination of trend and training image is consistent with the values along the single well. This question can be reformulated by looking for combinations such that the following conditional statement holds:

$$P(Tr, TI|\text{well data}) \neq 0 \tag{II.8.1}$$

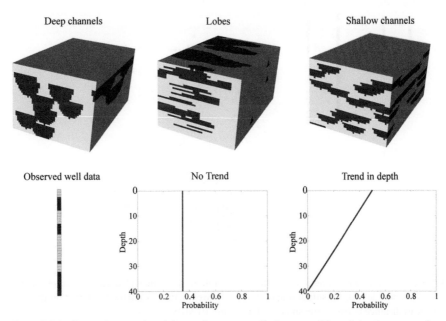

Figure II.8.1 Illustration case involving as data: one well, three possible training images, and two possible trend models.

where the variable Tr is a binary indicator for the choice of trend model; and the TI is a categorical variable modeling the choice between the three training images. Given that in this case the trend models represent large-scale variability, whereas the training image provides local variations, the following hierarchical decomposition is assumed for Equation (II.8.1):

$$P(Tr, TI|\text{well data}) = P(Tr|\text{well data})P(TI|Tr, \text{well data}) \qquad \text{(II.8.2)}$$

A total of 180 unconditional realizations are generated (30 realizations for each combination of trend and training image). By reading off values at the vertical well location, a set of 180 simulated (synthetic) well data (or a set of 181, if counting the actual well data) is generated. Consider first modeling $P(Tr|\text{well data})$. We first define a measure of similarity between any two well data in the set of 181 that is reflective of the trend difference. A simple smoothing is applied to each well datum, and a Euclidean distance calculated; see Figure II.8.2(a–d). Multidimensional scaling is used to visualize these differences; see Figure II.8.2(e). The actual field data are also plotted based on the distance between the trend in the field data and the simulated (synthetic) data. It is clear, based on the cloud of points, that the well data are only consistent with the model without trend. A similar procedure can be followed to estimate $P(TI|Tr, \text{well data})$; however, now a different measure of similarity needs to be used reflecting the local variation in the binary sand–shale sequence. To establish this measure, we use the MPH and

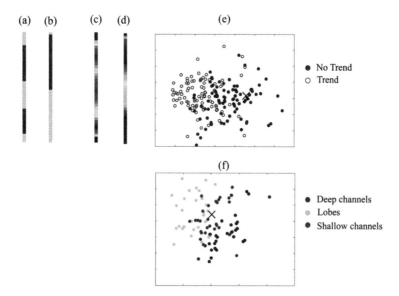

Figure II.8.2 (a) Single synthetic well datum; (b) another synthetic well datum; (c) smoothing of the first well datum; (d) smoothing of the second well datum; (e) MDS plot based on the Euclidean distance between the smoothed data; and (f) MDS plot based on the Jensen–Shannon distance of the MPHs for each well datum. Black crosses are the field well data.

calculate differences in MPH for the 181 well data using a Jensen–Shannon measure of divergence. An MDS plot corresponding to the Jensen–Shannon distance is shown in Figure II.8.2(f). Kernel smoothing (see Chapter III.1 for details on this procedure) can be used to estimate $P(TI|Tr, \text{well data})$ based on this distance. For this case, we find:

$$P(\text{Deep Channel}|\text{No trend, well data}) = 0$$

$$P(\text{Lobes}|\text{No trend, well data}) = 0.15 \qquad \text{(II.8.3)}$$

$$P(\text{Shallow Channel}|\text{No trend, well data}) = 0.85$$

The procedure outlined here is quite general because it focuses on the end product of the modeling exercise, namely, the realizations generated. Still, the method remains computationally involved because it requires generating realizations explicitly and would be applicable if this can be done fast. The procedure also illustrates that, in making comparisons between data and model, one cannot escape the definition of statistical similarity. This definition will be different depending on the nature of the data but also based on the intended application of the modeling exercise. For example, if in the above example one is interested is only figuring out trend, then the training image can be seen as a confounder, where by chance, well data reflecting a definite trend can be generated from a model without any trend.

8.3   Posterior quality control

8.3.1   Using summary statistics

After MPS realizations have been generated, quality control is needed. In general, a particular statistical property of a realization is considered to be a "summary statistic." This property is analyzed for a number of realizations and compared with the same property of the training image. The choice of the statistical properties of interest is often application-dependent and a choice of the modeler.

Comparisons of histograms (global proportions for categorical variables) and semivariograms (indicator semivariograms for categorical variables) are common ways of validating realizations against a training image. In many applications, it is expected that realizations represent the marginal distribution of the training image (global proportions in the categorical case, or histograms in the case of continuous variables). Although some deviation may be tolerated, too large of a deviation will alter the nature of the patterns of the training image. In continuous-variable cases, the extreme values of the realizations cannot be larger than what is found in the training image because of the resampling nature of MPS algorithms.

For binary images, it is often useful to transform the image into its geobodies equivalent, also known as the set of connected components. A geobody is a collection of adjacent grid nodes of the same category. Figure II.8.3 shows how a map of labeled connected components was obtained from remote sensing images of rainfall accumulation on the entire planet Earth. Based on such transformation into geobodies, several additional statistical properties can be established that allow characterizing complex features (Renard and Allard, 2013). Some of

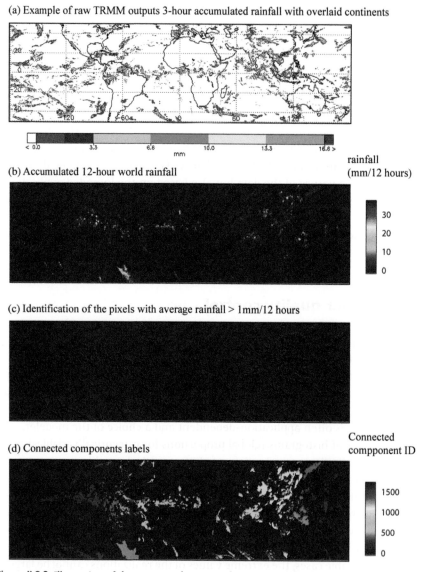

(a) Example of raw TRMM outputs 3-hour accumulated rainfall with overlaid continents

(b) Accumulated 12-hour world rainfall

rainfall (mm/12 hours)

(c) Identification of the pixels with average rainfall > 1mm/12 hours

Connected compponent ID

(d) Connected components labels

Figure II.8.3 Illustration of the concept of connected components based on the Tropical Rainfall Measuring Mission (TRMM) data, shown in (a). The continuous variable representing accumulated rainfall in (b) is converted to the binary variable (c) on which a connected component analysis is applied (d); 1793 connected components are found for the category corresponding to rainfall >1 mm/12 hours.

these properties are based on a statistical measure computed on the geobodies, such as the distribution of their size, which is similar to the granulometry curves used in sedimentology to characterize the distribution of the size of particles in a given sediment sample. Other summary statistics that can be extracted from the connected-components map are statistics on the aspect ratio of the connected components, their surface-to-area ratio, or the shape of their convex hull (AghaKouchak et al., 2011; Oriani and Renard, 2014).

One type of summary statistics often used in the context of MPS is the connectivity function (Pardo-Igúzquiza and Dowd, 2003). Based on the connected-component map, the probability for two locations A and B separated by a distance **h** to be connected is determined. This means a continuous path of nodes of the same category exists between them. A characteristic of the connectivity function is that it considers what occurs between A and B, whereas indicator semivariograms only take into account the category value at A and B, regardless of how these two locations are connected. The existence of a connection between two locations is straightforward to establish once the connected components map is available: two nodes are connected if they belong to the same geobody, otherwise they are disconnected. There exists one connectivity function for each category considered.

Connectivity functions are shown in Figure II.8.4, where the connectivity functions corresponding to the Tropical Rainfall Measuring Mission (TRMM) rainfall example (Figure II.8.3d) are displayed. Separate connectivity functions are displayed for each direction, and also for each category. The general shape of

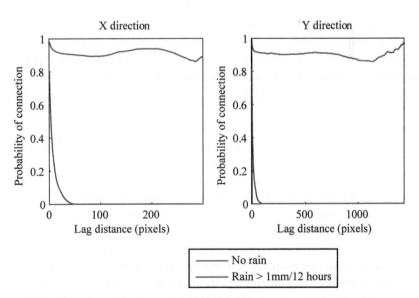

Figure II.8.4 Illustration of the connectivity function based on the connected component analysis of Figure II.8.3.

a connectivity function is a high probability of connection for short lags, which decreases with increasing lag. However, this may not always be the case, in particular when the shape of the connected components is complex, leading sometimes to an increase in probability. In this example, the connectivity functions are very different for the rainy and dry categories, and the connectivity of the rain cells also varies whether one considers the X or Y direction. In general, areas without rain are much more connected than rain cells. Intermittence in rainfall is more common than intermittence in dry spells (i.e., on average, rain spells are short-lived, whereas dry spells tend to have a longer duration).

Connectivity functions are often believed to provide a relatively complete characterization of categorical variables, and therefore they are often preferred to semivariograms when comparing realizations obtained with training images. They, however, cannot be used to compare realizations or training images with point sets, because computing connected component maps requires a fully informed image. The usual procedure is therefore to compute the connectivity function of the training image as well as the connectivity function for each realization, and, if those functions are similar, to consider this as a good indication that realizations contain similar connectivity patterns as the training image.

The concept of connectivity functions can be extended to continuous variables by using a varying threshold and determining how the connectivity properties vary with the different thresholds. This results in characteristic curves that can be used to compare complex variables in the same way as connectivity functions. These characteristic curves are described in detail in Renard and Allard (2013).

8.3.2 Avoiding verbatim copy

We define "verbatim copy" as instances where portions of the training images are found identically in the simulated realization. This typically occurs when the spatial constraints provided by the training image are too strong, resulting, for a given neighborhood, in a probability of 1 for a given value and a zero probability for all other values.

This may occur when the training image is small but complex, resulting in all counters in a pattern database being equal to 1. Then, with a given conditioning data event, only one pattern (for pattern-based methods) or one value (for pixel-based methods) is available for selection, rendering stochastic simulation deterministic. As a result, the multiple-point simulation ends up reproducing the training image to some extent, or even, in extreme cases, the simulation is an exact copy of the training image.

The occurrence of verbatim copy is not desired because often the goal of geostatistics is the modeling of spatial uncertainty. This problem is most prominent with training images that contain certain idiosyncrasies that are not repetitive enough and therefore appear unchanged in the simulations. Such idiosyncrasies are unfortunately common, and they are present in most training images that contain some amount of complexity. A good example is the "islands" in the classical channels training image of Strebelle (2002). In many realizations, these islands are reproduced to be almost identical to the islands in the training image (see Figure II.8.5a). Verbatim copy mostly occurs for unconditional simulations, because the conditioning data generally cannot be reproduced by using an exact copy of the training image.

One way of quantifying the occurrence and amount of verbatim copy is to use coherence maps (Lefebvre and Hoppe, 2005). Coherence maps are obtained by indexing the location of each node in the training image, and then by keeping track of this index during the simulation. As a result, the original location of each simulated node is known. Locating these indices in the simulation domain then constitutes the coherence map. If the index of a simulated node is only one increment away from the index of its neighbor, then this is an indication of verbatim copy.

Some amount of verbatim copying is inevitable and may even be desired: some patterns of the training image have to be directly borrowed, and the idiosyncrasies need to be copied to some extent because they are a constitutive part of the unique complexity being modeled. Therefore, an important validation step is for the modeler to display coherence maps and to make a call on whether the amount of verbatim copy is acceptable, knowing that verbatim copy is inversely proportional to the variability between realizations.

Ideally, the coherence map should consist of pure noise, meaning that no verbatim copy occurs. The realizations are then the product of a statistical sampling process rather than a sophisticated patching of the training image. In most cases, however, some verbatim copy is difficult to avoid. This observation has been used to argue that a training image–based spatial model (except for MRF) cannot be a substitute for a random function model as traditionally defined (as discussed in Emery and Lantuéjoul, 2014).

Figure II.8.5 illustrates different cases of verbatim copy. In Figure II.8.5a, the variable is categorical, which generally limits the occurrence of verbatim copy because a given data event can occur at several different locations in the training image, resulting in a spatial mixing of the training image patterns. The example of Figure II.8.5b is continuous. It is obtained by using direct sampling with a distance threshold of 0 and a neighborhood consisting of the 20 closest nodes. As a consequence of such a low threshold value, some verbatim is inevitable. Because the variable is continuous, there is a high likelihood for only one single location to exist in the entire training image corresponding to the data event searched for (note that, in this case, verbatim copy is artificially increased for demonstration purposes – it is, in general, not recommended to use a threshold of 0 with continuous variables). As a result, a number of compact areas are visible on the coherence map, corresponding to patches copied "as is" from the training image. In Figure II.8.5c, the same setting is used but now with a neighborhood consisting of the 60 closest nodes. This results in increased

Training image   Realization   Coherence map

(a) Categorical case

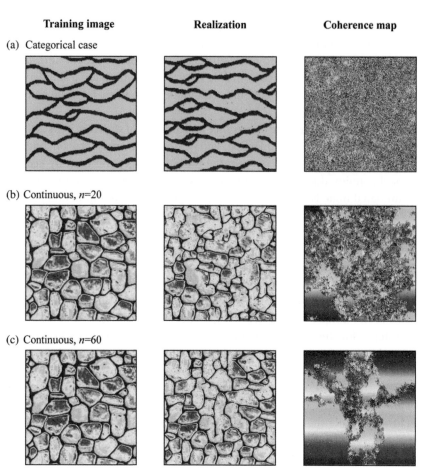

(b) Continuous, n=20

(c) Continuous, n=60

Figure II.8.5 Illustration of coherence maps for three test cases. All cases are unconditional DS realizations, with fraction of training image, $f = 1$, threshold $t = 0$. (a) A categorical variable with $n = 40$ neighbor nodes. (b) A continuous variable with $n = 20$ neighbor nodes. (c) The same continuous variable with $n = 60$ neighbor nodes. The color codes in the coherence maps represent the IDs of the nodes in the training image.

verbatim copy, with large patches copied from the training images and new patterns only occurring at the junctions between these patches.

Verbatim copy occurs more frequently with methods using raw storage of the training patterns (see Section II.2.4.1). When patterns are stored as probabilities of occurrence (e.g., in a tree or list), due to the finite size of the pattern database, counters with more than a single occurrence exist, limiting the occurrence of verbatim copy.

In general, avoiding verbatim copy can be achieved by parameterizing the simulation algorithms such that there is a possibility to create new patterns. With tree-based or list-based methods, setting a large enough minimum number of replicates (such as 10) generally suffices to avoid verbatim copy. In the case of direct sampling, the use of a threshold of zero, a fraction of scanned training image of 1, or very large neighborhoods should be avoided. Another solution is to use invariant distances, which produce a large number of patterns from a single idiosyncrasy. With patch-based algorithms such as cross-correlation simulation (CCSIM), the size of the patches should be kept relatively small, and the tolerance on the overlap error needs to allow for more than a single candidate patch to be selected. Some patch-based methods, however, allow breaking the verbatim copy of the training image by allowing for new patterns to be created, such as by means of image quilting.

8.3.3 Checking trend reproduction

Next to quality control on the pattern reproduction, in cases of nonstationary modeling, trend reproduction needs to be verified (Boisvert et al., 2010).

In evaluating the trend reproduction, one has to consider first the nature of the trend variable. Often, one has to distinguish between a constraint that applies to all realizations and a constraint that applies to each realization (Boisvert et al., 2010). For example, an aerial proportion map is a constraint that specifies, *in each realization,* the proportion of a category when the realizations are averaged over the vertical. The probability given in each grid cell of a probability cube is a constraint that acts *over many realizations.* In the absence of any other constraints, this probability indicates the frequency, over many realizations, of the occurrence of a category at that grid cell location. Often, proportions and probabilities are confused, partially because they both have values in the range [0,1].

8.3.4 Comparing spatial uncertainty

Most of the above methods focus on one or a set of summary statistics for comparing realizations with the training image. With such an approach, one has to be specific about the particular statistics used. In this section, we discuss the method of Tan et al. (2014), which does not require any particular input from the user, but rather relies on properties native to most MPS models. The general observation is that any MPS simulation algorithm creates two types of variability:

- A within-realization variability: the variation of patterns within a single realization and how they compare with the variation of patterns in the training image.
- A between-realization variability: the variation that exists between realizations. In geostatistics, this has been termed the spatial uncertainty (Chilès and Delfiner, 1999).

It can be argued that it is desirable that the within-realization variability is small (i.e., a good comparison with the training image) but that at the same time the between-realization variability is large (i.e., one doesn't just create copies of the training image) (see the discussion on verbatim copy in this chapter). Methods for exploring or comparing the spaces of uncertainty generated with MPS algorithms are therefore useful. They can be used for studying new algorithms, assessing the effects of input parameters on such spatial uncertainty, and determining the effects of conditioning (see the next section). Consider the example in Figure II.2.20, with a set of 50 realizations generated with two MPS algorithms (CCSIM and DISPAT) and a variogram-based SISIM method (Deutsch & Journel, 1992). The question here is on how to quantitatively compare the realizations generated by these algorithms.

A quite general framework for addressing such questions is the analysis of distance (ANODI) methodology. This method uses a distance as the basis for comparison. Two types of distance are proposed: a distance between the realizations and the training image (within variability), and a distance between realizations (between variability). The distance can be of any nature; for example, it could be the distance between various summary statistics calculated from realizations and training image. A more generally applicable approach is to rely on the realizations themselves and recognize that they generally contain various scales of variability. Multiscale pyramids can be used to represent the multiscale variation in any single realization (or training image). An example of such a pyramid is shown in Figure II.8.6.

For each resolution in the pyramid, one can then calculate the MPH or CHP, whichever is deemed more appropriate. This means that training image **ti** and realizations $\mathbf{re}^{(\ell)}$, where ℓ identifies one specific realization among an ensemble,

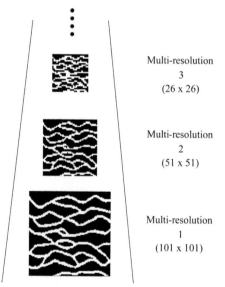

Multi-resolution
3
(26 x 26)

Multi-resolution
2
(51 x 51)

Multi-resolution
1
(101 x 101)

Pyramid of one single image

Figure II.8.6 Illustration of multiscale pyramids of a single model.

can be summarized using a pyramid of MPH or CHP. For example, in the case of the CHP:

$$\{\mathbf{ti}_1, \mathbf{ti}_2, \dots, \mathbf{ti}_G\} \xrightarrow{\text{summarize}} \{\mathbf{CHP}(\mathbf{ti}_1), \mathbf{CHP}(\mathbf{ti}_2), \dots, \mathbf{CHP}(\mathbf{ti}_G)\}$$

$$\left\{\mathbf{re}_1^{(\ell)}, \mathbf{re}_2^{(\ell)}, \dots, \mathbf{re}_G^{(\ell)}\right\} \xrightarrow{\text{summarize}} \left\{\mathbf{CHP}\left(\mathbf{re}_1^{(\ell)}\right), \mathbf{CHP}\left(\mathbf{re}_2^{(\ell)}\right), \dots,\right.$$

$$\left. \mathbf{CHP}\left(\mathbf{re}_G^{(\ell)}\right)\right\} \quad \ell = 1, \dots, L, \quad \text{(II.8.4)}$$

where CHP is a vector containing the counting distribution for each resolution, for example for the training image:

$$\mathbf{CHP}(\mathbf{ti}_g) = \{CHP_1(\mathbf{ti}_g), \dots, CHP_{c_g}(\mathbf{ti}_g), \dots, CHP_{C_g}(\mathbf{ti}_g)\} \quad \text{(II.8.5)}$$

where C_g is the number of clusters retained for that resolution g. Two distances can now be calculated: a distance between any realization and the training image, and a distance between any two given realizations. Any of the distances mentioned in Section II.2.5.5 can be used:

$$d\left(\mathbf{re}_g^{(\ell)}, \mathbf{re}_g^{(\ell')}\right), \quad \ell, \ell' = 1, \dots, L \quad g = 1, \dots, G$$

$$d\left(\mathbf{re}_g^{(\ell)}, \mathbf{ti}_g\right), \quad \ell = 1, \dots, L \quad g = 1, \dots, G \quad \text{(II.8.6)}$$

Summarizing the distance for one subresolution g, we obtain:

$$d_g^{between} = \frac{1}{L(L-1)} \sum_{\ell=1}^{L} \sum_{\ell'=1}^{L} d\left(\mathbf{re}_g^{(\ell)}, \mathbf{re}_g^{(\ell')}\right)$$

$$d_g^{within} = \frac{1}{L} \sum_{\ell=1}^{L} d\left(\mathbf{re}_g^{(\ell)}, \mathbf{ti}_g\right) \quad \text{(II.8.7)}$$

To obtain a single distance, we need to sum these subresolution distances over all resolutions. Not all resolutions should contribute equally to this total distance because the lower resolutions contain less information than the higher resolutions in the pyramid. A proposal is therefore to weight lower resolution less as follows:

$$d^{between} = \sum_{g=1}^{G} \frac{1}{2^g} d_g^{between}$$

$$d^{within} = \sum_{g=1}^{G} \frac{1}{2^g} d_g^{within}. \quad \text{(II.8.8)}$$

The ratio:

$$r = \frac{d^{between}}{d^{within}} \quad \text{(II.8.9)}$$

can then be regarded as a single quantitative assessment or "score" attributed to the set of realizations $\mathbf{re}^{(\ell)}, \ell = 1, \dots, L$. This score can be used to compare, for example, two sets of realizations, each generated with a different algorithm or generated with the same algorithm but with different parameters.

8.3.5 Consistency checks for conditioning

In this section, we discuss a general framework for assessing how MPS algorithms perform in terms of conditioning to hard-point data, which was more theoretically discussed in Section I.5.2.4. Point conditioning entails more than merely honoring the point data at their locations; otherwise, one could simply replace whatever simulated value at the data location with the available point data. Instead, it is expected that some form of regional effect is induced when generating multiple realizations constrained to the same point data. In semivariogram-based methods, in particular within Gaussian theory, this effect is exactly known. For example, in the stationary case, the ensemble average of a large number of Gaussian realizations is equal to the simple kriging estimate, and the ensemble variance equal to the kriging variance. In other words, the posterior mean and posterior variance or covariance are known, and realizations generated from any conditional Gaussian simulation algorithm can be verified against these conditions. This is no longer the case for MPS realizations because there is no analytical or expression for such posterior statistics.

The potential pitfall is that many conditioning tricks can be implemented without any verification whether biases are introduced near the data. A similar problem occurs in Boolean modeling: conditioning can always be artificially achieved, but doing so may lead to artifacts near the data and artificial reduction of uncertainty. It is therefore useful to test any algorithm for such biases. To set up such consistency checks, consider first the same algorithm without conditioning data. If a large set of unconditional realizations is generated, then such a large set could be seen as representing an unspecified unconditional distribution $f_{algo}(\mathbf{re})$. Consider now extracting, from each realization, a set of point conditioning data with the same configuration of locations over the domain. Because each unconditional realization is different, each point data $\mathbf{pd}$ set will be different, and each such set can be seen as a sample of an unspecified distribution $f_{algo}(\mathbf{pd})$. Consider now generating, with the same algorithm, a set of conditional simulations with each point data set. For a given point data set $\mathbf{pd}$, the set of conditional simulations now represent a conditional distribution $f_{algo}(\mathbf{re}|\mathbf{pd})$. It is an immediate consequence of Bayes' rule that the total set of all conditional realizations generated by varying the point data set obtained from unconditional simulations is:

$$\int_{pd} f_{algo}(\mathbf{re}|\mathbf{pd}) f_{algo}(\mathbf{pd}) d\mathbf{pd} = \int_{pd} f_{algo}(\mathbf{pd}|\mathbf{re}) f_{algo}(\mathbf{re}) d\mathbf{pd}$$

$$= f_{algo}(\mathbf{re}) \int_{pd} f_{algo}(\mathbf{pd}|\mathbf{re}) d\mathbf{pd} = f_{algo}(\mathbf{re})$$

$$(II.8.10)$$

In other words, the set of conditional realizations generated this way should be equivalent to the set of unconditional realizations. This makes sense because we randomize over the point data values, where the randomization is obtained from the unconditional models. It is Bayes' rule that suggests some form of consistency should exist between conditional and unconditional models.

The above procedure is useful in designing and testing conditioning. However, it requires generating many unconditional and conditional realizations. A faster way is to tailor the above "global" ANODI method into a "local" version. The ANODI procedure is global in nature in the sense that patterns are extracted over the entire modeling domain. To verify point conditioning, ANODI has to be performed considering only patterns near the conditioning data. Consider, as an illustration of the global and local variety of ANODI, the case presented in Figure II.8.7. Two algorithms are compared: CCSIM, which is known for reproducing well training image patterns, but as a patch-based technique has difficulty with conditioning to point data, and SNESIM, which is known to condition well to point data but has difficulty reproducing long-range pattern connectivity. The pattern reproduction aspect of the comparison is clearly visible when comparing realizations (Figure II.8.7c and II.8.7d) and in the global ANODI analysis. Figure II.8.8 shows the MDS plot generated with the distances (Equation (II.8.6)), where the Jensen–Shannon divergence between CHPs is used as a measure of distance. The local ANODI consists of repeating this exercise, but now a box of 50×50 is placed around each conditioning well, and distances are only based on what occurs within these two boxes. The MDS plot of Figure II.8.8a shows that SNESIM performs better

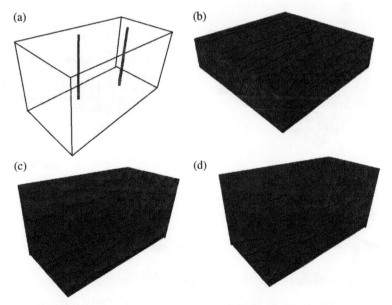

Figure II.8.7 Case with a two wells (a) and simple binary 3D stationary training image (b). Realization of both CCSIM (c) and SNESIM (d) from a set of 50 realizations with each method.

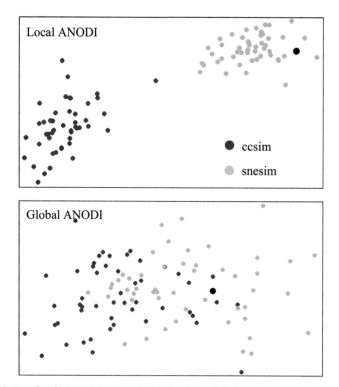

Figure II.8.8 Top: local ANODI. Bottom: MDS plot from global ANODI.

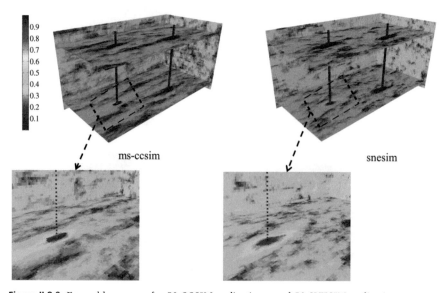

Figure II.8.9 Ensemble average for 50 CCSIM realizations; and 50 SNESIM realizations.

than CCSIM in this regard (patterns are closer to the training image patterns, and hence they are not distorted by any artifacts in conditioning). Figure II.8.9 shows the ensemble averages of each set where one notices an asymmetry near the conditioning well for the CCSIM case, and hence a distortion of patterns. This asymmetry is typical for methods that use raster paths instead of random paths (e.g., SNESIM).

References

Aghakouchak, A., Nasrollahi, N., Li, J., Imam, B. & Sorooshian, S. 2011. Geometrical characterization of precipitation patterns. *Journal of Hydrometeorology*, 12, 274–285.

Boisvert, J. B., Pyrcz, M. J. & Deutsch, C. V. 2007. Multiple-point statistics for training image selection. *Natural Resources Research*, 16, 313–321.

Boisvert, J. B., Pyrcz, M. J. & Deutsch, C. V. 2010. Multiple point metrics to assess categorical variable models. *Natural Resources Research*, 19, 165–175.

Caers, J. 2005. *Petroleum Geostatistics*, Society of Petroleum Engineers, Richardson, TX.

Caers, J. & Zhang, T. 2004. Multiple-point geostatistics: A quantitative vehicle for integrating geologic analogs into multiple reservoir models. *In:* Grammer, G. M., Harris, P. M., & Eberli, G. P. (Eds.), *Integration of Outcrop and Modern Analog Data in Reservoir Models, AAPG Memoir 80*, American Association of Petroleum Geologists, Tulsa, OK.

Chilès, J.-P. & Delfiner, P. 1999. *Geostatistics – Modeling Spatial Uncertainty*, John Wiley & Sons, Inc., New York.

Deutsch, C. & Journel, A. 1992. *GSLIB: Geostatistical Software Library*, Oxford University Press, New York.

Emery, J. & Lantuéjoul, C. 2014. Can a training image be a substitute for a random field model? *Mathematical Geosciences*, 46, 133–147.

Journel, A. G. 1999. The abuse of principles in model building and the quest for objectivity. *In: Geostatistics Wollongong 96 – Proceedings of the Fifth International Geostatistics Congress, Wollongong, Australia, September 1996*, 3–14.

Lefebvre, S. & Hoppe, H. 2005. Parallel controllable texture synthesis *ACM Transactions on Graphics*, 24, 777–786.

Mardia, K. V. 1970. Measures of multivariate skewness and kurtosis with applications. *Biometrika*, 57, 519–530.

Mirowski, P. W., Tetzlaff, D. M., Davies, R. C., McCormick, D. S., Williams, N. & Signer, C. 2009. Stationarity scores on training images for multipoint geostatistics. *Mathematical Geosciences*, 41, 447–474.

Mood, A. 1940. The distribution theory of runs. *The Annals of Mathematical Statistics*, 11, 367–392.

Oriani, F. & Renard, P. 2014. Binary upscaling on complex heterogeneities: The role of geometry and connectivity. *Advances in Water Resources*, 64, 47–61.

Pardo-Igúzquiza, E. & Dowd, P. 2003. CONNEC3D: A computer program for connectivity analysis of 3D random set models. *Computers & Geosciences*, 29, 775–785.

Perez, C., Mariethoz, G. & Ortiz, J. 2014. A training image selection program for multiple-point geostatistics. *Computers & Geosciences*, 70, 190–205.

Renard, P. & Allard, D. 2013. Connectivity metrics for subsurface flow and transport. *Advances in Water Resources*, 51, 168–196.

Strebelle, S. 2002. Conditional simulation of complex geological structures using multiple-point statistics. *Mathematical Geology*, 34, 1–22.

Tan, X., Tahmasebi, P. & Caers, J. 2014. Comparing training-image based algorithms using an analysis of distance. *Mathematical Geosciences*, 46, 149–169.

CHAPTER 9

Inverse modeling with training images

9.1 Introduction

Thus far in this volume, the conditioning to data in the presented MPS methods has been direct and noniterative. In this chapter, we present methods where such direct conditioning is not feasible and involves formulating and solving inverse problems. Our focus will be on the subsurface for the simple fact that most methods have been developed within that context. However, most of the presented methodologies are quite general and could be applied to other inverse problems.

In the subsurface context, two major data sources often require inverse modeling: dynamic data related to subsurface flow and geophysical data. Flow and pressure data change in time and are modeled using flow equations and simulators. Geophysical data can have both a time and spatial component. Geophysical forward models cover a wide spectrum ranging from simple convolution models to solutions of wave equations. Unlike dynamic data measured at a few well locations, geophysical data often cover the target volume exhaustively. They can be repeated in time, leading to 4D measurements. In some cases, these types of dynamic "time data" can be turned into static "space data"; for example, a well test can be translated into an averaged permeability in a circular region around the well. Seismic data can be calibrated into a soft probability. The translation makes conditioning to such data more straightforward; however, the conditioning in such cases remains only partial. For example, if one forward simulates the well test on such conditional models, a mismatch may still be present between the time-varying drawdown at the well and the simulated response from the model. In other words, the statistical conditioning in geostatistical models does not necessarily lead to physical plausibility or accuracy.

In this chapter, we focus on conditioning to data by means of physical models through the solution of inverse problems. Virtually any problem involving data and a model can be formulated as an inverse problem. Our aim is not to be exhaustive in the treatment of these problems, but to focus on those types of model formulations where a training image has been provided that quantifies additional information about the nature of the spatial model.

Multiple-point Geostatistics: Stochastic Modeling with Training Images, First Edition. Gregoire Mariethoz and Jef Caers.
Companion website: www.wiley.com/go/caers/multiplepointgeostatistics

9.2 Inverse modeling: theory and practice

9.2.1 Formulation

Tarantola (Tarantola, 1987, 2005) as well as his prior papers (Tarantola and Valette, 1982) provide perhaps one of the most general formulations of the inverse problem. The formulation is non-Bayesian, relying on more general concepts of conjunction of information sources. Consider the following nomenclature, which will be used throughout this chapter:

- Model variables **m**: a list of variables describing the model; also termed "model parameters." In the spatial, geostatistical context, this often involves the grid, with one or more properties (rock type, porosity, saturation, clay content, etc). Model variables can be static (porosity) or dynamic (saturation). We will, however, assume some form of discretization in space and time (the discrete inverse problem). In the context of inverse problems, model variables are taken as those variables that have no direct observation.
- Data variables **d**: those variables of the model description that are observed: porosity measured at a well, a pressure, or saturation over time. The measurements are then termed $\mathbf{d}_{obs}$.

The data variables are not treated differently from model variables; the only difference is that data are observed, whereas model variables are not. Tarantola and Valette (1982) formulate the general discrete inverse problem through a conjunction of states of information. Three such states of information are considered:

Information from theory: a physical theory is postulated on the joint distribution of (**m**,**d**). Such theory can be in the form of simple laws or partial differential equations, which can then be solved with simulators. The theory adds information on the nature of the relationship between **d** and **m**.

Information prior to theory: prior to formulating any theories, one may already have some information on these variables (**m**,**d**). The measurements are part of this prior, hence the approach taken here is non-Bayesian. In a Bayesian context, the measurements enter the likelihood, and the prior is formulated prior to obtaining such measurements. Bayes follows a reasoning based on information and knowledge streams, whereas Tarantola reasons from physics and how physical experiments work. Tarantolas' prior could include anything as long as it is obtained independently of a theory formulated on the joint (**m**,**d**).

Information on the representation of the variables: the fundamental definition of variables contains information about them, in particular on the coordinate system in which they are defined. Accounting for this information ensures that the inversion results are invariant to changes in the coordinate system (e.g., using polar coordinates versus Cartesian coordinates); see Tarantola (1987).

A general way to express each of these three sources of information is by means of a joint density. Information from theory is represented by $f_{theory}(\mathbf{m},\mathbf{d})$, and information prior to considering such theory as $f_{pre-theory}(\mathbf{m},\mathbf{d})$. Information on the nature of the variables is provided by a metric density $f_{metric}(\mathbf{m},\mathbf{d})$. The term "metric" reflects the nature of the space within which variables are defined. This density depends on the metric tensor of that space, hence the name "metric density." A mass m would preferably be represented by a Cartesian space (Euclidean), but a coordinate (φ, θ) in terms of latitude and longitude would not. Tarantola uses conjunctions of these states of information to derive the following general expression of the combined state of information:

$$f(\mathbf{m},\mathbf{d}) = K\frac{f_{theory}(\mathbf{m},\mathbf{d})f_{pre-theory}(\mathbf{m},\mathbf{d})}{f_{metric}(\mathbf{m},\mathbf{d})} \qquad \text{(II.9.1)}$$

Note that this formulation is symmetric in $\mathbf{m}$ and $\mathbf{d}$. This general formulation can now be simplified by making additional assumptions and more classical formulations such as Bayesian inverse problems deduced. The most common assumption made is:

$$f_{theory}(\mathbf{m},\mathbf{d}) = f_{theory}(\mathbf{d}|\mathbf{m})f_{metric}(\mathbf{m}) \qquad \text{(II.9.2)}$$

which now introduces an asymmetry in the formulation by first considering $\mathbf{m}$, then $\mathbf{d}$. Making such an assumption requires, however, that the relationship between $\mathbf{d}$ and $\mathbf{m}$ is only mildly nonlinear: $\mathbf{d}$ has to become "predictable" given $\mathbf{m}$ (this is not the case for chaotic problems). This assumption works reasonably well for the kind of subsurface physics (flow and geophysics) we are interested in. These types of assumption then lead to the following "a posteriori" density on the model variables:

$$f_{posterior}(\mathbf{m}) = K f_{pre-theory}(\mathbf{m}) \int_d \frac{f_{pre-theory}(\mathbf{d})f_{theory}(\mathbf{d}|\mathbf{m})}{f_{metric}(\mathbf{d})}d\mathbf{d} = K f_{pre-theory}(\mathbf{m})L(\mathbf{m})$$

$$\text{(II.9.3)}$$

where $L(\mathbf{m})$ represents a general definition of a likelihood function. There are differences here with Bayesian formulations from a theory point of view, and we refer to discussions in Tarantola (2005) as well as Mosegaard and Tarantola (1995). Apart from the inclusion of the metric density, these differences, for the type of problems discussed in this book, are, however, minor, so we will use the more traditional Bayesian formulation (Jaynes and Bretthorst, 2003) as a product of prior and likelihood as well. We will generally ignore the use of the metric density, and hence assume it to be uniform. Two major challenges now remain: (1) how to specify the various distributions on the right-hand side, in particular the prior; and (2) how to sample from the posterior.

9.2.2 A problem of prior

Most of the problems we consider use a nonlinear forward model:

$$\mathbf{d} = g(\mathbf{m}) \qquad\qquad (\text{II}.9.4)$$

where $\mathbf{m}$ is the spatial model entering the simulator, and $\mathbf{d}$ is the responses cal-culated (output) from the simulator. The likelihood in Equation (II.9.3) now depends on errors in the data and also errors in the model. The latter could reflect the use of proxy, surrogate, or upscaled models or reflect the limitations of any physical model in describing reality, but we will not treat that topic in this book. As a consequence, we will consider that the likelihood is a probabilistic measure of uncertainty in the data only. More specifically, we will use a simple statistical model to describe noise in the measurements using an exponential model:

$$L(\mathbf{m}) \simeq \exp(-O(\mathbf{m})) \qquad\qquad (\text{II}.9.5)$$

where the norm O describes some desired mismatch between the measurements $\mathbf{d}_{obs}$ and the simulated data responses $\mathbf{d}$. In that context, a misfit function is intro-duced as:

$$O(\mathbf{m}) = d(g(\mathbf{m}), \mathbf{d}_{obs}) \qquad\qquad (\text{II}.9.6)$$

where d is some user-specified distance between the forward simulated response and the observed data (e.g., a squared difference possibly weighted by a noise variance or covariance).

The main difficulty often lies in the specification of the prior, and this will be the focus in this chapter. A common assumption is to use a multivariate Gaus-sian distribution (a Gaussian prior), hence, only the mean and the covariance or variogram need to be specified. The Gaussian prior has been used frequently in the past, and various arguments are offered. In the Gaussian case, $f_{prior}(\mathbf{m}) > 0$; the strict positive means that any solution $\mathbf{m}$ is possible; the appeal here is that, regardless of the nature of the data, a solution will be reached as long as enough computation time is allowed. Second, a parametric problem is formulated where only two parameters need to be stated on the spatial model $\mathbf{m}$, namely, mean and covariance (which itself may require multiple parameters such as range, anisotropy, and nugget), making the specification of the density more manageable.

The influence of the prior on the posterior is dependent on the nature of the likelihood function. Should the data be very constraining, meaning that the likelihood is narrow, then the posterior will be narrow as well. In such cases, the prior choice should not matter much. If one reconsiders the Walker Lake data set of Part I, in particular Figure I.4.2 (i.e., two cases of hard-data information available: one case with 5% of the image informed by hard data, and a second case with only 0.1%), in the former case, the posterior solution is determined by the data; in the latter case, the properties of the multivariate Gaussian distribution become much more prevalent; and, in the case of a multi-Gaussian prior, they show up as destructured in the tail values.

However, the data considered in this chapter are not hard sample data; they are sparse and indirect (i.e., involving simulators of physics). Hence, most of the cases considered here fall in that latter category: the assumed multivariate distribution of the prior has a strong influence on the spatial characteristics of the posterior solutions. Most practical spatial inverse problems are of this nature, with the archetypical example in subsurface modeling being flow inversion. Flow in the subsurface is a function of a number of physical properties, more specifically: (1) the geological medium in which these fluids flow, (2) the fluid properties, and (3) the boundary conditions. The more uncertain one of these is often the geological medium. For example, any particle injected in a well traveling to another well follows an unknown path that is influenced by the three mentioned properties. When such particle arrives at the well, it only carries the information of arrival time; there is no direct record of the path or what happened along this path. Evidently, an enormous variety of solutions can be provided to such inverse problems, meaning that the assumed prior will have considerable impact on the spatial character of the posterior solution, as well as the uncertainty represented by such posterior.

9.2.3   Formulating a training image–based prior

In this chapter, we will formulate priors based on training images rather than on the multi-Gaussian distribution or any other parametric representation. We will focus on MPS algorithms to generate realizations using the specified training image(s). Hence, we do not assume any analytical or parametric expression of such prior, only a generating algorithm. The prior is thus explicitly expressed by the chosen algorithm(s) and training image(s):

$$f_{prior}(\mathbf{m}) = f_{\mathrm{algo},TI}(\mathbf{m}) \qquad\qquad (\text{II.9.7})$$

Unlike a Gaussian prior, we can expect now that for a large set of $\mathbf{m}$, $f_{\mathrm{algo},TI}(\mathbf{m}) = 0$ (i.e., that a large set of solutions is excluded a priori, i.e., those solutions that contain patterns not present in the training image(s)). This marked difference with parametric random field models (e.g., Gaussian and Markov random fields) has an upside and a downside when it comes to inverse modeling. The upside is that any expert or analog information can be explicitly enforced in the prior, and hence in the posterior. The downside is that now there exists a larger risk that the stated prior is inconsistent with the data. This will lead to nonconvergence of any sampling of the posterior should such an inconsistent prior enter into play (Equation (II.9.3)), a matter that will be discussed in the next section.

In practical field studies, it is rare that experts can state with certainty the single training image required to develop the prior. Instead, it is more likely, and also more effective, to work with several training images reflecting several sources of uncertainty. Based on past experience, and rules of deposition evaluated on a large amount of similar systems, one can speculate on the depositional environment leading to statements about the possible nature of such environments (e.g. alluvial, fluvial, deltaic). None of these various alternatives should be eliminated a priori. Within each system, one could have variations that are due to subclassifications within that system (see Jung and Aigner, 2012), or one can have variations in certain parametric descriptions of the system (e.g., proportions, orientation, or dimension of bodies). Databases (see Section II.7.9) can be consulted to amass information about the spatial distribution of rock-type bodies. From a modeling perspective, one may therefore have discrete variation (the style and the classification) and continuous variations within that style often described with parameters. Boolean methods could then be used to generate training images based on a selected style and on a set of sample parameters. Then, realizations can be drawn from that training image. In this way, a form of hierarchy is developed:

$$(\text{Depositional style} \rightarrow \text{parameters}) \rightarrow \text{realizations}$$

The first two are grouped because they are used to generate the training images from which realizations are created. Traditional statistical modeling would now proceed with formulating this as a hierarchical problem, for example using a two- or three-level hierarchical Bayes' problem (e.g., Banerjee et al., 2004; Gelman et al., 2004). The training image choice itself can then be seen as a nonnumerical categorical variable that could be part of the model parameters. One may have some prior beliefs stated on these training images. The models generated from a given training image then require another parameterization. Any parameterization of such models will need to reduce the dimensionality of the model considerably (see Section II.9.5). Statistical approaches dealing with these nonlinear, non-Gaussian priors, such as hierarchical Bayesian methods, rely on sampling by means of Markov chain Monte Carlo. This becomes rapidly prohibitive when the forward model takes more than a few minutes. In this chapter, we will therefore develop several alternatives that approximate this sampling, yet allow for solutions to large cases.

A priori screening through a set of many training images (created as a combination of style and parameters) is needed to evaluate which are consistent with the data and which are not. If training images are a prior in a real sense, meaning they have been created independently of any subsurface-specific data or any data to be inverted, then this inconsistency question needs to be addressed prior to considering any inverse sampling or solution method.

9.2.4 Prior-data inconsistency

Possibly the most important issue encountered in the actual practice of solving the type of inverse problem treated here is the prior-data inconsistency that

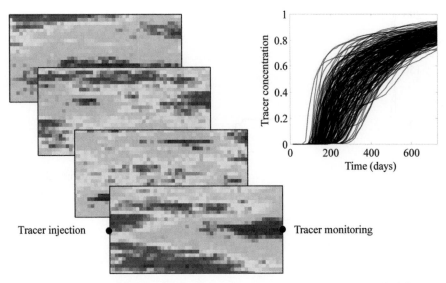

Figure II.9.1 A Gaussian prior and a simple tracer problem. A tracer is injected at the left location, and its arrival monitored at the right location. Various Gaussian models are sampled from which the tracer response is calculated.

occurs. We first revisit this issue with Gaussian priors, then discuss strategies for training image–based priors. In theory, the Gaussian prior is never inconsistent with any data, simply because it allows for all solutions. A Gaussian prior may, however, be practically inconsistent with data. Consider an example such as in Figure II.9.1, where a tracer experiment is presented. A Gaussian model with mean and covariance is specified, and tracer simulations are performed on each of its realizations; see Figure II.9.1. Now we pick one of those tracer simulations as the data to be inverted. We formulate the likelihood as a simple mismatch between the observed and simulated tracer concentration at some location on the right; see Figure II.9.1. The gradual deformation (GDM) of Gaussian realizations (see Section II.9.4.3) is used to construct models by iterative optimization. GDM iteratively perturbs Gaussian model realizations to achieve a match, within a specified error, to the data. The perturbations are gradual, which allows for an optimization method to be used in achieving a result, and GDM preserves the Gaussian prior model assumption (reproduces the mean and the covariance of the realizations). One notes that in the case of Figure II.9.2, iterations converge in a reasonable (finite) time; see Figure II.9.2. Consider now changing the data gradually to later breakthrough times. As shown in the iteration plot, at some point, convergence within a reasonable amount of time is no longer achieved. Even by allowing for uncertainty in mean and covariance, there is a practical range for which the Gaussian prior is consistent and a range where it is not consistent with the tracer concentration data.

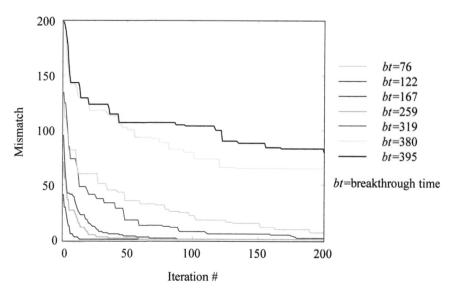

Figure II.9.2 Studying the convergence of the gradual deformation when changing the concentration curves to earlier and later breakthrough times.

One may be tempted to state the mean and covariance of a Gaussian prior based on the data itself or by calibration from it. For example, by trial and error, a few covariance models can be tried, and a few models can be evaluated through the forward model to see if at all any inverse solution can be achieved. However, such practice violates the theory formulated in Equation (II.9.3), where the following assumption was made:

$$f_{prior}(\mathbf{m}, \mathbf{d}) = f_{prior}(\mathbf{m}) \cdot f_{prior}(\mathbf{d}) \qquad (II.9.8)$$

meaning that the specification of prior on the model variables should be done independently from the way it is stated on the data variables. In other words, data should not be used to state the prior on model variables; otherwise, the product in this equation cannot be used; instead, the dependency would have to be explicitly modeled. The Bayesian formulations also rely on the same assumption; hence, any rigorous Bayesian approaches will require the prior on $\mathbf{m}$ to be obtained independently of the data that is being inverted. In practice, this rarely happens: data are used to interpret and state the prior, directly or indirectly. In many cases, training images are partly constructed from interpretations made on the data. Assuming independence when in reality it does not exist leads to reduced uncertainty (an overconfidence in knowledge). Therefore, "calibrating" priors with data may lead to a considerable reduction in uncertainty in the posterior distribution and, therefore, samples drawn from it.

One challenge in inverse modeling is to resist the temptation often observed in practice of building priors from data, whether directly or indirectly. The second challenge is to produce priors that are "informed". Meaningful priors in this context are charged with geological information and can be very specific and informative. If this path is taken, then the challenge is not a narrow prior, but a very wide prior, should one start from a possibly large set of training images. As a consequence, many of the training images in this prior will be inconsistent with the given data and one should investigate means to exclude them. Considering only the uncertain discrete choice of the training image, the full posterior is a function of both model parameters and training image choice $(\mathbf{m}, TI)$ which can be decomposed as follows:

$$f\left(\mathbf{m}, TI | \mathbf{d}_{obs}\right) = \sum_{k=1}^{K} f(\mathbf{m} | TI = ti_k, \mathbf{d}_{obs}) \, P\left(TI = ti_k | \mathbf{d}_{obs}\right) \qquad \text{(II.9.9)}$$

considering K training images have been proposed by the geological modeler. In other words, within the context of training images and MPS realizations, the modeling of the posterior distribution is decomposed into two parts: (1) modeling the posterior uncertainty on the training image, and (2) modeling the posterior uncertainty of $\mathbf{m}$ for each given training image.

In the context of the decomposition (Equation (II.9.9)), the question of inconsistency can be formulated as:

$$\text{find } ti_k : P\left(TI = ti_k | \mathbf{d}_{obs}\right) = 0 \qquad \text{(II.9.10)}$$

Several strategies can be employed to assess Equation (II.9.10):
- *Full sampling* (Section II.9.9.3) would invert for the joint $(\mathbf{m}, TI)$. The difficulty lies in the CPU cost and in the design of convergent McMC samplers.
- *Search methods* (Section II.9.9.4) can be employed to rapidly cover the space of joint $(\mathbf{m}, TI)$ and discover those TIs that can be excluded. Such methods are, however, only approximate to the full samplers, and hence the approximation made will need to be investigated.
- *Direct estimation* of $P\left(TI = ti_k | \mathbf{d}_{obs}\right)$ (see Chapter III.1) based on a few scoping runs. Any such method will, however, need to rely on a dimensionality reduction in the $\mathbf{d}$-variables to make such estimation feasible.

In the full sampling methods, one may be attempting to achieve too much at the same time: screening training images, and inverting for training image and model parameters in one single Markov chain sampling. If direct estimation of $P\left(TI = ti_k | \mathbf{d}_{obs}\right)$ is feasible based on a few scoping runs, then this approach is desirable. The training images with zero $P\left(TI = ti_k | \mathbf{d}_{obs}\right)$ can then be excluded and the nonzero probabilities carried forward in the further sampling of realizations.

9.3 Sampling-based methods

9.3.1 Rejection sampling

Consider a multivariate density $f(\mathbf{m})$ that is known up to its normalization. Consider the supremum of this density as S. Rejection sampling (e.g., Mardia, 1970) then proceeds as follows:

1 Draw a uniform sample $\mathbf{m}$.

2 Generate a random number p between 0 and 1.

$$if \ p \leq \frac{f(\mathbf{m})}{S} \ then \ accept \ \mathbf{m} \ as \ sample \ of \ f \ (\mathbf{m}) \qquad (\text{II}.9.10)$$

This procedure can only work for relatively small dimensions (five or less). However, we are not interested in a uniform $\mathbf{m}$, because such $\mathbf{m}$ is inconsistent with any training image. Rejection sampling therefore can proceed with high-dimensional $\mathbf{m}$ in the context of Equation (II.9.3), namely,

1 Draw a sample $\mathbf{m}$ from $f_{prior}(\mathbf{m})$.

2 Generate a random number p between 0 and 1.

$$if \ p \leq \frac{L(\mathbf{m})}{S_L} \ then \ accept \ \mathbf{m} \ as \ sample \ of \ f_{posterior}(\mathbf{m}) \qquad (\text{II}.9.10)$$

where S_L is now the supremum of the likelihood function, which is generally unknown and must therefore be set to 1, in which case rejection sampling on even low-dimensional problems quickly becomes unfeasible. The rejection sampler would only work if the prior model space is small and covers the posterior solution region, a considerable requirement. If the data do not fall in this region, meaning if models forward simulated with such prior have a large mismatch with data, then the rejection sampler cannot be applied to such cases (see the case in Figure II.9.1). The rejection sampler can be applied to wide priors derived from multiple training images. Chapter III.1 contains a real field case study where rejection sampling is actually achieved, albeit at the cost of considerable CPU time. Therefore, for most cases, it is (still) too inefficient for use in most practice and will therefore be used only as the benchmark sampler to evaluate other sampling methods.

9.3.2 Spatial resampling

Sampling from posterior distributions such as Equation (II.9.3) cannot be done directly, because the normalization constant cannot be computed. Markov chain Monte Carlo methods, such as Metropolis sampling (Metropolis et al., 1953) and Gibbs sampling (Geman and Geman, 1984), need to be used to sample from such distributions. Regardless of what Markov chain sampling method is used, all such techniques are iterative. A chain of geostatistical realizations is generated where the next realization is dependent on the previous one. The nature of the dependency between two successive realizations in the chain is therefore a controlling factor in McMC. A very simple way to inject such dependency in spatial models is to retain some part of the realization as conditioning data to generate the next

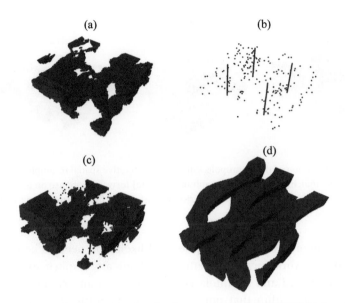

Figure II.9.3 An example of a new MPS realization generated with SNESIM (c), based on an existing one (a), using a spatial resampling of data points as well as constrained to existing conditioning data at four well locations (b). (d) Training image. The realization grid is of dimension 100×100×40, and the training image of dimension 150×200×80.

realization in the chain. This can be done in various ways; for example, one can retain a set of discrete locations (iterative spatial resampling (ISR); Mariethoz et al., 2010); see Figure II.9.3. Or one can return a compact block of locations (Hansen et al., 2012); see Figure II.9.6 later in this chapter. Other proposals are to retain only the edge of a block (Fu and Gomez-Hernandez, 2009). To generate the next realization in the chain, any MPS method then uses these hard conditioning data. The size of blocks or amount of resampled points retained allows controlling the amount of perturbation generated.

9.3.3 Metropolis sampling

Metropolis samplers (Metropolis et al., 1953) require evaluation of both the likelihood and prior density functions. In addition, the Metropolis requires a proposal mechanism able to sample the model space at random (a form of symmetry is imposed). Any of the above proposed spatial resampling methods can be used to generate a proposal $\mathbf{m}^{new}$ from some current model $\mathbf{m}^{old}$. The following algorithm then attempts to sample from the posterior:

1 Given a current model $\mathbf{m}^{old}$, generate a new proposal model $\mathbf{m}^{new}$.

2 Accept this proposal $\mathbf{m}^{old}$ with probability:

$$p = \min \left\{ 1, \frac{L(\mathbf{m}^{new})}{L(\mathbf{m}^{old})} \frac{f_{prior}(\mathbf{m}^{new})}{f_{prior}(\mathbf{m}^{old})} \right\} \qquad (\text{II.9.11})$$

3 If accepted, then $\mathbf{m}^{old} \leftarrow \mathbf{m}^{new}$.

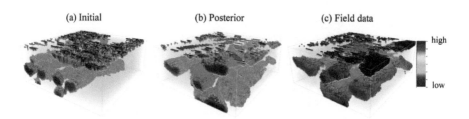

Figure II.9.4 Application of the Metropolis sampler with iterative spatial resampling to generate posterior sample matching the seismic amplitude data. (a) forward response on an initial realizations, (b) forward response on a posterior realization, (c) field seismic amplitude data.

In this traditional version, both prior and likelihood need to be evaluated. This would not work for MPS methods without any explicit expression of the prior. Mosegaard and Tarantola (1995) developed an extended version of the Metropolis algorithm that only requires an algorithm to sample a prior model:

1 Given a current model $\mathbf{m}^{old}$, generate a new sample from the prior, $\mathbf{m}^{new}$.
2 Accept this proposal $\mathbf{m}^{old}$ with probability:

$$p = \min\left\{1, \frac{L(\mathbf{m}^{new})}{L(\mathbf{m}^{old})}\right\} \qquad (\text{II}.9.12)$$

3 If accepted, then $\mathbf{m}^{old} \leftarrow \mathbf{m}^{new}$.

Note the omission of the ratio of priors in Equation (II.9.12) as compared to Equation (II.9.11). This extended version is clearly more applicable to large spatial problems with complex priors whose analytical form is not available, in particular to MPS-based priors. Figure II.9.4 shows an application of the Metropolis sampler with iterative resampling (Figure II.9.3). The data are the seismic response modeled using 1D convolution on the MPS realizations where seismic velocities differ considerably between each facies (channel and background). The results of the Metropolis sampler compare well with the rejection sampler; see Figure II.9.5.

9.3.4 Sequential Gibbs sampling

The above extended Metropolis sampler requires generation of samples from the prior $f_{prior}(\mathbf{m})$. This prior distribution need not be analytically known; it could be represented by the infinite set of MPS realizations generated from a given training image. One way of sampling from this prior is the Gibbs sampler, which when combined with sequential simulation can be turned into a sequential Gibbs sampler.

A Gibbs sampler is an iterative sampler that allows drawing from a multivariate distribution through iterative drawing from full conditional distributions.

(a)   (b)

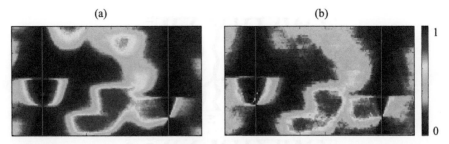

Figure II.9.5 A comparison of the Metropolis sampler (using iterative resampling) with rejection sampling. (a) Ensemble average using the rejection sampler; and (b) ensemble average using the Metropolis sampler. Shown is a slice through the 3D model at two wells locations (white lines) where facies conditioning data are present. The rejection sampler required about 100,000 forward model evaluations, whereas the Metropolis sampler took only 500 evaluations.

Consider $\mathbf{m}$ to be an MPS realization, and select randomly one m_i from that $\mathbf{m}$ for which we calculate the conditional distribution:

$$f_{m_i}\left(m_i|m_1, m_2, \ldots m_{i-1}, m_{i+1}, \ldots\right). \tag{II.9.13}$$

If we draw a sample from this distribution, we obtain a sample from the same prior $f_{prior}(\mathbf{m})$. The iterative version of this step is termed the Gibbs sampler. The Gibbs sampler requires only that a sample be generated; the actual specification of the full conditional distribution is not necessary. Secondly, generating only one single m_i at a time is not very efficient and may lead to long iterations; therefore, it may be more useful to generate a subset $\mathbf{m}_{i \in S}$ of $\mathbf{m}$ at one time (defined as either compact blocks or a set of points, as outlined above) requiring sampling from $f_{\mathbf{m}_{i \in S}}\left(\mathbf{m}_{i \in S}|\mathbf{m}_{i \notin S}\right)$.

In Hansen et al. (2012), the sampling of this distribution is done by sequential simulation and the SNESIM algorithm (see Section II.3.2.2). This sequential Gibbs sampler is then implemented in step 1 of the above extended Metropolis sampler to generate new samples from the prior. Figure II.9.6 shows an example of this resimulation of blocks to generate a chain of MPS realizations drawn from the prior.

9.3.5 Pattern frequency matching

The previous methodologies focus on sampling for obtaining multiple solutions to the inverse problem. In some practical cases, it may be useful to obtain just one solution, for example the solution $\mathbf{m}$ that corresponds to the maximum of $f_{posterior}(\mathbf{m})$, or a maximum a posterior (MAP) solution (Lange et al., 2012). Without loss of generality, the prior can be rewritten as

$$f_{prior}(\mathbf{m}) \simeq \exp\left(-h(\mathbf{m})\right) \tag{II.9.14}$$

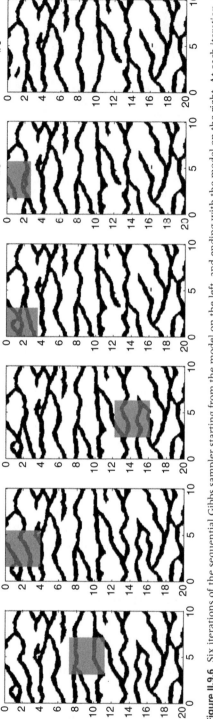

Figure II.9.6 Six iterations of the sequential Gibbs sampler starting from the model on the left, and ending with the model on the right. At each iteration, a 4×4 window is resimulated (from Hansen et al., 2012).

for some arbitrary bounded function $h(\mathbf{m})$. If the likelihood is also of the exponential form, for example

$$L(\mathbf{m}, \mathbf{d}_{obs}) \simeq \exp\left(-\alpha\, \ell(\mathrm{g}(\mathbf{m}), \mathbf{d}_{obs})\right) \tag{II.9.15}$$

with $\ell(\mathrm{g}(\mathbf{m}), \mathbf{d}_{obs})$ being some function measuring the difference between the observed data and forward-simulated data, then

$$f_{posterior}(\mathbf{m}) \simeq \exp\left(-\alpha\, \ell(\mathrm{g}(\mathbf{m}), \mathbf{d}_{obs}) - h(\mathbf{m})\right) \tag{II.9.16}$$

Then the MAP solution is defined as:

$$\begin{aligned} \mathbf{m}^{\mathrm{MAP}} &= \arg\max_{\mathbf{m}} f_{posterior}(\mathbf{m}) \\ &= \arg\max_{\mathbf{m}}\left(\alpha\, \ell(\mathrm{g}(\mathbf{m}), \mathbf{d}_{obs}) + h(\mathbf{m})\right) \end{aligned} \tag{II.9.17}$$

This equation is reminiscent of methods using regularization for solving inverse problems (Tikhonov and Arsenin, 1977): a mismatch term is coupled with a regularization term with the aim to induce some smoothness onto the solution. The weight α is used to regulate that smoothness. However, here the formulation is generalized to any arbitrary function of the model variables $\mathbf{m}$. This generalization provides an opportunity to inject regularization with respect to the training image. Instead of stating a degree of smoothness (e.g., by mean of covariance or derivatives), one could enforce the MAP solution to have patterns similar to the training image. This raises a question of how to define "similar." This is where the function $h(\mathbf{m})$ comes into play: $h(\mathbf{m})$ is defined to be a measure of dissimilarity between the training image patterns and the patterns of any realization generated from it. To obtain a summary of pattern frequencies, one can use for example a multipoint histogram (see Section II.2.4.4), which is a simple counting of the occurrence of pixel configurations within a given template or neighborhood. A chi-squared difference between both multiple-point histograms would then form such a function $h(\mathbf{m})$. Stochastic optimization methods such as simulated annealing can be used to find the maximum of Equation (II.9.17). Figure II.9.7 shows clearly the influence of the regularization, where MAP solutions without regularization deviate from the training image patterns and provide unrealistic-looking solutions.

9.4 Stochastic search

9.4.1 Aim

The previous sections of this chapter dealt with sampling methods within the inverse theory framework formulated in Equation (II.9.3). A posterior distribution is formulated through likelihood and prior, and the goal is to sample from this distribution. The main problem with these techniques is CPU time. McMC methods are notoriously slow, hence when a single evaluation of the forward

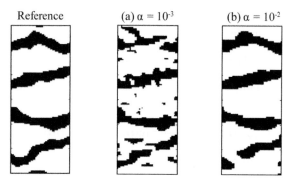

Figure II.9.7 A reference model and MAP solutions each with different values of α. The data inverted (not shown) consist of seismic travel times between boreholes located on each side of the model. The model parameters are the seismic velocity in each cell. The case with small regularization parameter (a) shows that such models are not geologically realistic, whereas adding the regularization leads to a solution (b) more consistent with the training image patterns (from Lange et al., 2012).

model (Equation (II.9.4)) takes a few hours of CPU time, they simply become impractical.

In this section, we discuss approximate methods. Their aim is still to sample, although such sampling is not rigorously done according to a theory; no posterior is explicitly defined by any theory, nor is there any sampling theory such as is the case for Metropolis or Gibbs sampling. Instead, the goal is to generate multiple solutions of the inverse problem and to make sure these solutions have certain properties; in the context of inverse modeling with training images, these are:

- In terms of spatial variability, the posterior realizations reflect the spatial patterns present in the training image. This also means that the posterior realizations and prior realizations share the same patterns.
- The data is matched up to a specified error ε. Objective functions are used to specify this mismatch error.

In addition, there is an aim of approximating the sample space of the theoretical formulation in Equation (II.9.3), although there is no guarantee, based on any theory, that this is achieved. A collection of models that follow the above bullet points often provide practitioners a false impression that one can use the variability of such a set for stating posterior uncertainty. However, if multiple clusters of models exist that lead to a good data fit, with different forecast implications, then that information in itself is practically relevant, without necessarily calling it "posterior uncertainty" in the rigorous sense.

One could compare the samples generated with these approximate methods with those obtained by sampling for simple cases, in order to study and research approximate methods, although no general conclusions can be reached from such study.

An illustrative and intuitive comparison between samplers and search methods is shown in Figure II.9.8. Samplers such as the Metropolis sampler (and iterative spatial resampling as a special case) rely on a "random walker" in the sampling space. A good sampler would move quickly to the areas of solution (located in the center) yet not too quickly so it can explore the posterior distribution uniformly. Search methods such as the probability perturbation or gradual deformation rely on search strategies and optimization, where likelihood decreases are always accepted; a uniform exploration of the posterior may not be achieved.

This approximate approach makes a lot of sense practically. The major subjectivity in Equation (II.9.3) is the prior, which is here formulated using a potentially large set of training images and any model realizations generated from it. One may therefore question, from a practical viewpoint, the need for rigorous sampling when the model formulation itself is highly subjective.

Instead of sampling, these methods propose to perform a stochastic search in the prior model space for those models that match the observed data up to a stated objective. Hence, any stochastic search makes a compromise between two important elements: CPU time (greediness) and coverage of the prior model space.

9.4.2 Probability perturbation method (PPM)

The PPM was the first method to solve inverse problems with training images (Caers, 2003). Compared to other stochastic search methods, it probably approximates best the sampling properties of the rejection or Metropolis sampler. Several tests, large and small, support this (Caers et al., 2006; Hoffman et al., 2006; Caers, 2007; Park et al., 2013). Probability perturbation works only with sequential simulation and was initially developed to solve inverse problems of flow where the prior models are generated with the SNESIM algorithm (Section II.3.2.2). Its basic version only applies to spatial models of discrete variables.

The stochastic search implemented in PPM was inspired by the gradual deformation method (see Section II.9.4.3) and uses a double-loop search. The outer loop considers a change in the random seed used in SNESIM, whereas the inner loop optimizes parameters linked to the perturbation of the current realization. The following algorithm summarizes the PPM for generating one realization:

1 Generate an initial realization $\mathbf{m}^0$ with SNESIM.
2 Outer loop: change the random seed.
  a Inner loop: optimize parameters $\mathbf{r}$.
    i Generate a new realization $\mathbf{m}^{new} = \mathbf{m}^{new}(\mathbf{r}, \mathbf{m}^0)$.
    ii Calculate the objective function $O(\mathbf{r}) = O(\mathbf{r}, g(\mathbf{m}^{new}), \mathbf{d}_{obs})$.
    iii Go to 3 when $O(\mathbf{r}) < \varepsilon$.
    iv Return $\mathbf{r}^{opt}$.
  b If $O(\mathbf{r}^{opt}) < O(\mathbf{0})$, then $\mathbf{m}^0 \leftarrow \mathbf{m}^{new}(\mathbf{r}^{opt}, \mathbf{m}^0)$.
3 Return $\mathbf{m}^0$.

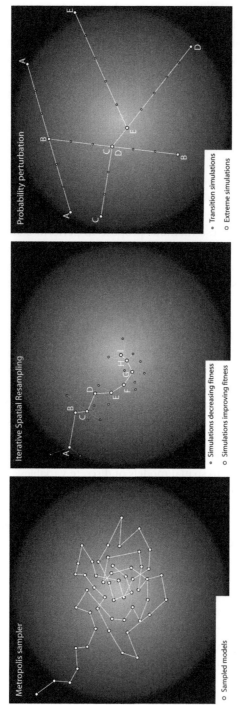

Figure II.9.8 Intuitive comparison of samplers versus search methods. The probability perturbation searches for an optimal model realization between two models drawn uniformly from the prior, whereas samplers rely on a random walker in the sample space.

The fundamental contribution of the PPM lies in step i: how to generate a new realization that is a perturbation of the existing realization and, at the same time, is also a realization generated from the prior (based on the single specified training image). Consider first the case with a single global perturbation parameter r and where **m** is a binary model. To achieve such perturbation, the current best realization is perturbed using a model of probabilities **p** defined on the same grid as **m**:

$$\mathbf{p}(r, \mathbf{m}) = (1 - r) \, \mathbf{m}^0 + r \, p_m \qquad (II.9.18)$$

where p_m is the marginal distribution, in this case simply the (global) probability of any m being 1 (global proportion). This probability model is then used as soft probability to generate a new realization with SNESIM. The tau model, with tau parameters equal to 1 (a form of conditional independence; see a sensitivity study in Hoffman, 2006), is used to combine probabilities in SNESIM. If $r = 1$, then the soft probabilities are all equal to the marginal. In that case, a completely new realization is generated. If $r = 0$, there is no perturbation because soft probabilities are either 0 or 1, namely, equal to $\mathbf{m}^0$. The value of r therefore regulates the perturbation from one prior model to another prior model, both of them generated with SNESIM; see Figure II.9.9. The double-loop optimization has therefore one loop that introduces a new realization (change of seed) and one loop that optimizes r. The perturbation is said to be global, because in probability, the same change is generated over the entire domain. To allow for more flexibility in the perturbation, regions, each with a different parameter r, can be introduced. This achieves a regional perturbation where some regions may change more than others to achieve a match. However, the perturbations do not show any region artifacts, simply because the SNESIM algorithm is not aware of any regions; they only enter in the soft probabilities. The nature of the sequential simulation assures that the perturbed realizations are consistent with the training image patterns. Hoffman (2006) shows that optimization using the polytope method with multiple regional parameters r is as simple as with a single parameter r.

Hu (2008) extended the PPM to any variable, whether continuous or discrete. Consider first a simple one-dimensional case, with one parameter m. The idea relies on a simple property, (Feller, 1971), namely, that given two samples m and m^0 from a given distribution $F(m)$, then any sample m^{new} drawn from

$$F(m|r, m^0) = (1 - r) \, i(m, m^0) + r \, F(m) \qquad (II.9.19)$$

with the indicator function defined as

$$i(m, m^0) = \begin{cases} 1 & \text{if } m \le m^0 \\ 0 & \text{else} \end{cases} \qquad (II.9.20)$$

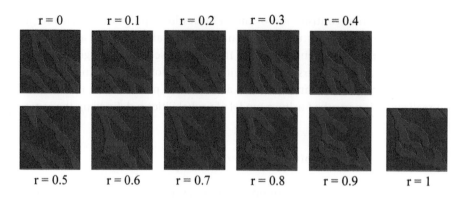

r = 0   r = 0.1   r = 0.2   r = 0.3   r = 0.4

r = 0.5   r = 0.6   r = 0.7   r = 0.8   r = 0.9   r = 1

Training Image

Figure II.9.9 Example of a perturbation from one MPS realization drawn from the prior into another prior sample by means of the probability perturbation method.

is also a sample of $F(m)$. Unlike in PPM, no assumptions are made on the nature of the distribution $F(m)$, allowing samples from, for example, Gaussian or uniform distributions to be gradually perturbed.

An alternative to PPM termed the "probability conditioning method" (Jafarpour and Khodabakhshi, 2011) uses the same idea, namely, a soft probability can be used to perturb realizations. In the case of PPM, that probability is updated using the ensemble Kalman filter (EnKf), which will be further discussed in Section II.9.5.3.

9.4.3 Gradual deformation of sequential simulations (GDSS)

The gradual deformation method (GDM) was initially developed to perturb Gaussian realizations (Hu, 2000). Consider two Gaussian vectors $\mathbf{y}^0$ and $\mathbf{y}$ as follows:

$$\mathbf{y}^{new}(r) = \sin(r)\mathbf{y}^0 + \cos(r)\mathbf{y} \qquad (\text{II}.9.21)$$

A simple proof shows that regardless of the value of r, the perturbed $\mathbf{y}^{new}(r)$ is also a Gaussian vector with the same correlation as vectors $\mathbf{y}^0$ and $\mathbf{y}$. This means that the covariance of the Gaussian realizations is preserved in the perturbation. A gradual deformation of Gaussian realization follows the same algorithm as given for PPM.

The GDM only works for Gaussian simulations. Adding up two binary images in whatever way will yield a realization with more than two categories. To gradually deform realizations generated using any other sequential simulation algorithm (such as SNESIM) a completely different approach needs to be taken. Any realization of a sequential simulation algorithm can be seen as a multivariate transformation $\mathbf{T}$ of a set of uniform random numbers $\mathbf{u}$ into a realization $\mathbf{m}$ exhibiting a nonrandom spatial pattern:

$$\mathbf{m} = \mathbf{T}(\mathbf{u}) \tag{II.9.22}$$

In sequential simulations, these uniform random numbers are used for two purposes: (1) to define the random path, and (2) to draw from conditional distributions. One cannot gradually deform a path to a new path, simply because the topology of these paths is very complex and a small change in the beginning of the path will cascade very rapidly and in an unstructured way. Hence, only the uniform random numbers for drawing from conditional distributions can be gradually deformed. Gradually deforming uniform random numbers is straightforward: one simply transforms them to independent Gaussian variables and use Equation (II.9.21) to perform the perturbation. The basic equation in step i of the above algorithm is therefore:

$$\mathbf{m}^{new}(r) = \mathbf{T}\left(G\left(\sin(r)G^{-1}(\mathbf{u}^0) + \cos(r)G^{-1}(\mathbf{u})\right)\right) \tag{II.9.23}$$

There is one considerable difference between GDSS and PPM, which lies in the nature of the path. In PPM, the random path changes with a change of the random seed, whereas in GDSS, the path is fixed: the path of the first realization is the same as the path used for generating the final inverse solution. As discussed in Section I.5.2.4, the path has considerable impact on the nature of the posterior uncertainty, when conditioning sequential simulations to data.

To study the effect of path, and hence the nature of posterior uncertainty covered by PPM and GDSS, consider the simple inverse problem presented in Figure II.9.10, with a 1D training image and a 1D model (m_1, m_2, m_3). Two types of data are present: (1) hard data informing the central grid location as being black, and (2) a second type informing that exactly two of the three grid locations need to be black, but it is not known which ones. Five prior model realizations consistent with the training image can be constructed (see Figure II.9.10), the probabilities of which can be exactly calculated by scanning the training image. Bayes' rule then provides posterior probabilities. For example, the posterior probability of the first node being black is:

$$\Pr(m_1 = 1 | m_2 = 1, m_1 + m_2 + m_3 = 2) = \frac{1}{3} \approx 0.33$$

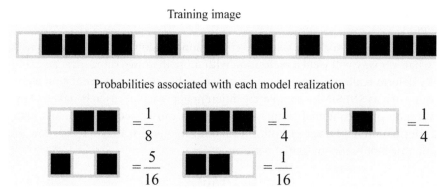

Figure II.9.10 A simple 1D case with a training image consisting of black ($m = 1$) and white ($m = 0$) cells. The model consists of three cells; the prior probabilities are extracted from the training image.

Consider solving this problem with PPM and GDSS. The estimated probabilities obtained using Monte Carlo are:

$$\hat{p}_{gdss}(m_1 = 1 | m_2 = 1, m_1 + m_2 + m_3 = 2) = 0.27$$
$$\hat{p}_{ppm}(m_1 = 1 | m_2 = 1, m_1 + m_2 + m_3 = 2) = 0.35$$

For this simple case, both methods approximate the exact posterior solution. But clearly the lack of permutation of the path in GDM has a considerable effect on the resulting posterior distribution.

9.4.4 Neighborhood algorithm (NA)

PPM and GDM rely on a perturbation of MPS realizations to search for solutions to inverse problems. The perturbations are constructed such that any new realization is still consistent with the specified training image; it is a sample of the training image–based prior model $f_{algo,TI}(\mathbf{m})$. An alternative is to first generate a large set of realizations of this prior initially, then simply search (efficiently) in that space of realizations for candidates that are likely to match the observed data. Although this would require that some models are at least close to obtaining a match, the advantage of doing this is that now, unlike PPM and GDM, multiple training images can be specified, from which multiple realizations are generated. This allows accounting for uncertainty in the training image, hence to consider a rich prior. It also allows assessing whether some training images are inconsistent with the data (see the above discussion). Once a few realizations have been found that are approximately matching the observed data, one could consider further refinement in the match by means of PPM, GDM, or possibly sampling methods.

A powerful search algorithm is the neighborhood algorithm (Sambridge, 1999a, 1999b). This algorithm is a stochastic search originally proposed for

Figure II.9.11 Sample space refinement in the neighborhood algorithm. An initial set of 10 realizations (represented by black dots) is refined progressively in areas with low misfit function (red areas).

seismic inversion problems and later applied to flow inverse problems (Demyanov et al., 2004; Subbey et al., 2004; Christie et al., 2006). The central idea of this method is to obtain an approximate misfit evaluation using previously evaluated misfit functions, and, based on the approximate misfit, identify multiple regions that achieve a minimum misfit. The approximate misfit surface is constructed by compartmentalizing the model space into Voronoi cells. Figure II.9.11 illustrates this idea at least conceptually. The NA can operate in any space defined by a distance. The axis in Figure II.9.11 could be two model parameters (e.g., object width and proportion), or it could be a projection (dimension reduction) from a higher-dimensional space (e.g., using PCA or MDS). The NA method searches iteratively and stochastically in that space by relying on a Voronoi tessellation of the space starting from a set of initial parameter choices or a set of models (10 initial guesses in Figure II.9.11). The goal is to refine the space in the high-likelihood (or low-misfit function) areas (the red areas in Figure II.9.11).

NA was originally designed for parameterized problems, where parameters could for example describe object shapes or layers, all listed as a vector. Such explicit parameterization is not necessarily possible when (1) dealing with multiple stochastic realizations and (2) dealing with multiple training images. Multiple MPS realizations represent spatial uncertainty that is not easily parameterized (see Section II.9.5.2). Secondly, the training images themselves may not be the result of parameter-based models; they may instead be constructed using nonparametric techniques or be directly borrowed from some analog data set.

Suzuki and Caers (2008) propose to "parameterize" the inverse problem, involving a prior model definition from multiple training images by means of distances. The use of a distance between model realizations (and not parameters) is motivated by the fact that algorithms such as NA only require the definition of a distance to move about and search for models sampled from the prior with low misfit. The focus now lies on the realizations and not on any parameters used to generate them. This idea therefore requires that a measure of dissimilarity (distance) needs to be defined between MPS realizations. Such a measure is inherently a subjective decision regarding what is deemed "similar," and several proposals for such measures have been formulated in Section II.2.5. To search for realizations that match the data, one then relies on the following assumption: for any stochastic search such as NA to be effective, the defined distance between any two realizations needs to statistically correlate with the difference in their forward model response. The higher that statistical correlation is, the more effective the search becomes. Should this assumption be reasonable, then the space within which one is searching becomes "structured": it is not a completely random function of the model realizations.

To illustrate this idea, consider a set of 81 training images representing alternative geological scenarios; see Figure II.9.12. A total of 405 facies realizations (five from each of 81 training images in Figure II.9.12) are simulated (using the SIMPAT algorithm; see Section II.3.3.1) conditional to facies observations at the wells depicted in Figure II.9.13, all penetrating channel sand. In this synthetic problem, background mud is understood as a sealing rock that does not have flow or storage capacity. The inner channel heterogeneity in petrophysical properties is constant and known. The spatial distribution of channel sands is considered the unknown in this inverse problem. The forward model response is the production performance or the pressure and water cut observed at producers. The flow data are shown in Figure II.9.13. It is not known a priori which one of the training images is in fact consistent (or not) with these data. Because we are dealing with discrete and distinct binary structures, the Hausdorff distance is used (see Section II.2.5.2), which has been shown for certain cases (Suzuki et al., 2008) to correlate reasonably with a difference in flow responses.

Once a distance is determined, NA can be applied because the division of the model space into Voronoi cells requires only the specification of a distance. The application of NA to this problem is illustrated in Figure II.9.14. The search is initiated by evaluating the objective function over some small number of selected MPS realizations, which are separated as far as possible in terms of the similarity distance (depicted as black dots in step 1 in Figure II.9.14). The remaining model realizations (white circles) are then associated to the nearest model. In the next

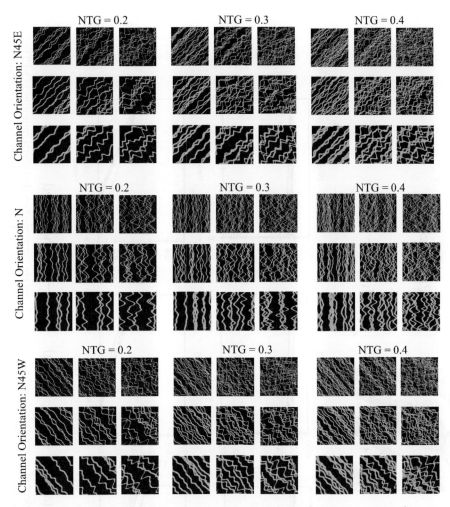

Figure II.9.12 A case study with 81 training images. Varied in such training images is the proportion of channel, the channel direction, and the channel sinuosity. Modified from Suzuki, S., & Caers, J. (2008). With kind permission from Springer Science and Business Media.

set, one of the Voronoi cells is selected for further processing through drawing from the following discrete distribution:

$$P(\text{selecting Voronoi cell } i) = \exp\left(-\frac{1}{T}\frac{O(\mathbf{m}_i)}{N_i}\right) \qquad (\text{II}.9.24)$$

where N_i is the number of prior models in Voronoi cell i; the parameter T is determined such that the total of the selection probability over all Voronoi cells equals 1.0. Within the selected Voronoi cell, another realization is picked at random, and its misfit function is evaluated; see Figure II.9.14. Given this model and its misfit function value, one can further compartmentalize the space and update the discrete selection probabilities. This procedure is repeated until one achieves

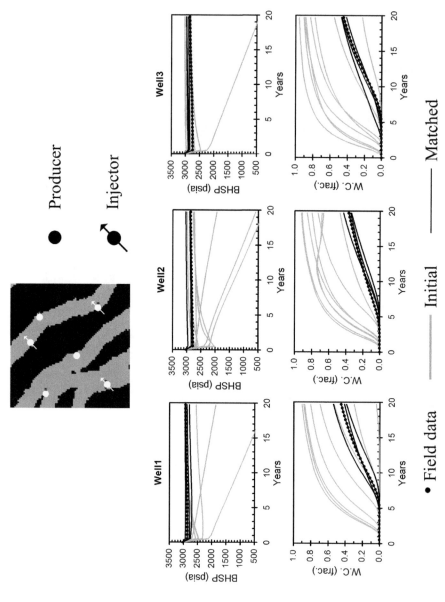

Figure II.9.13 A model setup with producing and injecting wells penetrating channel sand. Shown are the production data (dots) consisting of well bottom hole pressure (BHSP) and fractional flow of water. Gray curves are the forward simulated responses from a few initial models in the set, whereas black curves are those retained by NA. Modified from Suzuki, S., & Caers, J. (2008). With kind permission from Springer Science and Business Media.

Step 1   Step 2   Step 3

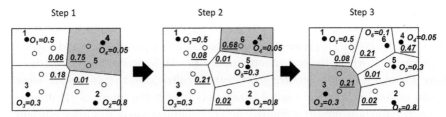

Figure II.9.14 The NA algorithm consists of an iterative refinement of the MPS model space based on the distance defined between MPS realizations. The misfit values (O_i) are turned into probabilities (underlined). The shaded polygon is the one selected by the NA algorithm. Modified from Suzuki, S., & Caers, J. (2008). With kind permission from Springer Science and Business Media.

a sufficiently low value of the objective function and hence fit to the data; see Figure II.9.15.

9.5   Parameterization of MPS realizations

9.5.1   Aim

Stochastic search methods search in the space of MPS prior model realizations for those that match the data up to a given precision. A search generates a walker in the space of realizations. Each method has a different walker. The search is often parameterized, enabling a structure in the search; PPM and GDM are examples of such parameterized searches.

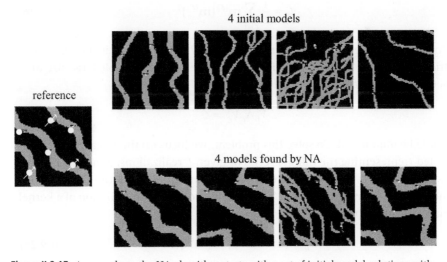

Figure II.9.15 A case where the NA algorithm starts with a set of initial model solutions with clearly different channel scenarios represented each by a different training image. The solution shows much less variability in terms of channel scenario. Flow responses of these models are shown in Figure II.9.13. Modified from Suzuki, S., & Caers, J. (2008). With kind permission from Springer Science and Business Media.

A different approach is to parameterize the realizations generated with MPS, then use optimization methods on these parameters to obtain a match to the data (Sarma et al., 2008). This approach is markedly different from what was discussed in Section 9.4, where the realization output from MPS algorithms was left as is. Now, the aim is to parameterize geostatistical realizations, that is, replace some model realization $\mathbf{m}$ with a set of parameters θ with the intent of (1) dimensionality reduction, $\dim(\theta) < \dim(\mathbf{m})$; and (2) optimization of θ is more efficient than searching for $\mathbf{m}$, for example through the use of gradients.

9.5.2   Kernel-based parameterization of MPS realizations

A successful parameterization would require obtaining a mapping between some small set of parameters θ and the geostatistical model realization $\mathbf{m}$:

$$\mathbf{m} = \mathbf{T}(\theta) \tag{II.9.25}$$

It was shown in the previous section that any MPS algorithm maps uncorrelated uniform random numbers $\mathbf{u}$ into a set of realizations. However, the size of $\mathbf{u}$ is usually equal to or larger than the size of the model $\mathbf{m}$. A dimension reduction therefore needs to take place.

One can start by considering a linear mapping, which is achieved by a linear principal component analysis (PCA). To construct such mapping, a number L of realizations $\mathbf{m}^{(\ell)}$ will need to be generated initially, from which the linear principal components are calculated by means of the experimental covariance:

$$C = \frac{1}{L} \sum_{\ell=1}^{L} \mathbf{m}^{(\ell)} [\mathbf{m}^{(\ell)}]^T \tag{II.9.26}$$

Note that the size of the covariance matrix is $N \times N$, where N is the number of grid cells in the MPS model $\mathbf{m}$. The problem here is the possible large size of C, hence the eigenvalue decomposition:

$$V_C \Lambda_C V_C^T = C \tag{II.9.27}$$

would be impractical. To solve this problem, we focus on the L realizations generated (representing the infinite set). For given L realizations, at most L positive eigenvalues exist. This eigenvalue problem has an equivalent formulation termed the "kernel eigenvalue problem." To that end, consider the definition of a kernel matrix K, with entries:

$$K_{ij} = (\mathbf{m}^{(i)} \cdot \mathbf{m}^{(j)}) = [\mathbf{m}^{(i)}]^T \mathbf{m}^{(j)} \tag{II.9.28}$$

An eigen-decomposition equivalent to Equation (II.9.27) can be formulated based on this kernel matrix as:

$$V_K \Lambda_K V_K^T = K \tag{II.9.29}$$

Note that the kernel matrix is of size $L \times L$, meaning that the decomposition is now computationally feasible for large grid sizes. The relationship between eigenvalues of C and eigenvalues of K is as follows:

$$\Lambda_K = L\Lambda_C \qquad (\text{II}.9.30)$$

and, equivalently for the eigenvectors:

$$\mathbf{v}_C^{(\ell)} = \sum_{\ell=1}^{L} \lambda_{K,\ell} \, \mathbf{m}^{(\ell)} \qquad \text{for } \ell = 1, \dots, L \qquad (\text{II}.9.31)$$

The eigenvalue decomposition can be used to determine a discrete Karhunen–Loeve (K-L) expansion of the model $\mathbf{m}$, which for covariance-based models is classically known as

$$\mathbf{m} = V_C \Lambda_C^{1/2} \theta \qquad (\text{II}.9.32)$$

where θ is a vector of uncorrelated random variables. Using the above kernel representation, the same decomposition can be expressed in terms of the computationally more accessible kernel matrix eigen-decomposition. Additionally, the size of the vector θ is at most L, hence a considerable dimension reduction can be achieved. If these uncorrelated variables are Gaussian, then so is $\mathbf{m}$.

At best, the linear PCA can only provide a Gaussian-type parameterization of the MPS realizations; Figure II.9.16 illustrates this. This would defeat the very purpose of using training images over covariances. What is, however, important about Equation (II.9.29) is the dimension reduction. In any real case, $L \ll N$; hence the size of the parameterization is at most L and could be chosen to be less, based on the number of significant eigen-components.

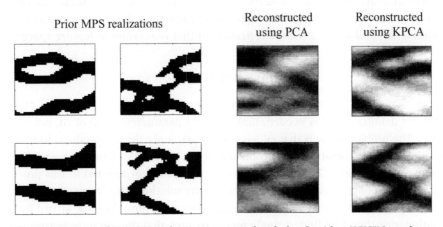

Prior MPS realizations · Reconstructed using PCA · Reconstructed using KPCA

Figure II.9.16 Four of 100 MPS realizations generated with the algorithm SNESIM are shown as well as two reconstructed images using the PCA-based parameterization and the KPCA-based parameterization. Modified from Sarma, P., Durlofsky, L. J., & Aziz, K. (2008). With kind permission from Springer Science and Business Media.

To extend the formulation for non-Gaussian models, such as MPS realizations, the above kernel formulation is extended to nonlinear, multiple-point representation of the decomposition (Equation (II.9.32)) that makes it much more amenable for parameterizing MPS models (Sarma et al., 2008). The dot product in Equation (II.9.28) can be extended as follows:

$$K_{ij} = \left[\varphi(\mathbf{m}^{(i)})\right]^T \varphi(\mathbf{m}^{(j)}) \qquad (\text{II.9.33})$$

where φ is some unspecified multivariate (multipoint) transformation of the MPS realization $\mathbf{m}$. In machine-learning language (see, e.g., Schölkopf and Smola, 2002), this transformation represents a mapping from model or realization space to a so-called feature space. Such mapping is used often to increase dimensionality, thereby aiming to obtain a better linear separability or increase in linearity of the models in that space, hence rendering linear techniques (e.g., linear PCA) more effective. In this context, such a transformation is used to capture better the higher-order features of the MPS realization that can evidently not be captured with the covariance in Equation (II.9.32).

In machine learning, this "kernel trick" is named after the observation that an analytical expression of φ (a highly nonlinear, high-dimensional transformation) does not need to be explicitly specified to produce such mapping; only a valid (positive-definite) dot product represented by a kernel function is required. Several such kernel functions have been proposed, such as:

$$\begin{aligned} &\text{the polynomial kernel: } \left(\mathbf{m}^{(i)} \cdot \mathbf{m}^{(j)}\right)^d \\ &\text{the radial basis kernel: } \exp\left(-\frac{\left\|\mathbf{m}^{(i)} - \mathbf{m}^{(j)}\right\|^2}{\sigma^2}\right) \end{aligned} \qquad (\text{II.9.34})$$

Once a kernel transformation has been chosen, linear PCA can be applied in the feature space, resulting in a kernel PCA (KPCA). The K-L expansion can be applied in that space as well, resulting in an expansion of the transformed realization $\varphi(\mathbf{m})$ and hence a parameterization of that realization in feature space:

$$\varphi(\mathbf{m}) = \theta^T \mathbf{T}(K, \mathbf{m}) \qquad (\text{II.9.35})$$

The mapping $\mathbf{T}$ is based on an eigen-decomposition of K (see Sarma et al., 2008, for details). What is needed, however, is a parameterization of $\mathbf{m}$, not of $\varphi(\mathbf{m})$. The problem now is that φ^{-1} is not known, only the dot product. In the machine-learning literature, this problem is better known as the pre-image problem (Kwok and Tsang, 2004). There is no direct analytical solution to this pre-image problem; moreover, the problem has many solutions. The kernel transformation is not bijective: many different $\mathbf{m}$ can be mapped onto the same $\varphi(\mathbf{m})$. By iteration, one therefore finds a single model $\mathbf{m}$ such that the distances between its projection and the expansion (Equation (II.9.35)) are as small as possible:

$$\tilde{\mathbf{m}} = \arg\min_{\mathbf{m}} \left\|\varphi(\mathbf{m}) - \theta^T \mathbf{T}(K, \mathbf{m})\right\| \qquad (\text{II.9.36})$$

It turns out that the Euclidean distance (Equation (II.9.36)) is only a function of the dot product of $\boldsymbol{\varphi}$ (hence, the kernel) and not of $\boldsymbol{\varphi}$ itself, making this optimization problem tractable.

Figure II.9.16 shows two KPCA reconstructed realizations. Clearly, the channel structure is better reproduced than in the PCA case. The parameterization, however, does not result in discrete structures such as in the original prior model realizations. The pre-image problem solutions are smooth solutions: the optimization in Equation (II.9.36) finds only those solutions that minimize a distance, and there is no guarantee that the solutions are samples of the original MPS-based prior. This problem of "smoothing" discrete spatial structures is present in most techniques relying on some form of dimension reduction or compression, regardless of whether one uses wavelets (e.g., Sahni, 2006) or compressed sensing (Khaninezhad et al., 2012a, 2012b). This is evidently a disadvantage if the goal is to be consistent with a training image–based prior (or any prior with discrete spatial structures). However, the smooth solutions do provide an advantage in finding inverse solutions through optimization of a misfit function because now gradients can be calculated and used to accelerate the optimization procedures.

9.5.3   Extensions on the ensemble Kalman filter applied to MPS realizations

The EnKf (Evensen, 1994, 2003; Houtekamer and Mitchell, 1998; Aanonsen et al., 2009) is a recursive filter operation where a mismatch in the data is used to adjust the model by a linear update operation. The recursive part refers to its consecutive application in time, as more data become available. This method of data assimilation is popular in meteorological and oceanographic forecasting, where large amounts of data are recorded in various time periods, and, hence, any forecast needs to be adjusted by including these new data. The theoretical formulation can be derived directly from Tarantola's formulation (Equation (II.9.1)), as well as any Bayesian formulations of the inverse problems. The main assumptions underlying the basic EnKf derivation are to (1) assume a multi-Gaussian distribution on the model and data variables $(\mathbf{d},\mathbf{m})$ and (2) a linear relationship between all the variables $(\mathbf{d},\mathbf{m})$ and (3) a metric density for $(\mathbf{d},\mathbf{m})$ that is uniform. In EnKf, the model variables are split into two parts, the static variables $\mathbf{m}_S$ and the dynamic variables $\mathbf{m}_D$. In the subsurface context, the static variables are typically porosity and permeability, and the dynamic variables are pressures and saturations. The couple $(\mathbf{d},\mathbf{m})$ or now triplet $(\mathbf{d},\mathbf{m}_S,\mathbf{m}_D)$ is now termed the state vector. In EnKf, an initial ensemble of various $(\mathbf{m}_S,\mathbf{m}_D)$ is generated. In the subsurface context, this would require generating a set of porosity or permeability models, including the initial saturations and pressures. Consider this initial ensemble at time $t = 0$. Next, a forecast step is made whereby a simulator forecasts the dynamic variables at the next time $t + \Delta t$, as well as provides a forecast of the observable variables $\mathbf{d}$ for all models in the ensemble. In the assimilation step, an update of all variables is made based on the difference between the forecasts and the actually observed **d** at that time $t + \Delta t$ using a linear filter as follows:

$$\mathbf{Y}_{t+\Delta t} = \mathbf{Y}_t + K_{t+\Delta t} \left(\mathbf{d}_{obs,t+\Delta t} - \mathbf{d}_{t+\Delta t} \right) \qquad \text{with } \mathbf{y} = \begin{bmatrix} \mathbf{d} \\ \mathbf{m}_S \\ \mathbf{m}_D \end{bmatrix} \qquad (\text{II}.9.37)$$

with the Kalman gain expressed as:

$$K = C_{\mathbf{Y}_{t+\Delta t}} H^T \left(H C_{\mathbf{Y}_{t+\Delta t}} H^T + C_{\mathbf{d}} \right)^{-1} \qquad \text{with } H = [0 \,|\, I] \qquad (\text{II}.9.38)$$

The covariance matrix $C_{\mathbf{Y}_{t+\Delta t}}$ is calculated from the ensemble itself. The matrix $C_{\mathbf{d}}$ represents the error covariance on the data variables. The original derivation involves a linear forward model:

$$\mathbf{d} = G\mathbf{m} \qquad (\text{II}.9.39)$$

The resulting posterior distribution in this linear case is also Gaussian and fully known through the specifications of the various covariance matrices (the original derivation of Kalman; see also Tarantola, 1987). The linear model can be replaced by a nonlinear forward model, but then the posterior distribution is no longer analytically known and, evidently, no longer Gaussian. Nevertheless, the filter (Equation (II.9.37)) can be applied to variables that are non-Gaussian. However, in such extension, several problems may occur:

- The variables are no longer linear (e.g., facies or permeability), hence metric densities are not uniform, and possibly unphysical values may be obtained (e.g., negative saturation or porosity; there is no preservation of the discreteness of the variables).
- The updated $\mathbf{m}_S$ are no longer guaranteed to preserve their original spatial structure; this is particularly true when the initial realizations are MPS realizations. As updates progress, the model becomes more Gaussian, even when the prior is distinctly non-Gaussian.
- One does not know what the posterior distribution of the ensemble is or how it compares with Metropolis or rejection sampling of the actual posterior distribution in the non-Gaussian and nonlinear case, as expressed through Equation (II.9.3).

Most extensions on this basic formulation that attempt to address these issues involve some form of transformation of the model variables. Clearly, the basic formulation cannot be applied to MPS realizations, but one could apply the EnKf in a space where it is more appropriate. For example, one can transform the realizations into a kernel space, as discussed above, and apply the EnKf in this space (Sarma and Chen, 2009). Alternatively, one could make a K-L expansion in this space, resulting in a model parameterization by means of a short Gaussian vector **y**. The EnKf can be applied to this Gaussian vector. The problem with the kernel approach lies in the pre-image problem (the back transformation), which remains largely unsolved, or it may involve solving another, equally difficult inverse problem. Transformations are only useful as long as back transformations are easily calculable.

Another line of approach is to apply the EnKf not directly to the MPS realizations but to some soft or auxiliary variable that controls the MPS realizations (Khodabakhshi and Jafarpour, 2011). Hu et al. (2013) propose applying the EnKf to the uniform random numbers used to generate the MPS realizations by means of a gradual deformation-based parameterization. Other approaches rely on transforming the non-Gaussian local distributions into Gaussian ones on which the EnKf then can be applied (Zhou et al., 2011).

An entirely different view on "fixing" the linear EnKf model is to abandon the idea of extending the linear filter altogether, yet remain (at least in principle) faithful to the idea of updating an entire ensemble of MPS realizations. In Zhou et al. (2012), a pattern-based search method is used in combination with direct sampling to directly make such updates. This approach relies on generating the initial set $(\mathbf{m}_S, \mathbf{m}_D)$ as before, then updating that set by searching the initial ensemble of realizations for a data set that matches the conditional pattern composed of models and observations.

References

Aanonsen, S. I., Nævdal, G., Oliver, D. S. & Reynolds, A. C. V., B. 2009. The ensemble Kalman filter in reservoir engineering: A review. *SPE Journal*, 14, 393–412.

Banerjee, S., Carlin, B. P. & A.E., G. 2004. *Hierarchical Modeling and Analysis for Spatial Data*, Chapman & Hall/CRC, Boca Raton, FL.

Caers, J. 2003. History matching under a training image-based geological model constraint. *SPE Journal*, 8, 218–226, SPE no. 74716.

Caers, J. 2007. Comparing the gradual deformation with the probability perturbation method for solving inverse problems. *Mathematical Geology*, 39, 27–52.

Caers, J., Hoffman, T., Strebelle, S. & Wen, X. H. 2006. Probabilistic integration of geologic scenarios, seismic, and production data – a West Africa turbidite reservoir case study. *Leading Edge [Tulsa, OK]*, 25, 240–244.

Christie, M., Demyanov, V. & Erbas, D. 2006. Uncertainty quantification for porous media flows. *Journal of Computational Physics*, 217, 143–158.

Demyanov, V., Subbey, S. & Christie, M. 2004. Uncertainty assessment in PUNQ-S3: Neighbourhood algorithm framework for geostatistical modeling. Paper presented at the 9th European conference on the mathematics of oil recovery, Cannes, France, August 31–September 2.

Evensen, G. 1994. Sequential data assimilation with nonlinear quasi-geostrophic model using Monte Carlo methods to forecast error statistics. *Journal of Geophysical Research C*, 5, 143–162.

Evensen, G. 2003. The ensemble Kalman filter: Theoretical formulation and practical implementation. *Ocean Dynamics*, 53, 343–367.

Feller, W. 1971. *An Introduction to Probability Theory and Its Applications*, John Wiley & Sons, Inc., New York.

Fu, J. & Gomez-Hernandez, J. 2009. A blocking Markov chain Monte Carlo method for inverse stochastic hydrogeological modeling. *Mathematical Geosciences*, 2009, 105–128.

Gelman, A., Carlin, J. B., Stern, H. S. & Rubin, D. B. 2004. *Bayesian Data Analysis*, Chapman & Hall/CRC, Boca Raton, FL.

Geman, S. & Geman, D. 1984. Stochastic relaxation, Gibbs distribution and the Bayesian restoration of images. *IEEE Transactions on Pattern Analysis and Matching Intelligence*, 6, 721–741.

Hansen, T. M., Cordua, K. S. & Mosegaard, K. 2012. Inverse problems with non-trivial priors: Efficient solution through sequential Gibbs sampling. *Computational Geosciences*, 16, 593–611.

Hoffman, B. T. 2006. *Geologically Consistent History Matching while Perturbing Facies*, PhD dissertation, Stanford University, Stanford, CA.

Hoffman, B. T., Caers, J. K., Wen, X. H. & Strebelle, S. 2006. A practical data-integration approach to history matching: Application to a deepwater reservoir. *SPE Journal*, 11, 464–479.

Houtekamer, P. & Mitchell, H. L. 1998. Data assimilation using an ensemble Kalman filter technique. *Monthly Weather Review*, 126, 796–811.

Hu, L. 2000. Gradual deformation and iterative calibration of Gaussian-related stochastic models. *Mathematical Geology*, 32, 87–108.

Hu, L. 2008. Extended probability perturbation method for calibrating stochastic reservoir models. *Mathematical Geosciences*, 40.

Hu, L. Y., Zhao, Y., Liu, Y., Scheepens, C. & Bouchard, A. 2013. Updating multipoint simulations using the ensemble Kalman filter. *Computers and Geosciences*, 51, 7–15.

Jafarpour, B. & Khodabakhshi, M. 2011. A probability conditioning method (PCM) for nonlinear flow data integration into multipoint statistical facies simulation. *Mathematical Geosciences*, 43, 133–164.

Jaynes, E. & Bretthorst, G. 2003. *Probability Theory: The Logic of Science*, Cambridge University Press, Cambridge.

Jung, A. & Aigner, T. 2012. Carbonate geobodies: Hierarchical classification and database – a new workflow for 3D reservoir modelling. *Journal of Petroleum Geology*, 35, 49–65.

Khaninezhad, M. M., Jafarpour, B. & Li, L. 2012a. Sparse geologic dictionaries for subsurface flow model calibration: Part I. Inversion formulation. *Advances in Water Resources*, 39, 106–121.

Khaninezhad, M. M., Jafarpour, B. & Li, L. 2012b. Sparse geologic dictionaries for subsurface flow model calibration: Part II. Robustness to uncertainty. *Advances in Water Resources*, 39, 122–136.

Khodabakhshi, M. & Jafarpour, B. 2011. Multipoint statistical characterization of geologic facies from dynamic data and uncertain training images. Paper presented at the SPE Reservoir Characterisation and Simulation Conference and Exhibition 2011, Abu Dhabi, UAE, October, SPE 146935, 206–217.

Kwok, J.-Y. & Tsang, I.-H. 2004. The pre-image problem in kernel methods. *IEEE Transactions on Neural Networks*, 15, 1517–1525.

Lange, K., Frydendall, J., Cordua, K. S., Hansen, T. M., Melnikova, Y. & Mosegaard, K. 2012. A frequency matching method: Solving inverse problems by use of geologically realistic prior information. *Mathematical Geosciences*, 44, 783–803.

Mardia, K. V. 1970. Measures of multivariate skewness and kurtosis with applications. *Biometrika*, 57, 519–530.

Mariethoz, G., Renard, P. & Caers, J. 2010. Bayesian inverse problem and optimization with iterative spatial resampling. *Water Resources Research*, 46.

Metropolis, N., Rosenbluth, A., Rosenbluth, M. & Teller, A. 1953. Equation of state calculations by fast computing machines. *Journal of Chemical Physics*, 21, 1087–1092.

Mosegaard, K. & Tarantola, A. 1995. Monte Carlo sampling of solutions to inverse problems. *Journal of Geophysical Research*, 100, 12, 431–447.

Park, H., Scheidt, C., Fenwick, D., Boucher, A. & Caers, J. 2013. History matching and uncertainty quantification of facies models with multiple geological interpretations. *Computational Geosciences*, 17, 609–621.

Sahni, I. 2006. *Multi-Resolution Reparameterization and Partitioning of Model Space for Reservoir Characterization,* PhD dissertation, Stanford, CA, Stanford University.

Sambridge, M. 1999a. Geophysical inversion with a neighbourhood algorithm – I. Searching a parameter space. *Geophyical Journal International,* 138, 479–494.

Sambridge, M. 1999b. Geophysical inversion with a neighborhood algorithm – II. Appraising the ensemble. *Geophyical Journal International,* 138, 727–746.

Sarma, P. & Chen, W. H. 2009. Generalization of the ensemble Kalman filter using kernels for non-Gaussian random fields. *In: Proceedings to the Reservoir Simulation Symposium, SPE, The Woodlands, TX, February,* 1146–1165.

Sarma, P., Durlofsky, L. J. & Aziz, K. 2008. Kernel principal component analysis for efficient, differentiable parameterization of multipoint geostatistics. *Mathematical Geosciences,* 40, 3–32.

Schölkopf, B. & Smola, A. 2002. *Learning with Kernels,* MIT Press, Cambridge, MA.

Subbey, S., Christie, M. & Sambridge, M. 2004. Prediction under uncertainty in reservoir modeling. *Journal of Petroleum Science and Engineering,* 44, 143–153.

Suzuki, S. & Caers, J. 2008. A distance-based prior model parameterization for constraining solutions of spatial inverse problems. *Mathematical Geosciences,* 40, 445–469.

Suzuki, S., Caumon, G. & Caers, J. 2008. Dynamic data integration for structural modeling: Model screening approach using a distance-based model parameterization. *Computational Geosciences,* 12, 105–119.

Tarantola, A. 1987. *Inverse Problem Theory: Methods for Data Fitting and Model Parameter Estimation,* Elsevier, Amsterdam.

Tarantola, A. 2005. *Inverse Problem Theory and Methods for Parameter Estimation,* Society for Industrial and Applied Mathematics, Philadelphia, PA.

Tarantola, A. & Valette, B. 1982. Inverse problems = quest for information. *J. Geophysics,* 50, 150–170.

Tikhonov, A. N. & Arsenin, V. Y. 1977. *Solution of Ill-Posed Problems,* Winston & Sons, Washington, DC.

Zhou, H., Gomez-Hernandez, J. & Li, L. 2012. A pattern search based inverse method. *Water Resources Research,* 48, W03505.

Zhou, H., Gómez-Hernández, J. J., Hendricks Franssen, H. J. & Li, L. 2011. An approach to handling non-Gaussianity of parameters and state variables in ensemble Kalman filtering. *Advances in Water Resources,* 34, 844–864.

CHAPTER 10

Parallelization

10.1 The need for parallel implementations

In multiple-point simulation, computational performance as well as memory demand present challenges. The first algorithm proposed (Guardiano and Srivastava, 1993) provided reasonable results in terms of reproducing complex spatial features. Despite these good results, even in comparison with more current techniques, it was initially not adopted mainly because of its unacceptable computational cost. Widespread adoption only came with the SNESIM algorithm, with the implementation of a tree structure for the storage of patterns effectively trading speed for memory (CPU for RAM). The tree structure preclassifies data events found in the training image, and this organized (and voluminous) database allows fast retrieval of such events during the simulation.

SNESIM has computational limits in terms of speed, and those come in addition to a possibly prohibitive memory cost. Therefore, with the advent of multicore processors and cluster computers, many researchers have sought to develop methods accelerating multiple-point simulations through the use of parallelization.

10.2 The challenges of parallel computing

Parallelization of an algorithm essentially entails that tasks are performed by several processing units simultaneously (i.e., processor cores), as opposed to a sequential execution where a single processing unit performs all steps of an algorithm.

When parallelizing an algorithm, several issues need to be considered. For example, some steps of an algorithm need to be completed before the next step can be started. One example in geostatistics is a sequential simulation (Equation II.2.44), where the value of each node is conditional to all previously simulated nodes. This means that it is in principle difficult to simulate several nodes at the same time without violating the underlying rules in calculating such conditional probabilities. The task of designing a parallel implementation involves constructing workarounds or finding approximations for such problems. As a

Multiple-point Geostatistics: Stochastic Modeling with Training Images, First Edition. Gregoire Mariethoz and Jef Caers.
© 2015 John Wiley & Sons, Ltd. Published 2015 by John Wiley & Sons, Ltd.
Companion website: www.wiley.com/go/caers/multiplepointgeostatistics

consequence, a parallel code may often be completely different than a sequential code, and sometimes the algorithms have to be redesigned for operation on parallel machines.

Another difficulty when designing or using a parallel algorithm is that the execution speed not only depends on the number of calculations performed (as is generally the case with a sequential code), but also is strongly affected by the time of communication between processors. When data are passed between different working processors, communication takes a certain amount of time. In addition, communication time depends on the hardware used. In this regard, two main types of computer architecture can be distinguished, shared-memory and distributed-memory systems:

- "Shared-memory systems" consist of a single memory unit that is simultaneously accessed by several processors. The architecture of multicore machines has now become more widely available. The communications between processors take place at the speed of reading or writing in the RAM, which is extremely fast. However, shared memory architectures generally have a limited number of processors.

- "Distributed-memory systems" consist of several independent computers that are connected, each with their own memory space. As a result, data must be communicated through a network of cables and switches. Several types of networks exist, some of them faster than others. An ethernet system is considered rather slow, whereas the InfiniBand system is considered fast. Most often, the communication speed is slower than in the shared memory case. However, such distributed memory architectures allow for a very large number of processors (up to a million cores for the largest existing clusters).

Depending on the type of architecture used, different implementation strategies can be considered, resulting in different codes. If the communication time is small, then it may be preferable to develop an algorithm that shares the computational load over a large number of processors. This may result in considerable "chatter" between the numerous processors. However, if communications are fast, the impact will not be substantial. In cases when one has to deal with slow communications, one may prefer using fewer processors than are available. In certain extreme cases, the overhead caused by communicating is not balanced by the gain provided by parallelization; parallel implementations may then be even slower than serial ones.

10.3 Assessing a parallel implementation

A common performance test for parallel algorithms is to compute the speed-up function, which is the ratio between the time taken by the algorithm to perform some computations in a serial way and in parallel. It is defined as

$$S(n_{CPU}) = \frac{t_{serial}}{t(n_{CPU})}, \qquad \text{(II.10.1)}$$

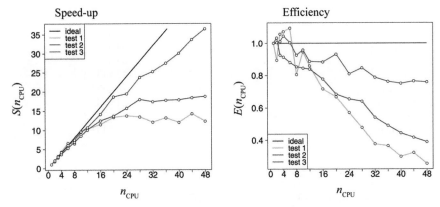

Figure II.10.1 Illustration of speed-up and efficiency curves for different simulation test cases using the IMPALA parallel algorithm. Modified from Straubhaar, J., P. Renard, et al. (2011). With kind permission from Springer Science and Business Media.

where t is the CPU time needed to solve a given problem with a certain number of processors and n_{CPU} is the number of processors used for a particular run. In the best case, the speed-up follows the ideal slope $sp(n_{CPU}) = n_{CPU}$ (Amdahl, 1967), meaning that using two processors results in half the time of a serial implementation, using three processors results in a third of the serial time, and so on. Such an ideal speed-up is only possible if the algorithm makes the best possible use of resources, resulting in no computational overhead for the management of tasks between the different processors. Another measure of performance of a parallel code is the efficiency curve, which is a normalized version of the speed-up curve:

$$E(n_{CPU}) = \frac{S(n_{CPU})}{n_{CPU}}. \tag{II.10.2}$$

In the ideal case, the efficiency curve should be constant at a value of 1. In practice, maximum speed-up cannot be reached, but approaching it means that the parallel implementation is efficient. Examples of speed-up and efficiency curves are shown in Figure II.10.1.

For a parallel algorithm to be efficient, the workload must be balanced between all processors. If large discrepancies are observed between the times used by different CPUs, then some processors work more than the others. This is usually assessed using load-balancing plots, such as is shown in Figure II.10.2. These curves show simultaneously the load of each processor, hence monitoring large CPU time differences between the processors used in the same run.

10.4   Parallelization strategies

Parallelization of geostatistical simulations is possible at three levels. The realization level is the easiest to parallelize. It consists in having each realization of a

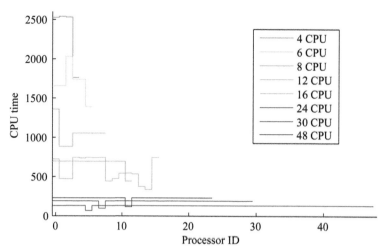

Figure II.10.2 Illustration of a load-balancing plot. Modified/Reprinted from Mariethoz, G. (2010). With permission from Elsevier.

Monte Carlo analysis computed by a different processor. As every realization is, by definition, independently constructed, no communication between processors is needed. The maximum number of processors that can be used with this strategy is equal to the desired number of realizations. This strategy is widely used and highly efficient, in particular when a large number of realizations need to be generated.

Parallelizing a simulation at the path level entails dividing the grid into zones and attributing a different zone to each processor. This strategy has been implemented by simulating groups of grid nodes at the same time (Dimitrakopoulos and Luo, 2004; Vargas et al., 2008) or by having a conflict management system ensure that only nonoverlapping patterns are simulated at the same time (Mariethoz, 2010). Path parallelization has also been applied in the context of simulated annealing (Peredo and Ortiz, 2011), by implementing a strategy called "speculative computing." The idea is to precompute, for each possible outcome of the node that is currently simulated, the value of nodes that are ahead in the path. For example, when simulating a binary variable, if the node currently simulated is path(i), the approach would consist of simulating in advance path $(i + 1)$ for both possible outcomes path $(i) = 0$ and path $(i) = 1$. Once the value of path (i) is determined, the precomputed outcome corresponding to this simulated value can be used without additional computation. The efficiency of speculated computing is, however, limited because only the computations corresponding to the actual outcomes of path (i) are used, while the precomputations for other prospective outcomes are lost CPU time.

The third level of parallelization is the node level. For example, the inversion of a large kriging matrix for SGS or the search for a data event in the multiple-points data events catalog can be shared among many processors. In this context, one example of an algorithm very difficult to parallelize is the tree structure for storing data events employed in SNESIM. As a result, SNESIM is efficient on single-core machines but cannot be scaled on multiprocessor machines. To address this issue, Straubhaar et al. (2011) redesigned the algorithm by using lists instead of trees. Searching a list takes more CPU time than searching a tree, but a list can be easily shared among several processors. As a result, each processor needs only to search its own sublist, which is fast because sublists are only a fraction of the size of the entire lists. In addition, each sublist takes significantly less memory than is the case with the entire search tree, and therefore memory issues are also mitigated. In the example of Figure II.10.1, it is shown that it is often possible to reach speed-ups of the order of 10 using lists. Therefore, if one considers tree-based storage to be approximately two times faster than list storage on a single core, the list storage can still be much faster when parallelization is considered, simply because it is not possible to parallelize tree storage.

These different strategies are not mutually exclusive. For example, the path can be distributed among different parallel machines, who themselves distribute the simulation of their individual nodes on local processors. In all of these cases, the efficiency of the parallelization is limited when a large number of processors is available, because the size of the problem to solve for each individual processor becomes small compared to the communication times between processors.

10.5   Graphical Processing Units

In recent years, a cost-effective alternative to multiprocessor clusters has emerged with the use of Graphical Processing Units (GPUs). A GPU is an assemblage of many (up to several hundreds) low-performance processors on a single chip, which are typically used for rendering operations in computer graphics. However, GPUs can also be used for other calculations, a practice called General-Purpose GPU (GPGPU). It has been shown that using GPGPU for geostatistical simulations bears a large potential for accelerating existing algorithms.

One limitation of GPUs is that all processing units on the GPU can only do the same operation in a single clock cycle. For example, if the operation is to read an integer value in memory, all processing units have to read an integer at the same time. If a single processing unit does another operation, all other units have to wait for it to complete its operation. As a result, *"if"* statements, for example, undermine the efficiency of a GPU implementation. Such a restriction means that existing algorithms may need to be redesigned for them to be adapted to this new generation of parallel hardware. Nevertheless, recent developments in multiple point simulation algorithms have brought GPU implementations of some algorithms described in Section II.2.3, such as SNESIM (Huang et al., 2013b), CCSIM (Tahmasebi et al., 2012), and direct sampling (Huang et al., 2013a).

References

Amdahl, G. Validity of the single processor approach to achieving large-scale computing capabilities. AFIPS Conference Proceedings, Thompson Books, Washington, DC, 1967, 483–485.

Dimitrakopoulos, R. & Luo, X. 2004. Generalized sequential Gaussian simulation on group size v and screen-effect approximations for large field simulations. *Mathematical Geology*, 36, 567–591.

Guardiano, F. & Srivastava, M. 1993. Multivariate geostatistics: Beyond bivariate moments. *In:* Soares, A. (ed.) *Geostatistics-Troia*. Kluwer Academic, Dordrecht.

Huang, T., Li, X., Zhang, T. & Lu, D. T. 2013a. GPU-accelerated direct sampling method for multiple-point statistical simulation. *Computers and Geosciences*, 57, 13–23.

Huang, T., Lu, D. T., Li, X. & Wang, L. 2013b. GPU-based SNESIM implementation for multiple-point statistical simulation. *Computers and Geosciences*, 54, 75–87.

Mariethoz, G. 2010. A general parallelization strategy for random path based geostatistical simulation methods. *Computers and Geosciences*, 36, 953–958.

Peredo, O. & Ortiz, J. M. 2011. Parallel implementation of simulated annealing to reproduce multiple-point statistics. *Computers and Geosciences*, 37, 1110–1121.

Straubhaar, J., Renard, P., Mariethoz, G., Froidevaux, R. & Besson, O. 2011. An improved parallel multiple-point algorithm using a list approach. *Mathematical Geosciences*, 43, 305–328.

Tahmasebi, P., Sahimi, M., Mariethoz, G. & Hezarkhani, A. 2012. Accelerating geostatistical simulations using graphics processing units (GPU). *Computers and Geosciences*, 46, 51–59.

Vargas, H., Caetano, H. & Mata-Lima, H. 2008. A new parallelization approach for sequential simulation. *In:* Soares, A., Pereira, M. & Dimitrakopoulos, R. (eds.) *geoENV VI – Geostatistics for Environmental Applications*. Springer, Dordrecht.

PART III
Applications

CHAPTER 1

Reservoir forecasting – the West Coast of Africa (WCA) reservoir

1.1 Introducing the context around WCA

1.1.1 Decisions carrying large financial risk

After the detailed review of existing simulation algorithms and modeling approaches of multiple-point geostatistics (MPS), we consider in this Part III different application areas where MPS and the use of training image are demonstrated. These case studies cover some of the domains where multiple-point geostatistics have been used and could be considered as templates for developing other areas of applications. These applications will be taken from the fields of petroleum engineering (reservoir characterization), mining (ore reserve estimation) and climate modeling (downscaling of regional climate models). The broader context of these applications is thoroughly introduced to provide the non-expert reader with a better understanding of the application area itself, thereby outlining why geostatistical models are used.

In this first chapter, we deal with subsurface reservoir modeling and forecasting, an area where geostatistics has had considerable attention (Kitanidis, 1997; Deutsch, 2002; Kelkar and Perez, 2002; Dubrule, 2003; Caers, 2005; Ringrose and Bentley, 2014). Subsurface reservoirs are important for several reasons: they contain hydrocarbons critical for the world's energy demands, are used to supply drinking water, provide storage (natural gas), or serve as sequestration of greenhouse gasses. Such reservoirs consist of a geological heterogeneous medium, either clastic or carbonate, although a few magmatic reservoir systems have been developed as well. This entails that engineering such systems is not simply a matter of "control" and "optimization" of a fully or almost fully known system (such as an airplane) but also a matter of mitigating the potential risk associated with the partial knowledge of such a medium. In this chapter, we will focus on a specific case study, namely, the planning of a new platform in deep water off the coast of West Africa. Due to the enormous financial risk associated with such operations (the average cost of a single oil platform in that area is around US$1–2 billion), energy companies need to understand the nature of financial risk, which is multiprong: economic factors such as oil price, local political climate,

Multiple-point Geostatistics: Stochastic Modeling with Training Images, First Edition. Gregoire Mariethoz and Jef Caers.
© 2015 John Wiley & Sons, Ltd. Published 2015 by John Wiley & Sons, Ltd.
Companion website: www.wiley.com/go/caers/multiplepointgeostatistics

the construction of offshore facilities, and, of particular concern in this example, the nature of the geological system.

Reservoir forecasting goes through various stages of development, starting with exploration when only limited seismic data (often 2D) and possibly one or two exploration wells are available to late field development and abandonment. The most critical phase is often the appraisal phase, when one typically has well logs from a few wells, 3D seismic data, and, if available, well-testing data. A team of reservoir experts consisting of geologists, geophysicists, and reservoir engineers collaborate on appraising the volume of oil in place, and the planning and design of wells. Such planning may go in several stages, but initially the recovery factor as well as optimizing locations of these wells are mostly targeted. Production of such reservoirs is possible due to in situ driving mechanisms: the oil may flow due to the in situ pressures in the reservoir or the support of an underlying aquifer system. However, nowadays most such (easy) oil has already been produced. What is increasingly common is the use of enhanced oil recovery systems to drive oil toward producers by means of injecting fluids (water) or even foams through injection wells. Due to the underlying geological heterogeneity, these fluids, useful as they are in enhancing oil production, make their way to producing wells, resulting in the need for installing costly facilities to separate the oil from water. A proper planning of wells would therefore attempt to minimize water production while maximizing oil production and at the same time keeping reservoir pressure high enough to avoid the dissolution of unwanted gas.

1.1.2   The West Coast of Africa case study

In this chapter, we will use a real-field deepwater turbidite offshore located on the West Coast of Africa (WCA). This reservoir is located in a slope-valley system below 500 m of seawater, resulting in large drilling costs and risks. Such systems comprise geological formations where steep submarine canyon cuts were formed. Petroleum reservoirs are found here less often than further down the slope where structures are less inclined, and when reservoirs are found they tend to have a more heterogeneous nature. The WCA reservoir is an amalgamated and aggradational channel complex that fills the canyon cut. As is typical with many geological systems, a large scale of variation of heterogeneity exists. Small-scale geological bodies such as channels can be grouped into larger complexes, which themselves can be grouped into even larger-scale channel systems (see Figure III.1.1). This hierarchy results in several levels of geological variability, but such nesting is much more complex than can be described with the traditional nesting of variogram structures. Due to the existence of wells and 3D seismic data, some of these large-scale features can be resolved, but often geological elements at the subseismic scale cannot be deterministically mapped, yet they may substantially impact production and recovery. Faults exist in this reservoir but are not the most critical element because they are not numerous, their geometry is reasonably simple, and they can be mapped relatively easily from

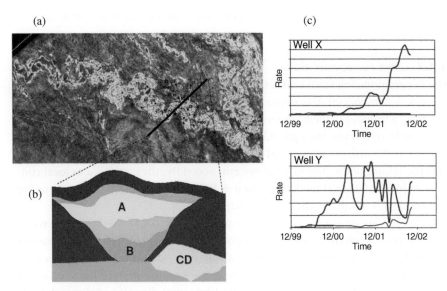

Figure III.1.1 (a) Interpreted seismic map of the upper layer in a WCA reservoir where a large-scale channel complex is visible; (b) conceptual depiction of the large-scale geobody definition within which the WCA reservoir is situated; and (c) early production data (water rate) from two wells. Data courtesy of Chevron.

seismic data. The most critical, impacting and uncertain reservoir elements are the petrophysical properties, permeability and porosity, whose spatial distribution is controlled through rock types.

Figure III.1.2 shows the structural framework and reservoir model grid viewed from the top. The question raised is: if one would develop a platform at the location of interest shown on the map in Figure III.1.2, what would be the oil production of the next decade from this platform? How much water will

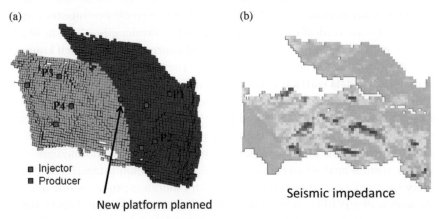

Figure III.1.2 (a) Map view of the structural framework of fault blocks; and (b) slice through the seismic impedance data used to constrain facies and petrophysical models.

be produced that will need to be separated from the oil? How uncertain are we about this? Is there a risk of no return on this large financial investment?

1.1.3   The reservoir data

The expert modeling team uses several data sources to resolve these critical questions. A first important data source is the detailed well logging and core data analysis from a number of wells. In terms of building geostatistical models, these data are used in two ways: they provide an insight into the vertical heterogeneity of the system (in case of only vertical wells) and hence aid in making interpretations into the nature of the depositional system. Secondly, they are used as local constraints in geostatistical modeling, often as hard data (although the exact nature of hard data is discussed later in this chapter).

A second important source of data is seismic: such data provide, unlike wells, an exhaustive yet indirect source of information. In geostatistics, it is used mostly as soft data for constraining petrophysical and rock type properties. Seismic data are of much lower resolution vertically than information derived from wells, but these data complement such wells in the horizontal direction. Seismic data do not measure or inform directly and uniquely rock type, permeability or porosity. The seismic waves are sensitive to an amalgamation of subsurface properties and features: layering, faults, the mineralogy, density, and porosity of rock, the rock acoustic and elastic properties and fluid composition. Geophysicists use inverse modeling tools to take the raw seismic measurements and turn them into an understandable "image of the subsurface". This image is presented either in terms of seismic amplitude migrated to the location where the wave reflected, or as an acoustic or elastic impedance cube. A slice through the impedance cube of WCA is shown in Figure III.1.2(b). These seismic volumes were mostly used for mapping the structures, faults, and layers, but they are increasingly used to inform the rock types, their porosity and what fluids they contain. One of the major differences between well-related data, such as cores or logs and seismic data, is the scale of such information. Seismic data inform reservoir heterogeneity at a much larger scale vertically (typically, one quarter wavelength, which in WCA coincides with 15 m). Horizontally, the scale of information is less well known, but in this case it would be around 50–100 m. The large-scale information provided by seismic data correlates better with large-scale rock type variations than with fine-scale petrophysical properties. The seismic data are therefore said to inform the system only partially, and probabilistic modeling is invoked to represent this partial information. Therefore, a probabilistic calibration of the information content of seismic information on rock type is often performed prior to any geostatistical analysis (see also Part I, relating elevation categories to a secondary variable). From the 3D seismic data, for calibration purposes, one retains only seismic observations made along the well paths that contain logs or have been cored. Next, one estimates the probabilistic relationship between rock type and seismic data from the pair of seismic and wells data. Various methods can be

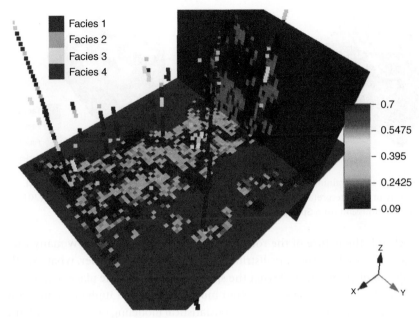

Figure III.1.3 Probability of facies 4 (channel sand) calibrated from seismic and well data. Lithological interpretations from logs are shown along the well bores.

used to determine the spatial probabilities in Figure III.1.3, some as simple as a co-located seismic-to-well calibration from a scatterplot, and some more complex involving neural networks and multivariate regression. At this early stage of the reservoir's life, one may already have a few years of production data, that is, data concerning the flow rates of oil and water (and the fraction of each) in producing wells. Production data provide very indirect information on the heterogeneity of the system because they are basically both integrated over space and time; they are also only available at specific locations and moreover influenced by other properties such as fluid properties (e.g., PVT properties and relative permeability) and the pressure or rate constraints imposed during production. Although these data may not inform very precisely rock types and/or petrophysical properties at that stage, they are very important indicators of reservoir connectivity (i.e., how fast injection water reaches producing wells). Figure III.1.1(c) shows 3 years of production data from some existing wells.

1.1.4 Geological interpretations

An expert geosciences team does not start with building directly geostatistical models using the various data discussed above. Geologists first interpret the nature of the depositional system. In doing so, they use spatial data, such as core–log and seismic data, to postulate on the depositional system. At the exploration or early appraisal stage, it is often clear to geologists what major type of system they are dealing with: fluvial, deltaic, carbonate reefs, or the like. What

Figure III.1.4 Three training images created with Boolean modeling techniques that reflect uncertainty in the interpretation of the depositional system. Modified from Park et al. (2013). With kind permission from Springer Science and Business Media.

is less clear is the nature of the rock type distribution and even how many rock types are present, how they are distributed in relation to each other, what exactly controls this distribution, and what the spatial trends are in the placement of the various rock types. Geologists are often more interested in understanding how the system was created, whereas most geostatistical modeling focuses on building numerical models of the subsurface. Critical to successful modeling is therefore the translation of this quest for understanding into a quest for quantitative model representation.

Specific to the WCA, geologists provided three different quantitative interpretations based on their qualitative understanding. The quantification is in terms of the geometry and spatial distribution of the various rock types. Boolean modeling and simulation are used to present these interpretations as visual appreciations in Figure III.1.4. Note that these are not reservoir models: they are clearly not on the same grid as Figure III.1.2 (i.e., they are not located in a real-world reference system), nor are they locally constrained to the well log, seismic, or production data. Each such conceptual interpretation is termed a "geological scenario." The first geological scenario has sand facies with low channel thickness, a low width–thickness ratio, and low channel sinuosity. The second geological scenario has sand facies with high channel thickness, a low width–thickness ratio, and low channel sinuosity. The third geological scenario has sand facies with low channel thickness, a low width–thickness ratio, and high channel sinuosity, including levees. In all scenarios, facies 2, 3, and 4 are sands, whereas facies 1 is nonreservoir shale. The petrophysical properties of all of these facies have different statistical distributions.

1.1.5 Summary of geostatistical modeling challenges

Reservoir modeling has seen considerable growth in the application of geostatistics from the mid-1980s to the present. The increased demand for energy is certainly not a stranger to this story. In addition, the easy access to oil has diminished rapidly, and energy companies are venturing in increasingly difficult and more unconventional environments, certainly in terms of geological heterogeneity. The time where reservoir engineers could forecast reservoir recovery from a simple decline curve analysis has mostly gone, and reservoir flow simulation is increasingly used to understand and manage these systems. Geological risk is now a common term.

Early modeling focused on layering resulting in so-called layer-cake models, which gave way in the 1990s to variogram-based models of rock types and petrophysical properties. The limitation of the variogram was rapidly understood by reservoir geologists, who then resorted to object-based or Boolean modeling (Deutsch and Wang, 1996; Holden, 1998; Skorstad et al., 1999; Lantuejoul, 2002) to better represent their understanding of the heterogeneity. Such Boolean models may represent geological heterogeneity fairly well, should such heterogeneity be presented as objects (as is the case for WCA), but they were (and still are) difficult to constrain to a wealth and variety of data. In addition, the emphasis on risk made uncertainty of the reservoir models and hence the flow responses calculated from them more important. Variogram-based models in the context of multi-Gaussian simulation were seen as providing too small uncertainty and as showing biases in flow predictions. Training images were introduced in the late 1990s as a way to convey an interpretation of geological heterogeneity. The aim was to address the lack of geological realism of variogram-based techniques, the difficulty of conditioning in Boolean models, and the increased importance of quantifying uncertainty. It is fair to state that petroleum geostatistics was the first major application of the MPS approach. Nevertheless, some geostatistical modeling challenges remain, which they are summarized as follows:

1 How does one quantify geological understanding of the genesis of the depositional systems into manageable quantitative numerical training images?
2 How does one verify and understand the relationship between such explicit geological interpretation and the reservoir data? Their possible inconsistency?
3 How can one generate rapidly reservoir models using these training images constrained to well log, seismic, and production data?
4 How can one account for the difference in scale between seismic and well data?
5 How does one rapidly update knowledge on both training images and models when new data become available?
6 How does one make this practical with a limited amount of CPU time?

The case study presented in this chapter is fairly typical for many reservoir forecasting problems where geostatistics has an impact: there is large spatial uncertainty about reservoir heterogeneity, yet at the same time there is considerable analog information in terms of interpretations from experts based on the reservoir data and outcrop studies. Few wells are drilled; and 3D seismic data are available, but of limited resolution. For that reason, Section III.1.2 will focus mostly on the more established procedure of multiple-point modeling in reservoirs. At the end, we will provide an overview of alternative workflows and how this field is still evolving.

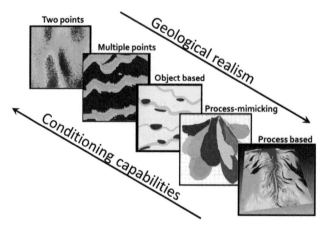

Figure III.1.5 Overview of various methods for quantitative reservoir modeling. A trade-off exists between conditioning to reservoir data from wells, seismic and production, and the realism of the description.

As a side note, we point out that MPS is not the only quantitative approach to geological modeling of these kinds of subsurface reservoirs systems (see Figure III.1.5); alternatives are summarized as follows:

- *Variogram-based geostatistics*: usually applied in the context of assigning petrophysical properties to fairly homogeneous layers or facies bodies.
- *Boolean or object-based geostatistics*: usually applied in cases where few well data are available and the facies geometries can be described by object shapes. Boolean methods are also used frequently to generate training images.
- *Multiple-point geostatistics*: applied in cases with conditioning that Boolean methods cannot handle, or when training images are available.
- *Process-mimicking models*: applied in cases with few wells data where a clear stacking pattern of sediments can be interpreted and modeled, as either erosion or depositional events (Pyrcz et al., 2005; Zhang et al., 2009; Bertoncello et al., 2013).
- *Process-based models*: these are rarely applied as reservoir models directly because they cannot be constrained easily to data (Sun et al., 1996, 2001; Michael et al., 2010).

1.2 Application of MPS to the WCA case

1.2.1 Validation of geological interpretations with reservoir data

Important in many geostatistical studies is not only the application or choice of the particular method of estimation or simulation, but the preparation of data as well as validation of the model choices prior to running the algorithms, procedures, or workflows. The "data" that enter geostatistical procedures in reservoir modeling are in fact not raw data themselves; they are already considerably processed or filtered. Geophysicists often term the output of their work "the geophysical model", which is used by geostatisticians as "the seismic data". Well logs need to be interpreted to provide sequences of sand–shale and porosity along the well trajectory; they also need to be positioned correctly within the modeling grid. Seismic data need to be tied to wells to ensure that layer markers that are accurately interpretable in wells are coinciding with those from a more error-prone seismic interpretation.

In variogram-based reservoir modeling, the variogram of rock type and petrophysical properties is calculated from the well data. Due to the folding and faulting, a coordinate transform is needed to make this meaningful. This matter is discussed in more detail here, along with the discussion of the grid for geostatistical modeling. Obtaining a meaningful experimental variogram is often difficult for subsurface reservoirs, and this is certainly the case for WCA. Most commonly, one may obtain a reasonable vertical variogram, due to the fact that many exploration or pre-production wells are vertical. However, with large inter well spacing, because of the heterogeneity of the system and the existence of many different facies each with different porosity and permeability characteristics, the experimental horizontal variograms often appear noisy, despite the presence of a strong geological structure, such as channels.

In such cases, the modeler often has to "invent" horizontal variogram characteristics such as range and anisotropy ratios. One may be tempted to consider seismic data to complete the variogram modeling because such data have good coverage and reasonable resolution in the horizontal directions. However, due to the filtering effect of such data, any variogram properties extracted may be biased (Mukerji et al., 1997).

Even if variogram parameters can be established, the variogram remains quite uninformative about the exact nature of the subsurface heterogeneity. Consider, for example, the three training images of WCA in Figure III.1.4. The vertical as well as omnidirectional horizontal variograms are calculated for these three training images. Despite the obvious visual differences between these training images, the variograms are all reasonably similar; see Figure III.1.6. Moreover, the vertical variograms of the conceptual images are not contradicting the variogram of the well data; see Figure III.1.6. This does not prove any of the training images as "correct"; it simply validates them with the wells data. Additional validation studies for WCA have been published in Boisvert et al. (2007).

Geoscientists often do not consider in great deal the production data of the reservoir, should such data be available. Instead, they focus on generating geological understanding and interpretations based on analog outcrops and principles of deposition, genesis, and process. Production data cannot be directly interpreted in terms of geological properties, such as channel geometries, proportions, or the spatial distribution of petrophysical properties. Such data require an entirely different expertise that is not part of the vocabulary of the geoscientist or geostatistician. For this reason, one of the major challenges in reservoir forecasting is to merge two different disciplines: descriptive geology and

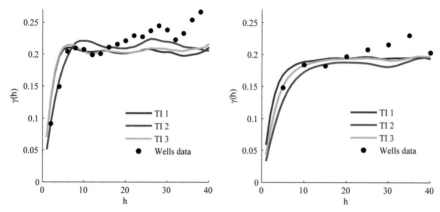

Figure III.1.6 (Left) Vertical indicator variogram of the channel facies (highest permeability facies) of the training images versus the well data (right) variogram of the training image along the channel direction versus the omnidirectional horizontal variogram of the well data.

time-varying flow in porous media data. If the modeling of reservoir properties turns out to be completely inconsistent with the production data, then all efforts put into geostatistical modeling are in vain. The major problem is that there are many factors that can create inconsistencies between the interpretations made on reservoir heterogeneity, such as those expressed in the training images, and the field production data. In fact, such inconsistency is often the norm. The result of this is that often the geostatistical modeling effort is abandoned and the model is forced to match the production data by simple manual adjustment. An extreme example of this is shown in Figure III.1.7. Evidently, such models have little use in the actual purpose: forecasting. It is therefore often important to discover whether any inconsistency between geosciences–geostatistics and the flow data of the reservoir occurs prior to building the model.

Any validation between conceptual training images and production data cannot escape from running some flow simulations. Such flow simulations require additional information such as relative permeability, properties of the fluids, boundary conditions and well constraints. It will also require building some reservoir models from the conceptual training images, which is the topic of Section III.1.2.2. After having treated the topic of geostatistical model building, we will therefore revisit the question of training image validation with production data.

1.2.2   Constructing 3D numerical reservoir models
1.2.2.1   The grid

Most commercial or industrial flow simulators rely on finite difference methods, requiring a grid with grid cells and volumes. The grid is of considerable importance in most reservoir forecasting methods, but we will focus on some important geostatistical aspects that often go unmentioned. There are two main

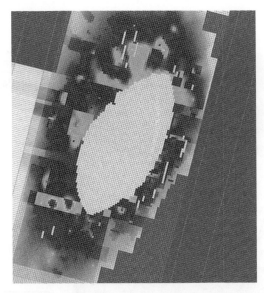

Figure III.1.7 An example of a reservoir model generated using variogram-based geostatistics, then manually adjusted (the various boxes and bars) to match the production data. The variable presented here is permeability.

differences between the grid concept in flow modeling and geostatistical modeling: the first pertains to the volume of the cells, and the second to the regularity of the cells.

It should be clear now that all multiple-point simulation methods outlined above, with the exception of ENESIM and direct sampling that can use flexible neighborhoods, require a regular Cartesian grid. Real reservoir traps are not contained in regular boxes; instead, they consist of complex layering and faults. Modelers build flow simulation grids in this physical domain, as shown in Figure III.1.8, where distances should be measured along layers and not with a simple Euclidean distance. For this reason, one first transforms the physical grid into a depositional grid. This grid is regular and flat; it represents the geological layers "at the time of deposition", although the latter should be considered only qualitatively. Geostatistical modeling takes place in this grid after all data (well data and seismic-derived probabilities) have been ported in this grid as well. Because of the grid spacing, well data may not fall onto the grid; see Figure III.1.8.

A second important difference between flow and geostatistics is that of volume, or, in geostatistical jargon, support. If the well data are considered as hard data, then geostatistical modeling proceeds by simulating values of the same volume as these hard data; see Figure III.1.9. Otherwise, by its very definition, hard data would not be deemed "hard". Figure III.1.9 shows this conceptual support difference between geostatistical grids and flow grid, even when no upscaling is performed. Flow grids contain cells with volumes that are neighboring each

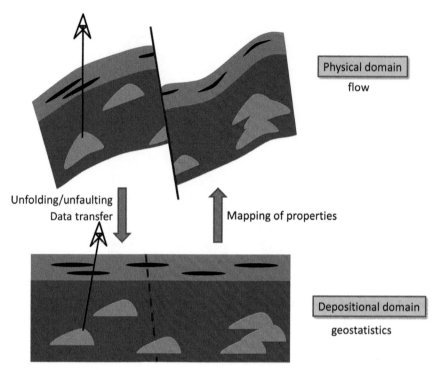

Figure III.1.8 Geostatistical modeling, including MPS, takes place in the depositional domain, whereas actual reservoir forecasting (including flow modeling) takes place in the physical domain (from Caers, 2011).

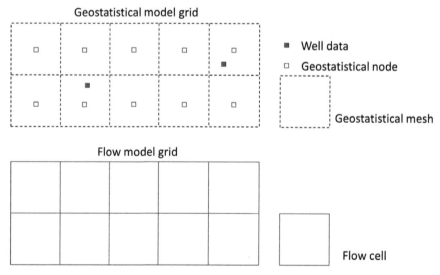

Figure III.1.9 Difference in representation between a geostatistical grid and flow grid, even in the case when no upscaling is used (a 2×5 grid is shown). The geostatistical algorithms simulate volumes located at the center nodes of a mesh of the same support as the hard data. The flow model contains larger volumes covering the entire mesh.

other. This is needed in the finite difference formulation because mass will be transferred through physical equations from one grid cell to the next. Geostatistics operates on a mesh. At the center of the mesh nodes sits a volume support that is equal to the volume of the hard data (the volume of a core). In all practice, however, the volume at the mesh node is assigned to the entire grid cell. This implicit "volume change" constitutes an upscaling termed "missing scale." The latter means that the geostatistical grid cell is considered homogeneous and that the single small volume is representative for that entire cell. Any pixel plot of a geostatistical model realization is therefore a bit deceiving; it shows the volume of the mesh colored with simulated values, while in reality we have much smaller volumes (typically, core volumes) that are represented at nodes of the geostatistical grid.

1.2.2.2   Choice of geostatistical methods and algorithms

Once the grid has been determined and the various data are assigned to that grid, geostatistical simulation can proceed. Geostatistical modelers now face a choice of algorithms. One of the goals of the case studies presented in this book is to provide some qualitative guidelines based on the author's experience and sensitivity studies performed for the various fields of application. In a reservoir context, the choice of an algorithm can be motivated by the following:

- What is the ultimate goal of the geostatistical models being created? In reservoir modeling, one can often distinguish four different purposes: (1) reserves estimations, (2) recovery estimation, (3) well location planning, and (4) well control planning. MPS plays a more important role in (2) and (3) when either no wells are drilled, or additional wells need to be planned. In such cases, the connectivity of high-permeability flow facies is often important. What are the amount and nature of the data available? Certain algorithms will perform better as the data become denser, whereas others perform optimally under very sparse data (one well). Some algorithms are ideal for multivariate problems with multiple soft data.
- What are the CPU requirements of the study? If a flow simulation takes 12 h, then it may not matter what is the CPU time of the chosen geostatistical algorithm, as long as it stays within the 10 min mark.
- What is the nature of the conditioning data? Dense well cases are better addressed with pixel-based modeling methods.
- What is the nature of the training image? The complexity of the training image may provide possibly the most important factor in deciding between multiple-point algorithms. Some algorithms do not perform very well with complex patterns consisting of strongly curving, thin features with low proportion over the entire domain. Other algorithms require stationarity of the training image, meaning that any trend needs to be imposed through soft-data variables.

In the WCA case study, the SNESIM algorithm (see Section II.3.2.2) is used for the following reasons:

- The soft data are a probabilistic representation of the seismic information.
- The well conditioning is important, and the wells consist of complex facies successions.
- The goal is to simulate categorical properties, not continuous ones.
- The training images are complex, but not overly complex: channel sinuosity exists but channels do not meander, nor are they extremely thin. All facies proportions are above 10%.
- The CPU requirements of SNESIM are on the order of 10 minutes, while flow simulation takes 4 h.
- In petroleum reservoirs, it is very important to have a control on the final proportions of the facies because there is often a desired target (see discussion later in this chapter), and this strongly controls the total amount of oil in place.

1.2.2.3 A note on marginal proportions

The marginal distribution of the rock types – or, simply, global facies proportions – has gone unmentioned so far. In the context of reservoir modeling, these proportions are important because they quantify the amount of reservoir rock containing fluids (shale does not contain producible fluids unless fractured), and hence impact not just recovery but also the modeled oil in place (reserves). What source of information is used to determine these proportions? Wells are often drilled in preferential areas or planned systematically based on seismic data, hence a considerable bias may occur when lifting proportions from wells. Training images are conceptual in nature; they primarily depict patterns deemed realistic from a geological point of view. Geologists may not (yet) pay considerable attention to proportions when creating 3D training images. Therefore, in practice, the target proportions are often imposed on the model. The only source of information that has large and exhaustive coverage is seismic data, yet such data do not directly inform proportions. The target facies proportions are therefore determined from a careful calibration between wells and seismic data as follows:

- For each facies (rock type), one determines, based on the wells data and co-located seismic, the likelihoods f_{well} (seis|facies i exists) and f_{well} (seis|facies i does not exists).
- Bayes' rule is applied to determine the posterior probability

$$P(\text{facies } i|\text{seis}) = \frac{f_{well}(\text{seis}|\text{facies } i)\, P(\text{facies } i)}{\sum_{all\, i} f_{well}(\text{seis}|\text{facies } i)\, P(\text{facies } i) + f_{well}(\text{seis}|\text{not facies } i)(1 - P(\text{facies } i))}$$

$$(\text{III.1.1})$$

This requires specifying the prior P(facies i), which is often based on the company's database of similar cases or simply some initial wide statement of proportion using a triangular distribution.

- This posterior probability is a function $\phi(seis)$ of the seismic and hence can now be applied to all grid locations (seismic is exhaustive), not just the well locations in the grid.
- Over the reservoir volume of interest, the sum is taken over all these probabilities.
- This sum is used as an estimate of global proportion for that facies.

Evidently, as with any reservoir property, uncertainty on facies proportions exists and the described calibration procedure has many sources of uncertainty: the difference in scale between well and seismic, the noise in seismic data, the calibration procedure, the seismic-to-well tie, as well as the game of chance played when drilling wells. Analytical methods (Biver et al., 1996; Haas and Formery, 2002) as well as spatial bootstrap (Caumon et al., 2004) are used to state uncertainty on facies proportions through pdfs. By drawing from these pdfs, the previously fixed proportion values can be varied, and hence each conditional simulation can exhibit such simulated proportions values. The conditional simulation generated by SNESIM can be constraint by these proportion values through the servo-system mechanism.

The most common way to integrate these probabilities is via the tau model (see Section II.2.8.3.2). In the WCA case, however, the seismic-derived probabilities did not provide strong constraints in terms of local variations in facies distribution.

1.2.2.4   Conditional simulation

In generating conditional simulation constrained to well data, we take the common two-step approach: first, perform conditional simulation to rock types (facies); and then fill those facies with petrophysical properties. Because the latter is typically performed with variogram-based geostatistics, we will not discuss it in this chapter. The conceptual 3D images of Figure III.1.4 are now used as 3D training images for the SNESIM algorithm. An equal number of conditional simulations are generated with each 3D training image. Figure III.1.10 shows a conditional simulation for training image 1 constrained to well data and seismic-derived probability cubes.

1.2.3   Quality control of the results

Posterior to generating models, various quality control checks are desirable. The same would be needed in traditional geostatistical methods: a check of the univariate distribution (marginal histogram) and of the variogram need to be performed. In Chapter II.8, we presented several techniques for quality control. Such quality control should focus on both conditioning and pattern reproduction.

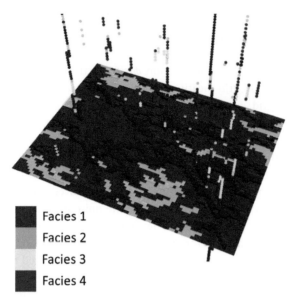

Figure III.1.10 Conditional simulations (horizontal section in a depositional grid) constrained to the facies data at wells locations for training image 1. Model shown in depositional coordinates (Cartesian box).

The quality of the conditioning can be easily assessed by looking at the ensemble averages of all conditional models per training image, in particular near wells. This is shown in Figure III.1.11.

1.2.4 Production data
1.2.4.1 Sensitivity analysis

After some production, such as in WCA, the reservoir models generated using the above methodology (constrained to wells and seismic) need further constraining with production data. As mentioned above, production data represent a data source considerably different from the geosciences data sources. The successful integration of these data sources into modeling is often termed "history matching", although this term gives the false perception that what only needs to be done is "matching" (Figure III.1.7). Matching simply means that the reservoir model, when subjected to flow simulation, matches the historical production data at the location of the wells (usually producer wells). Typically iterative or inverse procedures (ranging from simple manual trial and error to optimization and inverse problems) are used to obtain such matches. At each iteration, the reservoir properties (facies, structure, and petrophysical properties) are modified. However, since production data are rather ambiguously informative about reservoir properties, a possibly infinite collection of models can be created that match these data, some of these models make no sense geologically, or the

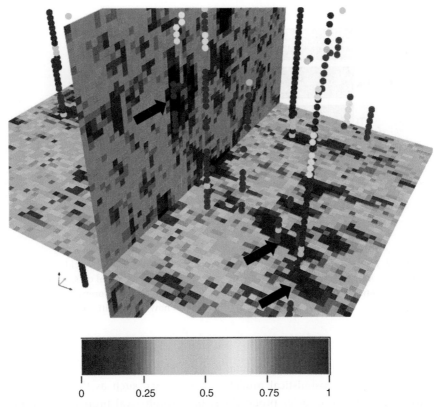

Figure III.1.11 Ensemble average of 50 realizations of the facies 3 binary indicator, representing the probability of occurrence of that facies given all conditional realizations. The black arrows near the conditioning wells indicate that the well conditioning data are achieved.

models do not match the previous well and seismic data any longer. Geologically realistic history matching is therefore desired to enhance forecasting.

In WCA, a number of factors have been determined to possibly affect flow, hence any forecast, including the simulated response of the historical production. Some of these are:

- The relative permeability curves: these curves quantify how two fluids (oil and water) jointly flow through the reservoir.
- The training image: in most cases at that stage of the reservoir life, several training images may be defined due to the uncertainty in geological interpretation.
- The fluid properties such as residual oil saturation: this is immobile oil saturation that cannot be produced with water displacement.
- The petrophysical properties: in particular, the ratio between horizontal permeability k_h and vertical permeability k_v. It is expected in clastic systems such as WCA; fluids flow easier in the horizontal than in the vertical, due to the presence of layering.

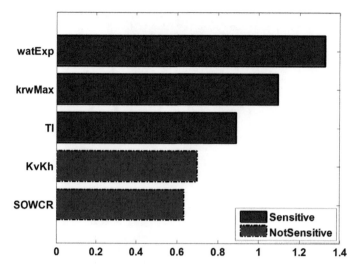

Figure III.1.12 Pareto plot listing the most important parameters affecting total field recovery in WCA determined using a sensitivity analysis. SOWCR = residual oil saturation; krwMax = maximum water relative permeability value; watExp = water Corey exponent; and TI = training image. See Fenwick et al. (2014) for details on how this sensitivity analysis is performed.

Prior to considering production data, it is therefore recommended to determine whether the geostatistical modeling parameters, such as the nature of the training image, the facies proportions, or petrophysical properties, are in fact impacting production, and hence historical production data. Figure III.1.12 shows a Pareto plot ranking from high to low the importance of some of the parameters mentioned: it is clear here that the k_v–k_h ratio is not importantly impacting the forecasts, whereas the training image has an impact on forecasting production responses. In this book, we do not focus on issues of relative permeability, but solely concentrate on facies models and training images.

1.2.4.2 Updating training image beliefs

Training images are constructed without considering any production data. Two main reasons can be stated for this: (1) geological interpretations often precede production, because such interpretation is needed at the appraisal stage; and (2) there is no directly discernible geological information in production data beyond rather qualitative and local assessments of connectivity (e.g., water injected from injector well I does not reach producer P).

In Chapter II.9 (Equation II.9.9), we therefore decomposed the problem into two parts: (1) updating the prior belief on each training image, and (2) inverse modeling with each individual training image. We now focus on explicitly modeling the likelihood term $f(\mathbf{d}_{obs}|TI)$, where $\mathbf{d}_{obs}$ denotes the production data.

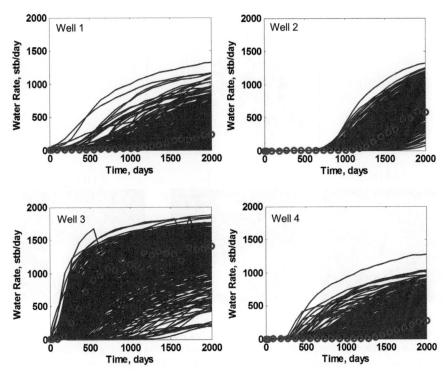

Figure III.1.13 Forward simulated production data on all 180 geostatistical models constrained to well log and seismic data (blue lines). The field production data are represented by the red dots. Modified from Park et al. (2013). With kind permission from Springer Science and Business Media.

This can be achieved in various ways; here, we first reduce the dimension of the response variable **d**, then apply kernel smoothing in that low-dimensional space to directly estimate the likelihood density. To reduce dimension, we first need to create a few scoping runs from the prior. To this end, 180 geostatistical reservoir models were generated based on wells data, seismic data and the three training images. All 180 models are run by the flow simulator. Figure III.1.13 compares the forward simulated production data on the geostatistical models, showing that the reservoir models created do not match the production data. Multidimensional scaling (Section II.2.4.5) is used to reduce the dimension of **d** to nine components (explaining 99% of variance), of which the first two are plotted in Figure III.1.14. For each training image, the density $f(\mathbf{d}|ti_k)$ is estimated from this scatterplot. Then, Bayes' rule:

$$P(TI = ti_k | \mathbf{D} = \mathbf{d}_{obs}) = \frac{f(\mathbf{d}_{obs}|ti_k)P(TI = ti_k)}{\displaystyle\sum_{k=1}^{K} f(\mathbf{d}_{obs}|ti_k)P(TI = ti_k)} \qquad (\text{III.1.2})$$

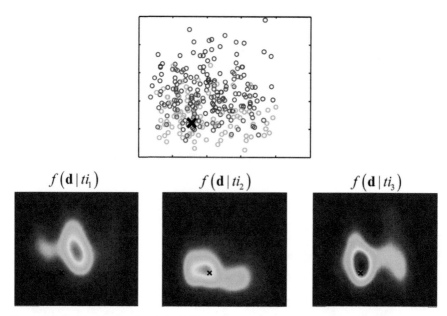

$f(\mathbf{d}\,|\,ti_1)$ $f(\mathbf{d}\,|\,ti_2)$ $f(\mathbf{d}\,|\,ti_3)$

Figure III.1.14 (Top) MDS plot of the field production data and the forward simulated production data for each training image (green = ti_1; red = ti_2; blue = ti_3); and (bottom) 2D projection of the kernel density estimation from 9D space showing the actual field production data marked with X. Modified from Park et al. (2013). With kind permission from Springer Science and Business Media.

allows calculating the updated beliefs as probabilities for each training image. Given prior probabilities of $P(TI = ti_1) = 0.50; P(TI = ti_2) = 0.25; P(TI = ti_3) = 0.25$, one obtains:

$$P(TI = ti_1 | \mathbf{D} = \mathbf{d}_{obs}) = 0.01; \quad P(TI = ti_2 | \mathbf{D} = \mathbf{d}_{obs}) = 0.39;$$
$$P(TI = ti_3 | \mathbf{D} = \mathbf{d}_{obs}) = 0.60$$

These probabilities are then used to determine how many history-matched models should be generated with each training image. The topic of history-matching models for each training image is presented next.

1.2.4.3   Probability perturbation with regions

For each training image that has received nonzero probability in the updated beliefs, history matches are generated. Because we are dealing with facies (categorical) models, including wells, and probabilistic seismic data constraints on such models, we employ the probability perturbation method (PPM; Section II.9.4.2). We will use a particular form of PPM termed "regional PPM" (Hoffman and Caers, 2005). In regional PPM, the field is decomposed into regions, either statically based on geological consideration or dynamically based on streamline geometry that delineates the drainage regions of the producing wells

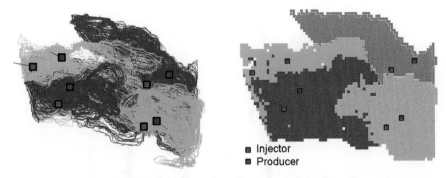

Figure III.1.15 Regions created using streamline simulation. Modified from Park et al. (2013). With kind permission from Springer Science and Business Media.

(Figure III.1.15). The advantage is that perturbation is varied per region, resulting in faster convergence.

Based on the updated belief $P\left(TI = ti_k|\mathbf{D} = \mathbf{d}_{obs}\right)$, 29 posterior (history-matched) models were generated, 12 from ti_2 and 17 from ti_3. Figure III.1.16 shows a few posterior models together with the matched production data. The nature of the regional PPM algorithm is such that, although regions are used for model perturbation, they do not affect the reproduction of channel geometries or create artifacts at the region boundaries. The total computational effort is 508 (PPM) + 180 (scoping) = 688 runs, or ~ 24 flow simulations per generated facies model.

1.2.4.4   Comparison with rejection sampling

Although requiring considerable CPU, for the sake of this case study and for validating what has been done, we apply the following implementation of the rejection sampler on both a TI variable and facies model:

1 Draw randomly a TI from the prior.
2 Generate a single geomodel **m** with that TI.
3 Run the flow model simulator to obtain a response **d**=g(**m**).
4 Accept the model if RMSE($\mathbf{d}_{obs}$, g(**m**)) < t.

RMSE is a root mean square difference between field and model response. A large number of tries – namely, 7236 flow simulations – are needed to obtain the same amount of models as PPM. The outcome of this rejection sampler is twofold: (1) a set of 29 history-matched models, and (2) updated TI probabilities. For those, we obtain:

$$P(TI = ti_1|\mathbf{D} = \mathbf{d}_{obs}) = 0.03\,(1\text{ model})$$

$$P(TI = ti_2|\mathbf{D} = \mathbf{d}_{obs}) = 0.34\,(10\text{ models}) \qquad\qquad \text{(III.1.3)}$$

$$P(TI = ti_3|\mathbf{D} = \mathbf{d}_{obs}) = 0.63\,(18\text{ models})$$

The updated TI probability obtained by rejection sampling is very similar to the one obtained by kernel smoothing in metric space.

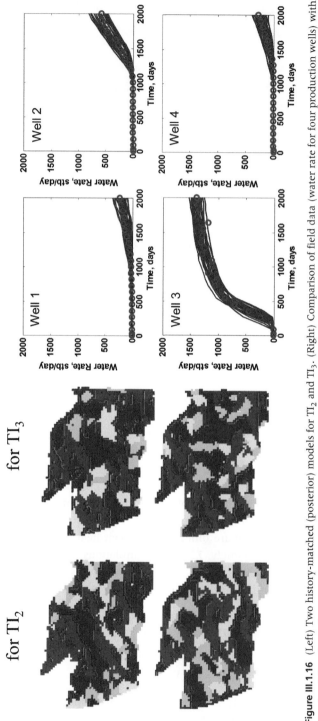

Figure III.1.16 (Left) Two history-matched (posterior) models for TI$_2$ and TI$_3$. (Right) Comparison of field data (water rate for four production wells) with simulated flow from the posterior models (modified from Park et al., 2013).

1.3 Alternative modeling workflows

The presented case study, WCA, is fairly typical for most applications of multiple-point geostatistics to reservoir modeling: the training image is generated using a stationary Boolean model and nonstationarity in facies proportion and variability is mostly enforced through a 3D probability model derived from seismic data. Without relying on their limited conditioning capabilities, Boolean models are great candidates for modeling cases such as the WCA. The modeler provides a geometrical description of each depositional facies, for example in map view and cross-section shape, and specifies certain prior distribution on object parameters such as length and thickness. Most of this information comes from well log data or is borrowed from a company's databases or outcrop studies. If indeed some of this information is borrowed from wells, the generated training images are consistent with that information. In general, the unconditional Boolean simulation that serves as a training image is stationary.

This basic workflow performed well for WCA, covering a relatively small geographical area. Based on what has been discussed in previous chapters, alternative workflows can be envisioned where the WCA modeling approach may not fully apply, or where different circumstances are present:

- Cases where the training image consists of very complex, thin, low-proportion curvilinear patterns (shale drapes and thin meandering channels) that need to be handled with transformed variables (see Section II.2.5.4).
- Cases where the reservoir heterogeneity is not easily decomposed into rock types, and hence modeling continuous variable training images is needed.
- Cases where geologists have difficulties when creating stationary training images simply because their understanding and interpretation of the system require them to think of trends as an integral part of such understanding. Specific methods then have to be applied to create nonstationarity models (see Chapter II.5).
- Cases where the reservoir heterogeneity cannot be easily interpreted in terms of objects, and hence other means than Boolean models, such as process-based models, need to be used for generating training images (see Chapter II.7).

One of the greatest challenges in quantitative reservoir modeling lies in merging two different worlds: geological sciences related to depositional systems, and quantitative and numerical geostatistical modeling. Although discussion on what algorithm (MPS or others) or combination of modeling approaches to use for what situation is important (from a pure geostatistical point of view), the overarching difficulty lies in the difference in language and purpose between geology and geostatistics. The primary purpose of the reservoir geoscientist lies in understanding the nature of the depositional system, to understand what factors contributed to the creation of the system in place. Geologists reason through genesis, process, and analogs, sometimes quantitative but most times qualitative. The geostatistician is mostly interested in what is current or present with a focus on geometry and topology: what is the quantitative description of those geometries as well as spatial variation affecting reserves and flow? If these geometries are known, then it almost does not matter how they got there in the first place. Yet, such a perfect situation rarely arises, and knowledge of processes may help constrain the understanding of the current geometrical features. In this context, the creation of relational database systems, relying on a geostatistics-focused classification of geological systems, combined with training image–generating algorithms has made great strides into merging these two worlds. The training image can be seen as a communicating language bridging two fields of science. There is not one single unique MPS technique or algorithm that fits all problems in reservoir modeling. This chapter provided an application-driven overview of how the toolbox can be applied in many circumstances.

References

Bertoncello, A., Sun, T., Li, H., Mariethoz, G. & Caers, J. 2013. Conditioning surface-based geological models to well and thickness data. *Mathematical Geosciences*, 45, 873–893.

Biver, P., Mostad, P. F. & Guillou, A. 1996. An overview of different techniques to quantify uncertainties on global statistics for geostatistical modeling. *In:* Baafi, E. Y. & Schofield, N. A. (eds.) *Geostatistics Wollongong '96.* Kluwer, Wollongong, Australia.

Caers, J. 2005. *Petroleum Geostatistics*, Society of Petroleum Engineers, Richardson, TX.

Caumon, G., Strebelle, S., Caers, J. K. & Journel, A. G. Assessment of global uncertainty for early appraisal of hydrocarbon fields. In: *SPE Annual Technical Conference and Exhibition, Houston, Texas, USA*, 2004. 685–692.

Deutsch, C. 2002. *Geostatistical Reservoir Modeling*, Oxford University Press, Oxford.

Deutsch, C. & Wang, L. 1996. Hierarchical object-based stochastic modeling of fluvial reservoirs. *Mathematical Geology*, 28, 857–880.

Dubrule, O. 2003. *Geostatistics for Seismic Data Integration in Earth Models*, SEG, Tulsa, OK.

Fenwick, D., Scheidt, C. & Caers, J. 2014. Quantifying asymmetric parameter interaction in sensitivity analysis: application to reservoir modeling. *Mathematical Geosciences*, 46, 493–511.

Haas, A. & Formery, P. 2002. Uncertainties in facies proportion estimation, I-theoretical framework. *Mathematical Geosciences*, 34, 679–702.

Hoffman, B. T. & Caers, J. 2005. Regional probability perturbations for history matching. *Journal of Petroleum Science and Engineering*, 46, 53–71.

Holden, L. 1998. Modeling of fluvial reservoirs with object models. *Mathematical Geology*, 30, 473–496.

Kelkar, M. & Perez, G. 2002. *Applied Geostatistics for Reservoir Characterization*, Society of Petroleum Engineers, Richardson, TX.

Kitanidis, K. 1997. *Introduction to Geostatistics: Applications in Hydrogeology*, Cambridge University Press, Cambridge.

Lantuejoul, C. 2002. *Geostatistical Simulation: Models and Algorithms*, Berlin, Springer.

Michael, H., Boucher, A., Sun, T., Caers, J. & Gorelick, S. 2010. Combining geologic-process models and geostatistics for conditional simulation of 3-D subsurface heterogeneity. *Water Resources Research*, 46.

Mukerji, T., Mavko, G. & Rio, P. 1997. Scales of reservoir heterogeneities and impact of seismic resolution on geostatistical integration. *Mathematical Geology*, 29, 933–950.

Park, H., Scheidt, C., Fenwick, D., Boucher, A. & Caers, J. 2013. History matching and uncertainty quantification of facies models with multiple geological interpretations. *Computational Geosciences*, 17, 609–621.

Pyrcz, M. J., Catuneanu, O. & Deutsch, C. V. 2005. Stochastic surface-based modeling of turbidite lobes. *AAPG Bulletin*, 89, 177–191.

Ringrose, P. S. & Bentley, M. 2014. *Reservoir Model Design: How to Build Good Reservoir Models*, Springer, Berlin.

Skorstad, A., Hauge, R. & Holden, L. 1999. Well conditioning in a fluvial reservoir model. *Mathematical Geology*, 31, 857–872.

Sun, T., Meakin, P. & Jøssang, T. 2001. A computer model for meandering rivers with multiple bed load sediment sizes 2. Computer simulations. *Water Resources Research*, 37, 2243–2258.

Sun, T., Meakin, P., Jøssang, T. & Schwarz, K. 1996. A simulation model for meandering rivers. *Water Resources Research*, 32, 2937–2954.

Zhang, X., Pyrcz, M. J. & Deutsch, C. V. 2009. Stochastic surface modeling of deepwater depositional systems for improved reservoir models. *Journal of Petroleum Science and Engineering*, 68, 118–134.

CHAPTER 2

Geological resources modeling in mining

Coauthored by Cristian Pérez, Julian M. Ortiz,[1] & Alexandre Boucher[2]

[1] *University of Chile*

[2] *Ar2Tech*

2.1 Context: sustaining the mining value chain

Mining operations are usually constrained by the processing plant capacity. In order to maximize revenue, a decision must be made regarding what materials must be sent to the processing plant. This requires knowing the properties of the ore in as much detail as possible. The plant is fed to its maximum capacity, if possible, with the best material that is available at that time of production.

In order to satisfy this requirement, understanding geological variability in the ore body is of paramount importance. From a limited amount of information, a geological resources team needs to characterize the properties of ore and waste. This includes predicting grades of different elements, quantities of ore above different cutoffs, the energy required to crush and mill the material to achieve a given particle size, and other metallurgical properties of the materials. This prediction is based on information that comes from different data types. Geological mapping at the surface is complemented with the logging of drill holes during exploration. Several models are built to improve understanding the 3D geological variability of the deposit: a structural model to characterize faults and major discontinuities is required, as these structures may have controlled the deposition of the elements of interest; mineralogical, lithological, and alteration models are also generated to understand the genesis of the deposit and develop metallogenic interpretations based on it, which constitute a conceptual model explaining how this deposit was formed. Often, the succession of geological events and spatial distribution determines the quantity and quality of the resources.

During production, blast holes are drilled for rock breakage. These blasted rocks are also sampled; therefore, more detailed information on the distribution of grades and other properties is available at this stage. This information can be used to check the quality of the existing resource model. One of the most significant challenges is to build and update the models of the geological units, which are then merged into estimation units.

The geological resource model is converted into mining reserves through the planning stage, when the volumes to be extracted and the proportions of materials to be sent to the plant or to the waste dump are defined. This plan is created at different time scales. A long-term plan, usually at the scale of years, is built from the exploration information. This plan provides a quantification of tonnages and grades over large volumes, defining the sequence of extraction of the different material types (ore, high- and low-grade stocks, waste). A medium-term model is then generated based on the exploration data along with additional infill drilling and blast hole samples, to refine the estimate of the tonnages and grades for shorter periods usually quarterly or monthly. The short-term plan is based on the production information (blast hole samples) and allows estimating properties of small volumes on a daily basis. Because it is at the origin of the mine plan, the resource model is the first and most essential step to sustain the mining value chain.

2.1.1   Issues in geological model construction

One of the main challenges in geological resource estimation is that resources are controlled by estimation units, which are defined based on the geological and geo-metallurgical behavior of the different types of materials (rocks). The resource geologist studies how changes in geological properties such as alteration mineralogy, lithology types, or mineralization zones control the spatial distribution of grades or other geo-metallurgical variables. These controls must be consistent with the metallogenic interpretation of the deposit. The metallogenic model constitutes a simplified interpretation of different types of deposits. It reflects an understanding of the spatial distribution of zones with different geological characteristics (lithologies, mineralogies and alterations) and aids exploration geologists in identifying targets for drilling.

Modeling starts by making interpretations over cross-sections, longitudinal sections and plan views, using the information from an exploration campaign, usually a set of drill holes over a wide grid in a pseudo-regular mesh. These interpretations in two dimensions are then integrated into large, three-dimensional volumes. This process is complex due to a highly structurally controlled deposit containing intricate shapes, rendering the construction of cross-sections tedious and time-intensive. Volumes defined during this process are limited by solids and surfaces that may cross each other. These inconsistencies generate errors when volumetric calculations are done. An alternative is to use some type of implicit modeling to directly build the solids and surfaces defining the volumes that represent the estimation units. Then, instead of processing two-dimensional sections, the model is constructed directly in 3D using Boolean operations over the domain. This has proven to work better than the wire-framing approach; however, including geological knowledge remains difficult.

Both approaches are somewhat cumbersome: they require a subjective input of the geomodeler and do not account for any uncertainty associated with modeling these volumes. Local variability of the contacts is usually smoothed out in areas of scarce data, providing an unrealistic rendition of the deposit. Traditional geostatistics offers many tools for modeling these types of categorical variables; however, they are seldom applied because they often do not reproduce accurately trends and transitions between the different categories. Local changes in the mean values, variance, or orientation of the spatial continuity are hard to impose.

Once completed, models often need to be updated with new information coming from additional drill holes or from blast hole samples, requiring a repeat of the same tedious exercise of updating the wireframe in 3D, building the solids, and correcting for any inconsistencies. After a few of these updating procedures, geological realism may be lost, and the focus lies mainly on being consistent with the drill holes and cross-section data. This often results in the construction of 3D shapes that have little resemblance to realistic geological shapes or spatial variability.

Successful geostatistical modeling methods should address the need for rapid modeling of complex geological variability, where geological knowledge is incorporated effortlessly, and for rapid updating with minimal manual intervention. Yet, any automated procedure is only valuable as long as geological realism is maintained, modeling time is reduced, and supervision by an expert geomodeler ensured. In this chapter, we present multiple-point geostatistics (MPS) methods that address specifically the problem of improving an existing deterministic model, either by assimilating new data into it or by expanding the single deterministic model into multiple realizations.

2.1.2 MPS in mining applications

The use of MPS in mining, and perhaps geostatistics in general, is different from its use in petroleum and other application areas. One of the characteristics of mining is that data may be quite abundant. In this case, conditioning to hard data will control many aspects of the model (such as local proportions) that in other applications may require additional information such as geophysics.

Furthermore, in mining, dense production information obtained from blast holes can be used to validate the generated models, or it can be considered as training information from which to infer pattern statistics. However, the blast hole data used may belong to a different area of the deposit; hence, a strong assumption of stationarity must be made to import these pattern statistics into the model. Global statistics (such as proportion in the lithological model) may not match the target statistics of the simulation area. Adjustments must be made to render the training set consistent with the modeling domain (Ortiz et al., 2007).

The critical issue in mining is the reproduction of geological features and the ability to impose expert knowledge about the phenomenon by means of additional data. This is somehow similar to what is seen in remote sensing applications (Chapter III.3) where abundant data are available and many layers of information must be integrated to allow for a reasonable model.

2.2 Stochastic updating of a block model

2.2.1 Introduction

In this chapter, we discuss some of the implementation details of building a geological model using MPS for the Escondida Norte mine in northern Chile. The challenges focus mainly on dealing with significant amounts of data and using MPS at the mine scale, requiring models that account for many blocks. Furthermore, the main goal of this process is to obtain models that respect the geological aspects of the deposit, in addition to honoring the statistics imposed, namely, global proportions, transitions between categories, and patterns.

In the remainder of this section, we show a path toward obtaining a model that is simple to build and fast to update. The case study progresses from a basic application of MPS using SNESIM (see Chapter II.3.2), considering the current deterministic geological model as the reference training image, to a more sophisticated application of direct sampling (DS; see Chapter II.3.3) to build a model based on the drilling data. As a final step, the possibility of combining the data-driven construction of the model obtained by DS with a few relevant sections incorporating the geologist interpretation is discussed as a way of controlling the geological realism of the final model. Lastly, an alternative workflow for generating spatial uncertainty from a single deterministic model based on contacts resimulation is presented.

2.2.2 The Escondida Norte case study

Minera Escondida is the largest copper producer in the world. It produces copper concentrate and copper cathodes. The operation is located in the Atacama Desert in northern Chile, 170 km southeast of the city of Antofagasta at an altitude of 3100 m above mean sea level. Mining commenced in 1990, and since 2005 the company has also operated Escondida Norte, which is a second open pit located 5 km from the main pit. Annual production reached 1,250,000 tons of copper in 2008. Oxide and low-grade sulfide ores are leached, then a solvent extraction process and electro-winning are used to produce copper cathodes. Sulfide ore is sent to the plants (Laguna Seca or Los Colorados) where copper concentrate is produced. Two pipelines transport the copper concentrate to the Coloso Port, located in the southern area of Antofagasta, where it is filtered and exported. The cathodes are shipped by truck to the port of Antofagasta for subsequent export.

Given the size of the operation, Escondida poses many challenges in terms of data processing, modeling, and forecasting. For example, in 2010, over 1 million meters of exploration holes were drilled. This implies that, in addition to having many drilling rigs in the pits and surrounding areas, around 3000 meters of drill hole cores had to be logged, stored, prepared, and analyzed every day. The existence of different types of material in the mine and different options for processing requires making decisions with a considerable financial risk. Escondida carries a thorough geo-metallurgical program to characterize the types of materials, and to understand the performance of these materials at the different processes, in order to have the information necessary to optimize the destination of the different blocks of ore extracted from the mine. A systematic quality assurance and quality control program is in place to ensure that every process is within standard ranges of performance and that the best decisions are made at every stage.

There are essentially three types of data available for this case study: the topography of the Escondida Norte deposit sector, the current deterministic geological block model, and the drill hole data. The current geological model is a block model with blocks 25×25×15 m in size.

The case study focuses on the lithological model. Challenges may be twofold:

- With only scarce information, any kriging or estimation models are usually too smooth and do not represent properly the true underlying variability in the contacts between lithological units. A simulation model helps define the short-scale variability; however, with limited information, it is difficult to infer variograms, in particular for short distances. Analog models from an already mined area may, however, provide a training data set.
- When information is abundant, model building becomes easier; hence, the challenge often lies in updating the model with new information. An automated procedure can alleviate the burden of rebuilding the model every time new data are gathered.

2.2.3 Geological model simulation with SNESIM

A first approach to test the capacity of MPS to build lithology category distribution models for the Escondida Norte is to use SNESIM. Because many drill hole data are available, the data set is split into two subsets: one for model construction and another one for validation. These different data sets are defined by selecting complete drill hole data instead of composites (i.e., materials sampled in an interval along a drill hole), in order to mimic a drilling campaign with fewer drill holes in each case. The number of data for each subset built is shown in Table III.2.2. Performance is measured based on the prediction capacity over multiple realizations with SNESIM when compared to the validation subset. The exercise considers reducing the data up to one tenth of the available drill hole information to understand the value of information in this case, as measured by the percentage of match between models and validation data.

The study begins with an exploratory analysis of the data to assess the quality of the database, determine the number of categories, identify any missing or erroneous value, and determine the representativeness of the data. Escondida Norte's model accounts for seven lithologies (Table III.2.1), in addition to the category of "Air" in the model (above the topography).

Table III.2.1 The different categories in the Escondida Norte model

Lithology	Code given
Carmen porphyry	0
Feldespar porphyry	1
Rhyolitic porphyry	2
Andesite	3
Breccias	4
Granitic porphyry	5
Gravels	6

Table III.2.2 Number of drill holes in each data set

Conditioning drill holes (in %)	Conditioning data in amount	Validating drill holes (in %)	Validating data in amount
10	10,403	90	92,962
30	31,241	70	72,124
50	51,506	50	51,859
70	72,829	30	30,536
90	92,734	10	10,631

The block model is built considering 25×25×15 m³ blocks, and it comprises over 760,000 blocks. The simulation area is located between 2675 and 3305 m above mean sea level. Drill hole information contains the logged lithology in composites of 5 m in length. There are 103,650 composites from 1780 drill holes.

The declustered proportions of lithology categories using a nearest-neighbor approach are depicted in Table III.2.3 for each one of the data sets considered for the study. The construction of the subsets for modeling and validation preserves the proportions of the different categories with relatively small fluctuations.

Table III.2.3 Categories' proportions after declustering

Code	Percentage of conditioning data				
	10%	30%	50%	70%	90%
0	0.33%	0.57%	1.10%	0.54%	1.00%
1	22.37%	19.68%	22.36%	20.30%	19.49%
2	36.46%	35.73%	26.69%	29.89%	28.53%
3	32.38%	35.61%	39.03%	40.33%	41.94%
4	3.65%	3.38%	5.33%	3.42%	3.27%
5	1.46%	1.39%	1.75%	1.63%	1.81%
6	3.35%	3.64%	3.74%	3.89%	3.97%

In the Escondida case as well as in many mining applications, the most convenient training image is the existing deterministic model. Such models are heavily informed by a dense array of drill holes and incorporate site-specific characteristics such as nonstationarity. Training images are therefore not used in the traditional way, but rather to create a model of spatial uncertainty from a single deterministic model. The problem is then similar to using training images that have the same nonstationary characteristics as the domain to simulate (Honarkhah and Caers, 2012; Hu et al., 2014). Using single deterministic models as training images may, however, not be ideal because they may suffer from several drawbacks: (1) it is a representation of the true distribution of lithologies, but not the ground truth; (2) contacts between different categories will usually look smoother than in reality; and (3) there is only a single deterministic training image, therefore uncertainty in the conceptual model is not accounted for.

However, due to the abundant drill hole information, uncertainty in the spatial distribution of the geological rock types (facies) may not be as relevant as in other applications (see the petroleum application in Chapter III.1, where only a few boreholes are available). In this context, MPS is used to rapidly construct geological models that honor the interpretation and respect the data, avoiding cumbersome wire framing and solid generation.

For this case study, two training images are considered: the first one corresponds to the full interpreted geological model, and the second one corresponds to a subset of it. Two modeling decisions are made regarding nonstationarity: one type of model considers the enforcement of global proportions (stationary in global proportions) and a second model type considers five regions where local proportions are enforced.

Fifty realizations are built with each training image. One of the models is shown in Figure III.2.1. The SNESIM models were postprocessed with the maximum a posteriori selection (MAPS) algorithm proposed by Deutsch (1998) to enhance pattern reproduction. Quality controls include a comparison of reproduced global and vertical proportions with the existing data and a comparison of indicator variograms between models and training images, all of which were well reproduced.

Performance is measured by the success rate in matching the lithology logged in the validation subset. Results are presented in Table III.2.4.

Results show that the use of regions with imposed local proportions does not make a significant difference. Conditioning data, if in abundance, usually force the realizations to respect the local proportions. Another interesting observation from these results is that the choice of the training image only has a minor impact on model performance. This application shows that multiple-point simulation techniques can be applied successfully at the mine scale. The spatial continuity of the lithologies is well reproduced in the realizations, even when a small proportion of the original data is used for conditioning.

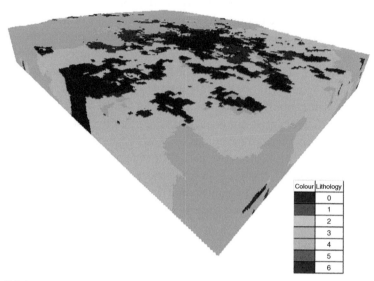

Colour	Lithology
	0
	1
	2
	3
	4
	5
	6

Figure III.2.1 One conditional realization using SNESIM.

Table III.2.4 Success rate in matching the validation set for various modeling choices

		Percentage of conditioning data				
		10%	**30%**	**50%**	**70%**	**90%**
Complete TI:	Maximum	61.67%	67.44%	70.71%	71.82%	74.73%
global	Minimum	57.74%	65.41%	68.28%	70.24%	70.82%
proportions	Average	59.18%	66.59%	69.48%	71.17%	72.56%
Subset TI:	Maximum	59.33%	66.14%	69.51%	71.34%	72.80%
global	Minimum	55.25%	63.61%	67.27%	69.12%	69.23%
proportions	Average	57.47%	64.80%	68.29%	70.09%	70.97%
Complete TI	Maximum	63.08%	67.95%	70.22%	72.29%	73.39%
with local	Minimum	59.49%	65.21%	67.50%	68.97%	70.09%
proportions	Average	61.37%	66.52%	68.88%	70.68%	71.71%
Subset TI	Maximum	61.93%	66.53%	68.47%	71.08%	72.19%
with local	Minimum	57.00%	64.21%	66.07%	67.80%	68.48%
proportions	Average	59.58%	65.43%	67.54%	69.72%	70.67%

Note: *Complete TI* refers to using the full deterministic model as TI; *local proportions* refers to using regions and enforcing local proportions; *subset TI* refers to using a subset of the full deterministic model as TI. The success in matching the validation data set is reported as the *average*, *minimum*, and *maximum* over a set of 50 realizations.

Table III.2.5 Presentation of conditioning and validation data sets

Conditioning drill holes (in % of total)	Amount of conditioning data	Validating drill holes (in % of total)	Amount of validation data
30%	26,926	70%	62,543
50%	45,008	50%	44,461
70%	62,453	30%	27,016

Because using the interpreted geological model seems to contradict the idea of avoiding its lengthy construction, an alternative approach for building stochastic models of the spatial distribution of these units is presented in Section 2.2.4.

2.2.4   Geological model simulation using direct sampling

Multiple-point simulation based on the DS approach (Chapter II.3.2.3) is computationally efficient and often more flexible than SNESIM. Because data are abundant, the "direct sample" may be directly lifted from the data set itself. This would avoid the construction of a training image altogether. In the Escondida Norte case, the generation of the geological model can be considered a reconstruction problem (Mariéthoz and Renard, 2010), where the simulated values are directly extracted from the same data used for conditioning. Using this approach, we redo the same validation exercise as above, evaluating the performance by means of validation data sets of different sizes or amounts.

In the application of DS, a smaller area is considered that contains just over 225,000 blocks. As before, the data contained within the model area are split into a conditioning subset and a validation subset. Sensitivity on the relevance of the available information is determined by studying three scenarios, as described in Table III.2.5. The proportions for all categories are well preserved within the subsets for all cases.

The main parameters for the simulation using DS are presented in Table III.2.6. In an attempt to incorporate local features and nonstationarity, coordinates are used as a secondary variable when accounting for the similarity measure to match pattern statistics. In this sense, coordinates are used as a control map, in a similar fashion as in the application of Chapter III.3.

Table III.2.6 Parameters used for the direct-sampling algorithm

Search radius X	35
Search radius Y	35
Search radius Z	15
Threshold	0
Maximum number of neighbors	30

Figure III.2.2 Top row: lithological models simulated with DS, bottom row: postprocessing of the top row models.

Fifty models are built in each of the three scenarios, with and without consideration of the local coordinates, which results in six cases. Realizations are postprocessed using MAPS. Figure III.2.2 shows individual realizations before and after applying the cleaning algorithm, built with increasing amounts of conditioning data (30%, 50%, and 70% of data are kept for conditioning, from left to right).

Models are assessed based on their capacity to reproduce global proportions as well as indicator variograms (not imposed), and by measuring the success rate in reproducing the lithologies logged in the validation subsets in each case. It should be noted that a 100% success rate is unachievable, because the support of the composites is smaller than that of the blocks (5 and 15 m, respectively). Also, the models are compared to the current deterministic geological model. Comparison of the final results with the validation subsets is shown in Table III.2.7.

When performance is validated against the current interpreted geological model, results improve slightly, as depicted in Table III.2.8.

For comparison, models are also built using sequential indicator simulation (SISIM) with the same conditioning sets. Indicator variograms are fitted and declustered proportions are used as targets. Results show success rates 1–4% lower when compared to the validation sets, and 2–5% lower when the comparison is made against the current deterministic geological model. All in all, the approach with MPS using DS shows a better performance and is easier to apply.

2.2.5 Improving the models using additional geological interpretations

The previous sections of this chapter show a natural progression from applying MPS simulation in a conventional way to adapting its use to the particular

Table III.2.7 Success rate when comparing realizations with the validation data set

	No secondary variable used					
	30% case		**50% case**		**70% case**	
	Not processed	**Processed**	**Not processed**	**Processed**	**Not processed**	**Processed**
Minimum	63.26%	68.08%	65.53%	70.41%	68.89%	71.49%
Maximum	64.64%	68.93%	66.89%	71.17%	70.29%	72.22%
Average	63.84%	68.58%	66.13%	70.73%	69.67%	71.86%
	Using secondary variable					
	30% case		**50% case**		**70% case**	
	Not processed	**Processed**	**Not processed**	**Processed**	**Not processed**	**Processed**
Minimum	63.49%	68.47%	65.43%	70.38%	69.05%	71.62%
Maximum	64.87%	69.40%	66.79%	71.21%	70.12%	72.66%
Average	64.01%	69.09%	66.25%	70.81%	69.51%	72.21%

Note: The processing refers to the postprocessing of the DS realizations. Postprocessing generally improves performance in terms of matching the validation results.

Table III.2.8 Success rate when comparing realizations with the current deterministic model

	No secondary variable used					
	30% case		**50% case**		**70% case**	
	Not processed	**Processed**	**Not processed**	**Processed**	**Not processed**	**Processed**
Minimum	64.68%	70.01%	65.71%	70.73%	69.92%	73.23%
Maximum	66.42%	72.32%	67.90%	73.06%	72.37%	75.66%
Mean	65.51%	71.11%	66.66%	71.75%	71.19%	74.48%
	Using secondary variable					
	30% case		**50% case**		**70% case**	
	Not processed	**Processed**	**Not processed**	**Processed**	**Not processed**	**Processed**
Minimum	64.77%	70.49%	65.35%	70.31%	69.92%	73.98%
Maximum	66.60%	72.82%	67.90%	73.06%	71.77%	75.23%
Mean	65.50%	71.33%	66.60%	71.72%	70.81%	74.35%

conditions of mining data. In general, we find that in mining applications, the amount of available information is considerable. This allows moving away from the use of a training image, which carries important assumptions of stationarity, and instead using the available dense conditioning data set as training data from which multiple-point statistics are borrowed, including the nonstationarity properties.

Although the training image may belong to the same deposit, a typical case is to use a training image that comes from blast holes in a mined-out area, where global statistics such as the mean and the variance are often different from those of the volume to be modeled. These changes, even when small, alter the frequencies of multiple-point patterns and may affect the quality of the final models.

A natural approach to further improve the model is to combine the use of a pure data-driven modeling using DS and inferring the statistics from the hard data, with some representative detailed cross-sections of the deposit, where detailed geological features can be interpreted, including all the transitions between categories. This is particularly important when there is evidence of features that are not easily observed in the drill hole data, such as discordant units cutting off others, a hierarchical ordering of sequences, small intrusions of categories within larger units, and so on.

The basic idea is then to replace the construction of systematic cross-sections—say, every 50 m—with a few representative sections that show the typical rock type distributions over the domain. These cross-sections can be used as conditioning information along with the hard data obtained from the drill holes.

2.3   An alternative workflow: updating geological contacts

The modeling method described above considers the entire deterministic model as a training image for generating multiple realizations or updating models. However, in facies models, certain locations may be known accurately, whereas other locations may be highly uncertain. It may be possible to update only the uncertain locations, therefore making the procedure very efficient in terms of both computation time and modeling effort. Here, we investigate a case where the main uncertainty is the location of the contacts between facies, and therefore the updating focuses only on these contacts (Boucher et al., 2014).

The contact updating workflow is illustrated using the Quadrilátero Ferrífero, which is one of the most important iron-ore mining districts in the world. It is located on the southeastern border of the São Francisco Craton, in the state of Minas Gerais, Brazil. Iron ores have been formed throughout the Quadrilátero Ferrífero by the enrichment of itabirites (a banded-quartz hematite also known as hematite schist). High-grade ores have iron contents higher than 64%, with a very low amount of contaminants.

The ore body considered is composed of discontinuous lenses of the iron-rich hematite interfingered with iron-rich soft itabirites. The deposit is approximately 800 m in length, 800 m in width, and more than 420 m in depth. The morphology of the deposit is partly controlled by several folding events. To capture this complex structure, a 2 millions cell stratigraphic grid was created to better represent geological units. Four major lithotypes were interpreted: waste, hard itabirite, friable itabirite, and hematite. Each lithotype has its own spatial characteristics (e.g., a histogram or variogram) of grade variation (similar to facies having different porosity or permeability variation in the WCA case of Chapter III.1). Based on the structural interpretation and the 1720 drill holes where lithology data are recorded, a single interpretation of the spatial distribution of these four lithotypes was built (see Figure III.2.3) using surfaces and wireframe generated with computer-aided design (CAD) software. As for the Escondida case, building such a single deterministic model is a tedious and time-consuming process. Modelers need to merge drill hole data with geological interpretations (often made at the surface). Any quantification of uncertainty by means of multiple realizations is impossible to achieve within a reasonable amount of time, and automation is therefore required. Here, we present an application of MPS to automatically perturb the single lithological model and yet remain consistent with the drill-hole data, thereby providing some measure of uncertainty consistent with data.

A first step in designing an MPS workflow to address this problem is to recognize that ore bodies such as in Figure III.2.3 consist of large zones with a single lithotype. From a spatial statistical viewpoint, Figure III.2.3 does not display a high degree of spatial variation in lithotypes; rather, they consist of massive bodies in contact with each other. In other words, the major uncertainty lies in the contact between lithotypes and how these contacts vary spatially. The focus of

Figure III.2.3 Single deterministic lithotype model.

modeling, therefore, will be on the perturbation of these contacts rather than on the direct perturbation of the lithotypes.

To state uncertainty on such contacts, the modeler has therefore to provide, in addition to the single deterministic model, an interpreted "zone of uncertainty" believed to represent the maximum fluctuation of that contact surface. This is similar to interpreting faults from seismic data (Chapter III.1), where the exact location of the fault is not known, yet interpreters may provide bounds on its location. Such zones can be drawn manually, or perhaps guided by statistical measures such as a local kriging variance. Figure III.2.4 shows such interpreted zones, which have an average extent of approximately 30 m in radius.

A contact is essentially a surface delineating two different lithotype volumes. In terms of a numerical model, this means that grid cells on each side of that contact will have different indicator (categorical) values. The 3D volume that encompasses the same type of transition (from lithotype X to lithotype Y) therefore contains patterns of contact spatial variability for that specific transition. With four lithotypes, there is a total of 16 possible transitions, which are reduced to 12 transitions because we do not take into account the transition between areas of the same lithotype. Out of these 12 transitions, 11 are possible in this model, meaning that the single deterministic model with its delineated zones of uncertainty is transformed into a volume with 11 categories, each indicating a different transition. This volume of contact types is now used as a training set or image for resimulating the contact areas and conditioning to the drill hole data. To achieve this, we use SNESIM with the search tree partitioning into smaller trees for each contact type (Boucher, 2009). After simulating contacts, the corresponding lithotypes can be retrieved.

Figure III.2.5 shows sections in the realizations of simulated waste, friable itabirite, and the hematite lithotypes, which can be compared with the single

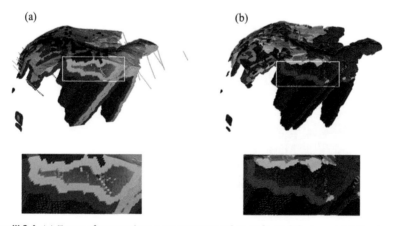

Figure III.2.4 (a) Zones of uncertainty near contacts; colors indicate lithotypes. (b) Contact type; the color indicates the 11 possible transitions; this volume is used as a training image.

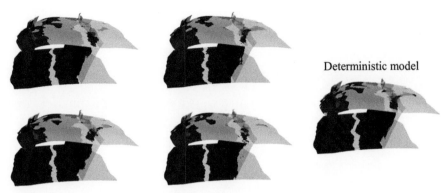

Figure III.2.5 Sections of four realizations of lithotypes conditioned to the drill hole data compared to sections of the single deterministic model of Figure III.2.3.

deterministic model. Given the complex interplay of hematite, friable itabirite, and waste, the top 100 m are in this case almost completely resimulated.

References

Boucher, A. 2009. Considering complex training images with search tree partitioning. *Computers and Geosciences*, 35, 1151–1158.

Boucher, A., Costa, J. F., Rasera, L. J. & Motta, E. 2014. Simulation of geological contacts from interpreted geological models using multiple-point geostatistics. *Mathematical Geosciences*, doi: 10.1007/s11004-013-9510-1.

Deutsch, C. V. 1998. Cleaning categorical variable (lithofacies) realizations with maximum a-posteriori selection. *Computers and Geosciences*, 24, 551–562.

Honarkhah, M. & Caers, J. 2012. Direct pattern-based simulation of non-stationary geostatistical models. *Mathematical Geosciences*, 44, 651–672.

Hu, L. Y., Liu, Y., Scheepens, C., Shultz, A. W. & Thompson, R. D. 2014. Multiple-point simulation with an existing reservoir model as training image. *Mathematical Geosciences*, 46, 227–240.

Mariethoz, G. & Renard, P. 2010. Reconstruction of incomplete data sets or images using direct sampling. *Mathematical Geosciences*, 42, 245–268.

Ortiz, J. M., Lyster, S. & Deutsch, C. V. 2007. Scaling multiple-point statistics to different univariate proportions. *Computers and Geosciences*, 33, 191–201.

CHAPTER 3

Climate modeling application – the case of the Murray–Darling Basin

3.1 Introduction

The third application area considered in this book concerns the downscaling of climate models. Climate models play an important role in assessing the impact of natural and human-induced forcing on environmental Earth systems. Global Climate Models (GCMs) consist of equations describing fluid flow and energy transfer discretized on a finite difference grid representing the entire Earth's surface. These models play a crucial role in policy making because they provide critical insight into future climate conditions based on known physical laws. However, such models present enormous computational challenges. With GCMs, it is important to represent the entire Earth in a single model because this allows avoiding the specification of lateral boundary conditions, reflecting the fact that every element in the system is interdependent with all other elements. To represent vertical fluxes, models need to be three-dimensional (3D). In addition to having parameters that vary horizontally, the system is stratified vertically into different atmospheric layers. Coupled ocean–atmosphere models include additional layers that model oceanic currents and related heat and moisture exchange processes. Hence, models have to be discretized into a large number of elements, and their resolution involves the coupling of several processes. As a result, running climate models is one of the most computationally intensive scientific applications.

Only a relatively small number of GCMs have been developed (24 models were mentioned in the fourth assessment of the International Panel on Climate Change), and for each model only a limited number of runs are available, with each run requiring times on the order of 10^5 CPU hours. Even then, these models need to be kept relatively coarse, with a single element representing a surface area on the order of 100×100 km. For several applications, this scale is too coarse for making useful predictions, and a finer-scale model resolution is needed. An example would be the prediction of future extreme temperatures or droughts around a particular city. Temperature and rainfall may present significant variability at a scale smaller than a GCM element, making detailed

Multiple-point Geostatistics: Stochastic Modeling with Training Images, First Edition. Gregoire Mariethoz and Jef Caers.
© 2015 John Wiley & Sons, Ltd. Published 2015 by John Wiley & Sons, Ltd.
Companion website: www.wiley.com/go/caers/multiplepointgeostatistics

predictions impossible. One approach usually adopted is the creation of models that focus on a smaller spatial area, called Regional Climate Models (RCMs). RCMs are local, higher-resolution models that are run within one or a few GCM cells, and they use the GCM outputs as boundary conditions (Wilby and Wigley, 1997; Maraun et al., 2010; Evans et al., 2012). Essentially, the aim of RCMs can be seen as imparting small-scale variability where computational limits impose restrictions on the resolution of GCMs. Although representing a significant computational gain compared to GCMs, RCMs are themselves still very computationally demanding.

The area considered in this case study corresponds to the Murray–Darling catchment in southeastern Australia, which is the region where most Australian agricultural production takes place. The economy of this region strongly depends on climatic conditions that present a large variability from one year to the next. Wet years allow for economic returns that compensate for the economic losses that occur during dry years. Moreover, agriculture heavily relies on artificial reservoirs (dams) to absorb the variability in rainfall input. Artificial reservoirs have a dual use because they provide water during dry periods but also absorb floodwaters that would otherwise be lost and cause damage. However, the control of dam water release needs to be planned in advance: one should maintain reservoirs at low capacity when floods are likely to occur, and conversely they should be kept relatively full during dry periods. In this regard, climate predictions are used to assess the likelihood of aquifers being replenished or depleted, which is critical for the water-trading schemes that are in place in Australia.

Climate predictions at the scale of small catchments or subcatchments offer some precious insights for applications related to water management. In the problem considered, the inputs consist of the results of a relatively coarse climate model of southeastern Australia informing temperature, soil moisture, and latent heat flux for the period from 1985 to 2006, and these data are available as seasonal averages. The resolution of this model corresponds to elements having a size of 50×50 km. We now place ourselves in the year 2005, and consider that we are planning hydrological management for year 2006. The 50 km resolution given by the climate model is insufficient; therefore, we need a higher-resolution prediction (10×10 km) for the year 2006. The coarse-scale prediction (50×50 km) for year 2006 that is given by the climate model needs to be obtained at a finer scale – or downscaled – in a physically realistic manner. The standard solution would be to set up an RCM with cells 10×10 km that takes the boundary conditions from the coarse model and use it to downscale the coarse grid using physical equations.

The alternative approach to RCMs is statistical downscaling, often using multivariate regressions (Hewitson and Crane, 1996; Raje and Mujumdar, 2011). Although computationally inexpensive, these methods only take into account collocated variables when estimating downscaled values. Neighboring values are generally not considered; therefore, these methods do not ensure that the downscaled values present the correct spatial variability (i.e., the spatial patterns observed at the 10 km scale for the period 1985–2005).

According to the concepts introduced in Section II.2.6, in MPS, the relationships between variables are inferred from training images, and relationships between scales can be considered similarly. Here, the training images we use to represent these multivariate and multiscale relationships are derived from past data. In this case, latent heat flux, temperature and soil moisture have been measured on the entire continent with relatively low uncertainty by combining satellite imagery and in situ measurements, and they are available for the years 1985–2005. Hence, both scales (50 km model output and 10 km data) are available exhaustively for the past period (1985–2005). The problem consists in determining the 10 km resolution values for year 2006, with the constraint that they should be coherent with the 50 km resolution model prediction. RCMs typically perform physically based downscaling, honoring known relationships between scales; however, such a procedure is very CPU demanding. The alternate route pursued here is to consider statistical downscaling: using observed statistical relationships between the different scales. The entire approach assumes that the past can be used to statistically represent the future, and also that the past data are rich enough to encompass future variability.

Such statistical downscaling is challenging, first because the relations between the 10 km data and the 50 km model output are typically nonlinear, hence the 50 km values are not a simple averaging of the corresponding 10 km data. Also, the spatial patterns can be complex due to the heterogeneity in the landscape features. Another difficulty is that the variables considered (temperature, soil moisture, and latent heat flux) are not linearly correlated with each other, as shown in Figure III.3.1. Moreover, the variables considered are nonstationary in both space and time. Spatial nonstationarity is caused by the study area comprising different climatic zones, ranging from coastal to desert, as well as different types of topography. Temporal nonstationarity is due to the different seasons considered, with magnitude of the values, spatial patterns and relations between the variables that are expected to differ for each season. In this case, the temporal nonstationarity is represented by the past data because it is caused by seasonal fluctuations occurring as yearly cycles. In the case of a global trend, the nonstationary would not be contained in the past data (Chu et al., 2010). Such can be the case, for example, with long-term climate change that results in trends, or fluctuations occurring at a low frequency compared to the period where data are available. In the case of the Australian climate, this would correspond to the Pacific Decadal Oscillation (PDO), which is a long-term climatic cycle in the sea surface temperature that has a 20–30-year period, not represented in the 20 years of available satellite imaging data. Therefore, in these cases, the past is not representative of the future, and the stochastic approach based on training images cannot be used directly.

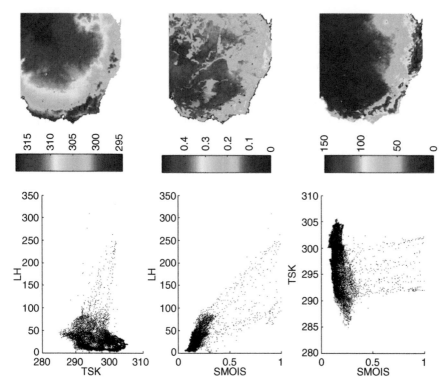

Figure III.3.1 Top: example of the type of patterns expected for the three main variables considered (temperature [K], soil moisture content [%], latent heat flux [W/m²]). Bottom: scatterplots representing the complex multivariate relationships (modified from Jha et al., 2013).

Modeling problems concerning several variables and nonstationarity have often been addressed in the context of traditional geostatistics, with co-kriging and co-simulation techniques that are based on linear correlations. Block kriging and block simulation are often used to model the relationships between scales, which assume that the variables are additive. In contrast, the challenge in this problem is to capture complex spatial features that cannot be characterized with covariance-based methods, across different scales, subject to multivariate nonlinear relationships (as shown in Figure III.3.1). As a consequence, a limitation of traditional statistical downscaling is that it tends to underperform compared to physical downscaling techniques that do account for complex relationships between variables across scales. In this regard, a statistical downscaling method including the nonlinear multivariate and multiple-point relationships can be closer to the physical ideal, yet be much less CPU demanding. This is the approach we present in this chapter; it was originally introduced by Jha et al. (2013).

3.2 Presentation of the data set

Here, we focus on downscaling climatic data for an area encompassing the Murray–Darling basin (MDB) in southeast Australia (Figure III.3.2). With a size of approximately 1 million km^2 and supporting a population of over 3 million people, the MDB is the largest and most economically productive catchment in Australia. Because of its importance in the Australian economy, the effect of climate change on the long-term productivity and sustainability of the basin is of high importance. The data set used in this study was derived from Evans and McCabe (2010), who evaluated the Weather Research and Forecasting (WRF) model RCM for the period from 1985 to 2009 over the MDB. WRF has been extensively tested over this region at different scales.

WRF model outputs at spatial resolutions of 50 km (coarse resolution) and 10 km (fine resolution) are used. In an initial step, all coarse and fine variables are interpolated on a simulation grid (161×171 nodes) using cubic spline interpolation. On this simulation grid, the cell size then corresponds to 0.09 angular degrees. It is important for all variables to be on the same resolution grid in order to define multivariate data events as discussed in Chapter II.2.3. The four main variables are considered are:

- TSK: surface skin temperature [°K],
- SMOIS: soil moisture [%],

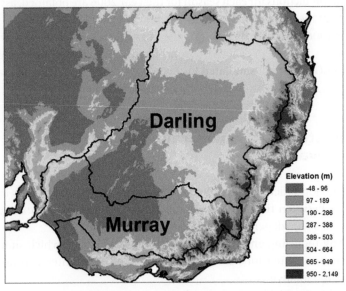

Figure III.3.2 The Murray–Darling basin in southeastern Australia (modified from Jha et al., 2013).

- LH: latent heat flux [W/m^2], and
- VEGFRA: vegetation areal fraction [%].

Land surface variables such as the surface skin temperature, soil moisture and latent heat flux are not linearly related, with multiple thresholds in the system affecting the strength, and even the sign, of their relationships. TSK is an estimate of the temperature of a very thin surface layer of the land or water, and it responds rapidly to changes in direct sunshine or shade. SMOIS plays a key role in estimating partitioning of the components of the water balance, such as infiltration, runoff and evaporation, and it can vary significantly due to large heterogeneities in land cover types, soil type, leaf area index and topography. The SMOIS data correspond to the moisture content in the top 10 cm of the soil and respond very quickly to precipitation and evaporation. LH describes the energy used for transporting the water from the land surface to the atmosphere as evapotranspiration, with the soil moisture condition directly influencing the evaporative flux (Kalma et al., 2008).

Twenty years of WRF-generated spatial data, corresponding to conditions in the years ranging from 1985 to 2005, are available for all variables at both 50 km and 10 km resolutions. Although the variables are available at a daily temporal scale, we consider here seasonally aggregated data, with the months of December, January, and February (DJF) for the summer; March, April, and May (MAM) for autumn; June, July, and August (JJA) for winter; and September, October, and November (SON) for the spring season (southern hemisphere seasons). These data are used as training images. Using the MPS approach, downscaled 10 km predictions for each season for the year 2006 are obtained, assuming that the 50 km resolution is known for that year. The actual 10 km WRF outputs for the year 2006, which are known, are used to validate the downscaling results. The vegetation fraction (VEGFRA) is considered to be constant, and therefore the 10 km VEGFRA is not simulated for 2006 but is known as being the same as in 2005.

3.3 Climate model downscaling using multivariate MPS

One of the major difficulties identified in climate model downscaling is that the relationships between the different variables considered are complex and nonlinear, as well as the dependence between the different scales.

As for any modeling endeavor, an initial step is to decide which simulation algorithm is best suited for the application at hand. The criteria to take into account are the following:

- The simulated variables are strongly nonstationary; therefore, appropriate handling of nonstationarity must be devised (see Chapter II.5).

- Several variables are considered simultaneously, and it is important to honor their respective relationships. This means that a simulation method allowing for multivariate modeling must be used (see Chapter II.6).
- All variables considered are continuous, ruling out simulation methods designed specifically for categorical variables.
- The climate model outputs provide local data at every location on the domain considered. This large amount of local information is better addressed with pixel-based methods rather than with patch-based methods.
- In this case, CPU time is not of paramount importance, because even long simulation times remain orders of magnitude smaller than when using RCMs.

The DS method (Section II.3.2.3) is chosen because it fulfills all those criteria.

One fundamental aspect of the MPS approach is that it relies on analog information. In geological applications, it can be challenging to construct training images. In the case of surface variables, however, remote sensing such as satellite imaging typically offers exhaustive coverage of a given domain. Here, data are available for 20 years at frequent intervals, meaning that there is a wealth of data usable as analogs. Such coverage is ideal in providing training images that are based on actual measurements rather than on interpretation. The challenge in this case is not to find an analog, but to deal with the wealth of analog information and to filter or extract the relevant information from it.

The past data embed the intervariable and interscale relationships; therefore, reproducing the properties of the analog should be sufficient to honor these relationships. The spirit of the approach adopted is to impair small-scale variability to the climatic variables using the past analog as a training image. Whereas RCMs use physical equations and classical statistical downscaling methods use regressions, the idea here is to use multiple-point statistical relationships inferred from the training images. The data corresponding to the past (years 1985–2005) are used as a series of training images for the simulation of the three variables in the future (year 2006). Additional variables are used to represent multivariate relationships and nonstationarity (in this case, the additional variables are control maps). Any number of variables could potentially be included in the model, for example known spatial attributes that have a significant relationship with the variables to predict (also known as external drift). In a similar fashion, the different scales present in the model are also modeled as additional variables. Therefore, the training images have to contain both coarse (50 km) and fine (10 km) resolution for all variables to have the ability to represent the relationships between both scales. Figure III.3.3 represents the DJF state for six main variables in a single year. These variables vary for each season of each year in the training data set. Three additional variables are static (i.e., they do not vary with time): the vegetation fraction and two control maps consisting of the longitude (X coordinate) and latitude (Y coordinate) of each grid node.

TSK SMOIS LH

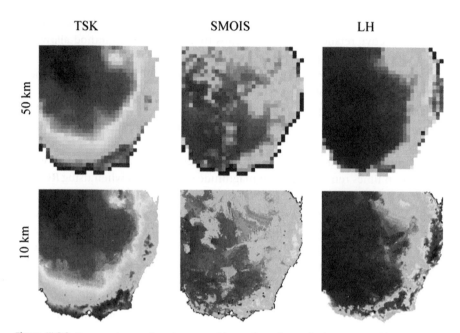

Figure III.3.3 Temperature, soil moisture, and latent heat flux at both coarse and fine resolution for a single December–January–February season (modified from Jha et al., 2013).

These nine variables are exhaustively known in the past; hence, they can serve as training images for the simulation of future time steps. In the future, only three variables are unknown: soil moisture, temperature and latent heat flux at a 10 km resolution (i.e., these are the variables to predict). The statistics and patterns of land–atmosphere values for each variable are known to differ depending on the season. To represent such seasonal changes in the modeled variables, a separate multivariate training image is built for each season. For each season, the training image comprises nine variables, and the time is represented in the third dimension. Therefore, in total, a training image consists of 180 maps (9 variables × 20 years), and there are four training images corresponding to each season.

Prediction for the 10 km variables is then performed separately for each season. In 2006, TSK, SMOIS, and LH are known at the 50 km resolution. In addition, VEGFRA and the control maps (longitude and latitude) are all known. These variables are used as conditioning data, and the direct sampling algorithm only simulates for a given season TSK, SMOIS, and LH at 10 km resolution. Regarding the parameterization of the simulation algorithm, the neighborhoods are defined as the 20 closest nodes for all variables, except for the latitude and longitude, for which a single neighbor (the central pixel of the data event) is enough to define the location of a point. For all variables, a Manhattan distance is used with a distance threshold of $t = 0.01$. All variables are given an equal weight in the distance calculation, except for latitude and longitude, which have a weight of one fourth of the other variables to allow sampling patterns across a reasonably broad area of the domain being considered. Note that the simulations consist of a single year and are therefore 2D, whereas the training images for each season are 3D (stacking of several 2D maps in the time dimension).

3.4 Results and validation

For validation, the downscaled results are compared against the actual 10 km WRF outputs for all the variables in the year 2006. Figure III.3.4, Figure III.3.5, and Figure III.3.6 present the coarse input data and the simulation outputs for TSK, SMOIS, and LH for all four seasons of 2006, along with the reference image. The coarse data, shown in the first column, display land features that are smoothed over large areas. Comparing the figures in the second and third rows, one observes that the simulations not only capture the overall variation across geographical locations but also preserve detailed land features. For example, in the summer situation for Latent Heat Flux (first row of Figure III.3.6), the 50 km data show that LH increases from west to east. There is an area on the east coast where LH has its highest magnitude. However, the corresponding simulation output, while maintaining those regions of high and low values of LH, also provides insight into regions with high LH in the south of the area, which were not visible in the coarse data. There are also specific features on all images at the edge of the domain, related to the boundary conditions of the RCM used. Because these features are present in the training image at both scales, they are also reproduced in the downscaled realizations, as expected.

The spatial distribution of the ensemble average of 50 realizations represents well the reference features, as shown in Figure III.3.7. However, locally reproducing the high and low values is not enough: an important validation criterion is that the relationships between the simulated variables, which are known to be complex and nonlinear, are also reproduced in the downscaled results. The scatterplots comparing the multivariate relationships of the simulated and reference variables are shown for the summer and winter seasons in Figure III.3.8 and Figure III.3.9.

The downscaled results from the multiple-point simulations show excellent agreement with the spatial distribution of WRF reference variables at a fine scale for TSK and LH across all seasons. Apart from providing an explicit capacity to represent uncertainty in the downscaled product, which is generally not possible with RCMs, the multiple-point simulation approach also has the advantage of reduced computational cost compared to RCMs. One full simulation of an RCM takes approximately 6000 CPU h, whereas DS takes about 48 h of CPU time for 50 realizations. This relatively important cost is due to the large number of variables and the size of the training images spanning 20 years. Despite this, MPS and DS are still orders of magnitude faster than physically based methods.

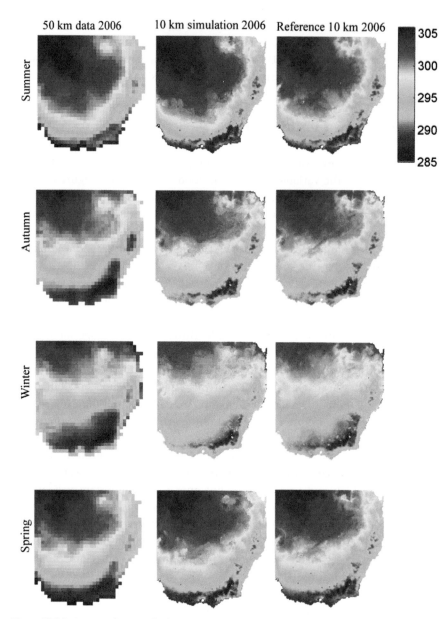

Figure III.3.4 Downscaling results for temperature (TSK) and comparison with reference (modified from Jha et al., 2013).

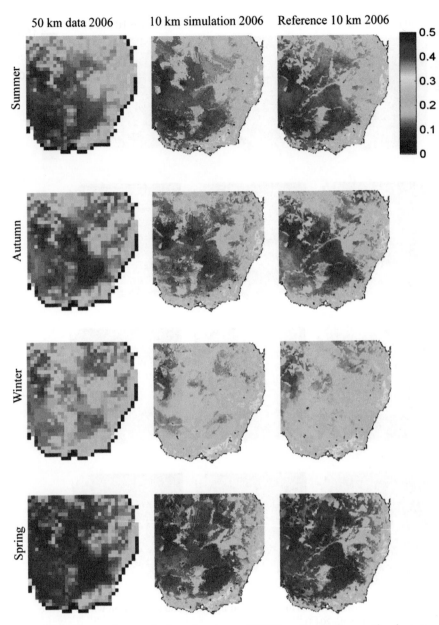

Figure III.3.5 Downscaling results for soil moisture (SMOIS) and comparison with reference (modified from Jha et al., 2013).

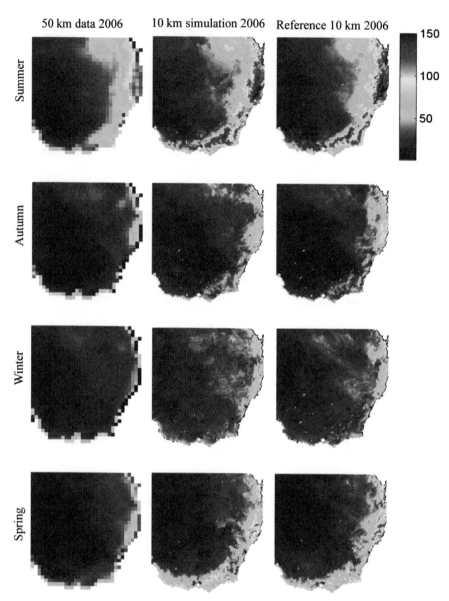

Figure III.3.6 Downscaling results for latent heat flux (LH) and comparison with reference (modified from Jha et al., 2013).

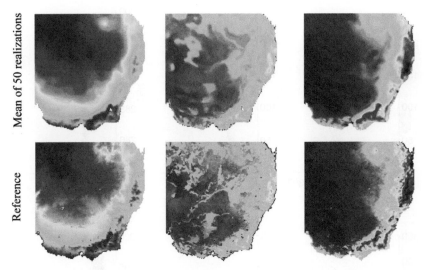

Figure III.3.7 Comparison of the mean of 50 downscaled realizations (top) and the reference (bottom) for the summer of 2006, for temperature, soil moisture and latent heat flux (modified from Jha et al., 2013).

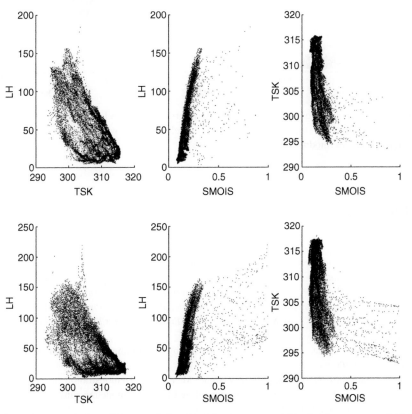

Figure III.3.8 Scatterplots of downscaled and WRF reference variables for the summer of 2006. Top row: scatterplots of values in the downscaled model. Bottom row: scatterplots of values in the reference model (modified from Jha et al., 2013).

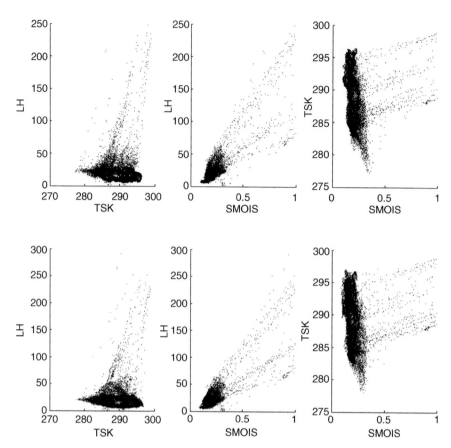

Figure III.3.9 Scatterplots of downscaled and WRF reference variables for the winter of 2006. Top row: scatterplots of values in the downscaled model. Bottom row: scatterplots of values in the reference model (modified from Jha et al., 2013).

References

Chu, J., Xia, J., Xu, C. Y. & Singh, V. 2010. Statistical downscaling of daily mean temperature, pan evaporation and precipitation for climate change scenarios in Haihe River, China. *Theoretical Applied Climatology*, 99, 149–161.

Evans, J. P. & McCabe, M. F. 2010. Regional climate simulation over Australia's Murray-Darling basin: A multitemporal assessment. *Journal of Geophysical Research D: Atmospheres*, 115.

Evans, J. P., McGregor, J. L. & McGuffie, K. 2012c. *Future Regional Climates. The future of the World's climate*, Elsevier.

Hewitson, B. C. & Crane, R. 1996. Climate downscaling: Techniques and application. *Climate Research*, 7, 85–95.

Jha, S. K., Mariethoz, G., Evans, J. P. & McCabe, M. F. 2013. Demonstration of a geostatistical approach to physically consistent downscaling of climate modeling simulations. *Water Resources Research*, 49, 245–259.

Kalma, J. D., McVicar, T. R. & McCabe, M. F. 2008. Estimating land surface evaporation: A review of methods using remotely sensed surface temperature data. *Surveys in Geophysics*, 29, 421–469.

Maraun, D., Wetterhall, F., Ireson, A. M., Chandler, R. E., Kendon, E. J., Widmann, M., Brienen, S., Rust, H. W., Sauter, T., Themes, L. M., Venema, V. K. C., Chun, K. P., Goodess, C. M., Jones, R. G., Onof, C., Vrac, M. & Thiele-Eich, I. 2010. Precipitation downscaling under climate change: Recent developments to bridge the gap between dynamical models and the end user. *Review of Geophysics*, 48, 3003.

Raje, D. & Mujumdar, P. P. 2011. A comparison of three methods for downscaling daily precipitation in the Punjab region. *Hydrological Processes*, 25, 3575–3589.

Wilby, R. & Wigley, T. 1997. Downscaling general circulation model output: A review. *Progresses in Physical Geography*, 21, 530–548.

Index

Multiple-point Geostatistics: Stochastic Modeling with Training Images, First Edition. Gregoire Mariethoz and Jef Caers.
© 2015 John Wiley & Sons, Ltd. Published 2015 by John Wiley & Sons, Ltd.
Companion website: www.wiley.com/go/caers/multiplepointgeostatistics

Printed and bound by CPI Group (UK) Ltd, Croydon, CR0 4YY

16/04/2025

14658552-0002